Second Edition

New Food Product Development

From Concept to Marketplace

Second Edition

New Food Product Development

From Concept to Marketplace

Gordon W. Fuller

CRC PRESS

Boca Raton   London   New York   Washington, D.C.

Library of Congress Cataloging-in-Publication Data

Fuller, Gordon W.
    New food product development : from concept to marketplace / Gordon W. Fuller.—2nd ed.
       p. cm.
    Includes bibliographical references and index.
    ISBN 0-8493-1673-1
    1. Food—Marketing. 2. Food—Research. 3. New products—Marketing. 4. Product
management. 5. Food industry and trade—Technological innovations. I. Title.

HD9000.5.F86 2004
664′.0068′8—dc22   2004045485

Visit the CRC Press Web site at www.crcpress.com

© 2005 by CRC Press LLC

No claim to original U.S. Government works
International Standard Book Number 0-8493-1673-1
Library of Congress Card Number 2004045485
Printed in the United States of America   3 4 5 6 7 8 9 0
Printed on acid-free paper

Dedication

This book is dedicated to my grandchildren:
Caitlin, William, Seaira, Nicholas, and Amy,
whose inventiveness and creativity at all things have
never ceased to amaze and amuse me.

Preface to the Second Edition

This revision of the 1994 edition has been completely reorganized with much material added. All new material had to meet one of the following criteria: (1) it had to make a substantial intellectual or technical contribution to the understanding of the problems of new food product development; or (2) it had to illustrate cautionary issues associated with the new product development process. In addition, the material had to describe the real world environment of product development and not describe what many authors wish the real world to be.

Some of the suggestions made by reviewers, colleagues, and readers have been included. Suggestions to include fictional examples of new product development, thus copying styles of presentations developed by many business schools, were studiously avoided. Fictional examples are too blatantly obvious in the elements contributing to the success or failure of the product situation they describe. In my opinion, such examples teach little about product development and provide insignificant understanding of the process. Real life is ever so much more educational. Success stories also leave one at a loss about what the crucial act or acts leading to a successful introduction were. What was the "Eureka" stage? Did one successful act in development outweigh all the stupid blunders that had been made, or were all phases of new product development carried out correctly?

Others suggested that I obtain permission from my clients to include real life examples of new food product development. All work with my clients is confidential and this shall remain so. I did, however, meet this suggestion halfway. I have identified more clearly than I did in the original edition what are anecdotes from my own experience. In these examples, neither my clients nor their products are clearly identified. If my clients do see themselves, they should be ashamed. The anecdotes illustrate particular misadventures in new product work. They are a reminder that work on a new product should be undertaken in an organized, thoughtful fashion with careful attention given to every aspect of development; that all testing should be completed before the launch; and that calling for advice, assistance, or guidance to solve problems arising after the product's introduction is too late. Unfortunately the real world of product development is not as it should be.

For those who have disparaged the age of some of my references I suggest they read Chapter 12. I am quite happy to use older literature that is clearly written so that principles are easily grasped.

I am deeply indebted to my wife, Joan, for reading and making many helpful suggestions on the text; to my son, Grahame, a senior technical writer for Softimage Co., a subsidiary of Avid Technology, Inc., for preparing many of the figures, for many interesting and stimulating discussions regarding the text, and also for reading and emending my notes on computers, communication technology, and the Internet; and to my son-in-law, Dr. David Gabriel, associate professor in the Department of

Physical Education and Kinesiology at Brock University, for suggestions and assistance with figures. To these members of my family, thank you.

Helpful discussions on Chapters 10 and 11 were had with Timothy Beltran, BScE, who at the time of our discussions was executive chef of the J. P. Morgan Chase dining room on Wall Street, in New York, and who has now formed his own catering company, Culantro Caterers, in New Jersey; and with Henry B. Heath, MBE, BPharm. (London), MFC, FRPharmS, FIFST (U.K.), and retired president of Bush Boake Allen Corp., Dorval, Quebec. To both of these gentlemen, my heartfelt thanks.

I want also to thank Eleanor Riemer, late of CRC Press, who gave me much encouragement and direction, and her assistant, Erika Dery, who answered my questions and provided much advice and guidance on style.

Gordon W. Fuller

The Author

Dr. Gordon W. Fuller has a wide variety of training, experience, and skills that he has used successfully in his consulting practice for over 30 years. He graduated with a BA (1954) and an MA (1956) in food chemistry from the University of Toronto and a PhD (1962) in food technology from the University of Massachusetts.

His work experience includes a stint as a research chemist with the Food and Drug Directorate in Ottawa, Canada, working on gas liquid chromatography of food flavors. In addition, he spent 2 years as a research food technologist working on chocolate products for the Nestlé Co. in Fulton, New York, and another 2 years in applied research and product development on tomato products for the H. J. Heinz Co. Fellowship at the Mellon Institute for Industrial Research, Pittsburgh, Pennsylvania.

Dr. Fuller served as associate professor in the Department of Poultry Science at the University of Guelph, Ontario, Canada, where in addition to teaching and research responsibilities in poultry meat and egg added-value products, he carried out extension work for food processors in southern Ontario. He also held a fellowship at the Food Research Association, Leatherhead, England, where he gained experience working with meat products.

Prior to forming his own consulting company, Dr. Fuller was for 8 years vice president, Technical Services, Imasco Foods Ltd., Montreal, Canada, where he was responsible for corporate research and product development programs at the company's subsidiaries in both Canada and the U.S. His consulting practice has taken him to the U.S., South America, Europe, and China. He has lectured on agricultural economics and food technology topics in North and South America, England, Germany, and The Netherlands. As an outside lecturer he presented courses at McGill University in agribusiness management and new food product development and was a guest lecturer at Concordia University for many years.

Contents

1 New Food Products and New Food Product Development in a Nutshell: The Mystique and Mythology of New Product Development

> … the production of a new food commodity might seem to be a trivial matter unworthy of serious consideration, this is not necessarily so. The technological expertise upon which any one item depends may require the full depth of scientific understanding.
>
> *Magnus Pyke (1971)*

I. INTRODUCTION

New food products are described by several sources as the lifeblood of food companies. Companies would eventually languish and die if they relied solely on their old bell-ringer products year after year. The rewards from a successful new product can be great, but if the products are failures, the company must absorb severe financial losses as well as endure a loss of face.

New food products and their development can be written about from many perspectives: that of the consumer, that of the company's management, that of a food journalist — often describing what went wrong — that of the technocrat involved in the development, and that of the marketer responsible for the consumer research and promotion materials. Each perspective contributes something to the study of new products and their development, but each also brings a bias reflecting the position of the contributor.

An exploration of the mystique surrounding new product development requires first a complete understanding of the terms used and, here especially, the terms as they will be used in this book. Agreement on these terms is important in describing and understanding the new food product development process.

TABLE 1.1

Characteristics of a New Food Product as Introduced by a Specific Food Company

Product has never before been manufactured by that company.

Product has never before been distributed by that company.

An old established product manufactured by a company is introduced into a geographically new area by that company.

An old established product manufactured by a company is introduced in either a new package or a new size or form.

An old established product manufactured by a company is introduced into a new market niche, that is, positioned as one with a new function.

II. DEFINING NEW FOOD PRODUCTS AND THEIR CHARACTERISTICS

A. NEW PRODUCTS

Very simply, a new food product should be one that is new; that is, the product has not been presented in the local marketplace or, indeed, in any marketplace, before. This is, of course, a rare occurrence. Most products advertised as "new" usually have an analogue, a comparable look-alike, produced locally by a competitor, produced elsewhere in the country, or imported.

A food product may be new with respect to a company that has not sold it before but is not necessarily new with respect to a marketplace. Therefore, the characteristics of a new food product that might be introduced by a company can be described in Table 1.1.

The simplest definition of a new product is:

> a product not previously manufactured by a company and introduced by that company into its marketplace or the presentation by a company of an established product perhaps in a new form or into a new market not previously explored by that company.

This definition should not be applied too rigidly. It becomes a little shaky if pushed. For example, no-name or store brand products, even those as famous as President's Choice®, do not fit some of the characteristics or categories.

No-name or store brand products are products purchased from a food manufacturer by a store, which then puts its own no-name or house brand label on them. The retailer has established purchasing standards for the product (often very high standards, cf. President's Choice) and has neither developed, manufactured, nor market researched the product. In effect, these products piggyback on the research and development work of the food manufacturer.

Tables 1.2 and 1.3 enumerate new food products, describe some general characteristics of each type, and provide some examples.

TABLE 1.2
General Characteristics of Classes of New Food Products

Types of New Product	General Characteristics
Line extensions	Little time or research is required for development.
	No major manufacturing changes in processing lines or major equipment purchases are required.
	Relatively little change in marketing strategy is needed.
	No new purchasing skills (commodity trading) or raw material sources are required.
	No new storage or handling techniques for either the raw ingredients or the final product are needed. This means regular distribution systems can be used.
Repositioned existing product	Research and development time is minimal.
	Manufacturing is comparatively unaffected.
	Marketing must develop new strategies and promotional materials to interpret and penetrate the newly created marketing niche.
	Sales tactics require reevaluation to reach and make sales within the new marketplaces.
New form or size of existing product	Highly variable impact on research and development.
	Highly variable impact on physical plant and manufacturing capabilities. Major equipment purchases may be required if manufacturing to be done in-house.
	Marketing and sales resources will require extensive reprogramming.
Reformulation of existing product	Moderate research and development required consistent with reformulation goal.
	Generally little impact on physical facilities.
	Generally little impact on marketing and sales resources unless reformulation leads to repositioning of product.
Repackaging of existing product	The novelty of the repackaging will dictate the amount and degree of research and development required.
	Slight impact on physical facilities. New packaging equipment will be required.
	Little impact on marketing, sales, and distribution resources.
Innovative products	Amount of research and development dependent on the nature of the innovation.
	Highly variable impact on manufacturing capabilities.
	Possible heavy impact on marketing and sales resources.
Creative products	Generally heavy need for extensive research and development; therefore a costly venture.
	Extensive development time may be required.
	May require entirely new plant and equipment. Degree of creativity may require development *de novo* or unique equipment.
	Will require total revision of marketing and sales forces. Creation of a new company or brand may be required.
	Risk of failure is high.

TABLE 1.3
Examples of Different Types of New Products

Type of New Product	Examples of Category
Line extensions	A new flavor for a line of wine coolers or for a line of flavored bottled waters
	New varieties of a family of canned ready-to-serve soups
	New flavors for a snack product such as potato chips
	New flavored breadcrumb coating
	A coarser or more natural peanut butter
Repositioned existing product	Oatmeal-containing products positioned as dietary factors in reducing cholesterol
	Soy-containing products repositioned as dietary factors combating cancer
	Soft drinks positioned as main meal accompaniments
New form of existing product	Margarine or butter spreadable at refrigerator temperatures
	Prepeeled fruit or sectioned grapefruit or oranges
	Fast-cooking products such as rice or oats
	Instant coffees, teas, and flavored coffees
	Dehydrated spice blends for sauces
Reformulation of existing products	Low-calorie (reduced sugar, fat) products
	Hotter, spicier, zestier, crunchier (e.g., peanut butter), smoother products
	All natural ("greener") products, organic products
	Lactose-free milk products
	High-fiber products
New packaging of existing product	Single serving sizes of, for example, yogurt
	Branded fruits and vegetables
	Pillow packs for snack food items
	Institutional sizes for warehouse stores
	Squeeze bottles for condiment sauces
	Pull-top containers of snack dips
	Use of thin profile containers
Innovative products	Dinner kits
	Canned snack food dips (see above)
	Frozen dinners
	Simulated seafood products
Creative products	Reformed meat cuts
	Extruded products
	Surimi- and kamaboko-based products, soy bean curd (tofu), and limed corn if these were discovered recently
	Short-chain fatty acid–containing products

1. Line Extensions

A line extension is a variant of an established line of food products, that is, one more of the same. Line extensions represent a logical extension of a family of similarly positioned products.

Some thought must be given to distinguish what are and are not line extensions, for example:

- Adding a canned three-bean salad product to a line of canned bean products involves a change in processing and quality control technology. Developers have gone from a high pH, low acid product to an acidified product. This is no longer a line extension; a different market niche is targeted.
- A snack manufacturer that extends its product line from potato chips to the manufacture of corn chips, corn puffs, roasted peanuts, or popcorn is not making simple line extensions. Such products have in common only the snack food element. They may be distributed through the same channels and displayed in the same section of a retail store, but purchasing philosophies, storage facilities, and manufacturing technologies have changed extensively for the manufacturer.

Line extensions are not to be confused with brand extensions. Brand extensions, particularly brand overextensions, can be a death knell for a brand and the products under its protective umbrella. One need only think of a favorite brand of food and picture that brand name extended to carry a line of women's lingerie or men's underwear. It has been tried with disastrous results.

Marketing programs are usually not affected by line extensions, but there can be some surprises. A manufacturer of a family of confectionery products originally positioned for children may be presented with some marketing difficulties if adult flavors are introduced. Children may not appreciate the flavors; adults may not readily accept the product if they do not realize that the product is flavored for them. Therefore, different promotions, advertisements, and store placements for the adult products must be considered.

2.　Repositioned Existing Products

A company can be very startled to find, either through consumer letters or surveys of their consumers' product usage, that their consumers have come up with a new use for an existing product, one that the company had never anticipated. These observations may allow a whole new market direction to be taken and give a new life to an existing product (a process called product maintenance). The classic example is ARM & HAMMER® Baking Soda finding a new niche as both a body deodorant and a deodorant for food odors especially in refrigerators.

These repositionings require extensive marketing development because they take the company into entirely new markets for which brand extension might not be suitable.

3.　New Form or Size of Existing Products

Putting an existing product into a new form, for example, a paste product converted to a tablet or a sauce to a powder, is a radical departure for customers and consumers alike to accept. Neither one may appreciate the so-called improvement in the modified product. Advantages or improvements of the new form over an old established product form must be perceived as an advantage by customers and consumers. For example, the advantages of a dried, sprinkle-on version of a previously wet condiment sauce may not be appreciated or preferred by users over the traditional form. Prepeeled, precut french fry–style potatoes were never successful over whole potatoes in the chilled food, retail market. (This added convenience of a new form may change; see below.)

4. Reformulation of Existing Products

The "new, improved" product is typical of this category. Reformulation may be necessary for any number of reasons:

- Some improvement, such as better color, better flavor, more fiber, less fat, greater stability, and fewer calories, is required in a product to match competitive products or to fit with trends.
- A raw material is unavailable or too costly because of a scarcity due to a poor harvest, trade embargo imposed on the source country, or strike-bound supplier. Reformulation is required to overcome the scarcity.
- New technologies bring improved processes or cheaper ingredients with vastly improved characteristics and properties that must be investigated to make an improved product.
- Reformulation is needed to lower the costs of a product to compete with cheaper products from competitors. This is often the reason for the new, improved product.
- Regulatory agencies have altered the legal status of an ingredient or an additive. This has happened in the past with several color additives and an artificial sweetener. More recently questions have arisen respecting the safety of some nutraceuticals, for example, kava kava and St. John's wort.
- Reformulation may also be necessary to create a new market niche for existing products, for example, one with fewer calories.

Reformulation can usually be accomplished comparatively inexpensively and within a relatively short developmental time. However, if the reformulation goal is fewer calories through the replacement or removal of fat or sugar, research and development may be extensive.

The new, improved product can present a marketing dilemma for a company. Customers and consumers are apt to wonder, if this new version is an improvement why was it not offered in the first instance?

5. New Packaging of Existing Products

In its simplest form, the packaging of bulk produce into unit packages typifies this category (see Figure 1.1). The brand labeling of packaged produce and meats gave these existing products a new life as new products of higher quality on which a

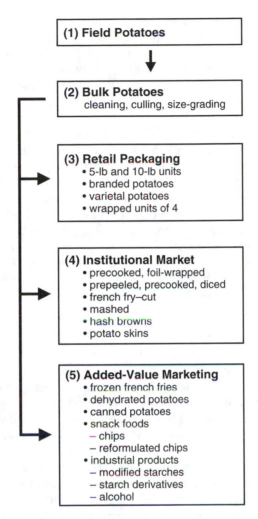

FIGURE 1.1 Adding value to the humble potato.

company is proud to put its name (Gitelman, 1986). For example, bananas, plums, pears, and many other fruits and vegetables were, until recently, all no-name foods; these now display stickers of well-known food companies. Packages of "speed scratch products," that is, mixed, washed salad greens and prepeeled fruits and vegetables, fit in here. Here, of course, development is minimal. Manufacturing involves purchasing, inspection, grading, cleaning, trimming, storage, weighing, packaging, and distribution. The key responsibility in making this new product a success rests mainly with marketing a brand, the quality of which is known and respected by customers and consumers. As adding convenience progresses, the amount of technology required to make these products a success and to make them safe increases.

New technologies such as modified atmosphere packaging and controlled atmosphere packaging have permitted the creation of a number of new products and provided existing products with an extended shelf life for both satisfying existing markets and allowing the opening up of new markets in a larger distribution area. Both technologies may require extensive research for safety and shelf life stability of the products (see Chapter 5, Section 7). Another packaging development is biobased materials, which offer a marketing ploy as environmentally sound features (Shahidi et al., 1999; Petersen et al., 1999). Here again, care must be taken over the safety of these packaging films.

To change a package, for example, from metal to glass containers, means a new packaging line. Even the changeover from steel cans to aluminum cans to save weight requires an extensive overhaul of the packaging operation. Similarly, the use of plastic squeeze bottles with snap-cap lids for dispensing mustard, ketchup, or other sauces is a major packaging changeover from glass containers.

To change from a cylindrical, conventional can to a thin profile container (the pouch and the semirigid tray) for thermally processed foods requires a reformulation. Such a change improves the quality (added value) of the product through faster heat penetration but requires recalculation of the thermal process. The original formulation was based on withstanding the more rigorous thermal process, and the formulation may not be suitable for the new heat process. In addition, extensive changes to the packaging line are required.

6. Innovative Products

The remaining two categories, innovative products and creative products, cannot properly be described without clearly defining what is meant by innovation and what is meant by creation. *Webster's Ninth Collegiate Dictionary* defines *innovate* as "make changes." *The Concise Oxford English Dictionary*'s definition is "make changes in." So, an innovative product is one resulting from making changes to an existing product.

Innovative products are difficult to categorize. According to dictionary definitions every new product description previously mentioned in this chapter is an innovation. Despite the dictionary definitions, none of these products were innovative. Generally, the more innovation (change) in a product, the more costly the marketing strategies of that novelty may become since the customer and consumer have to be educated on the value of the innovation. In short, the development of innovative products could be costlier and riskier than any of the other previously discussed paths to new products.

However, little research and development are required for a frozen food processor to put a stew, frozen vegetables, and a frozen pastry on a tray and call it a frozen dinner. Certainly there will be some production line changes. Likewise, putting a can of tomato sauce, a package of dry spaghetti sauce spices, and a package of dry pasta together to make a dinner kit requires little research and development effort. Yet both were remarkably successful innovative products that engendered many imitators.

In the nonfood category, Akio Morita, the inventor of the Walkman® (trademark of the Sony Corporation), claimed the Walkman involved no invention and no costly research and development, just expert putting-together of established inventions into a superbly marketed product (Geake and Coghlan, 1992).

New ingredients can form the basis for innovative products. Simulated crab legs, lobster chunks, shrimp, and scallops based on surimi technology have formed the basis for many seafood dishes (Johnston, 1989; Mans, 1992).

An ambitious program of innovation with all its attendant costs for a canner of commodities such as chick peas, navy beans, and other lentils would be to venture into the leisure foods arena to manufacture such added-value products as hot bean dips or ethnic dishes such as hummus tahini. The demographics of many cities indicate unique marketing opportunities for the adventurous in such ethnic dishes.

7. Creative Products

The dictionary sources noted above both define *create* as "bring into existence." A creative product is, therefore, one newly brought into existence: the rare, never-before-seen product.

Creative products are harder to define than innovative products and still more difficult to exemplify. Surimi, a fish gel developed several hundred years ago, is considered a creative product, as are kamaboko-based products, which were developed from surimi. Tofu, bean curd, and limed corn meal are also creative products. Today, one might consider reformed meat products as a creative development, and certainly extrusion to produce new puffed products is creative.

When creative products are successful, imitators rapidly flood the market with me-too products. They telescope the time and effort the developers of the creative product took to create an imitation and market it. In general, the more a product is a copycat of an existing product, the less development time will be required even if the product is creative. Development time may be only as long as it takes to get new labels printed. Similarly, development costs and costs of market entry will be minimal. Truly creative products have greater costs and development times that may be measured in years rather than weeks or months.

B. ADDED VALUE

Added value is a characteristic many new products are purported to have. The late Mae West had a memorable line in the film *She Done Him Wrong*: "Beulah, peel me a grape!" Beulah provided added value for a consumer; peeled ready-to-eat fruit possesses this characteristic. Added value, then, describes the degree of innovation or change that makes a product more desirable to customers and consumers. The novelty might be improved stability, improved functionality, better color, better texture, or better service or convenience. It could mean less preparation time (greater convenience) or less waste from preparation to dispose of. Whatever the value is, consumers want it. Meltzer (1991) rather unclearly defined value-added processing (the terms *added value* and *value added* are used synonymously) as "any technique that effects a physical or chemical change in a food or any activity that adds value to a product."

The concept of added value is depicted more concretely in Figure 1.1. I worked closely with a potato processing operation that produced prepeeled, french fry–cut potatoes (Stage 4). Not too many decades ago, only field grade potatoes were sold.

They were packed in bulk as 100-lb burlap sacks in grocery stores. Convenience was added by breaking these into bushel, peck, or half-peck baskets of potatoes — measures never seen in retail outlets today. There was no culling of bad potatoes, no size sorting, and no cleaning. One bought one's potatoes in the fall, stored them in the cellar, and used them throughout the winter. The quantity of potatoes purchased was not convenient for smaller families and was far in excess of the needs of the increasing number of apartment dwellers who had neither cellars nor storage cupboards. Many ethnic families did not want large quantities of potatoes, nor did such quantities reflect the changing food habits of consumers who did not want meals designed around potatoes.

Initially, added value was created by cleaning, culling, and size grading (Stage 2 in Figure 1.1), followed by more suitably sized packaging in 5- and 10-lb units (Stage 3). Smaller, more convenient unit packaging (foil- or glassine-packaged potatoes) introduced new market niches by targeting customers such as "live alones" and occasional potato users. Offering specific varietal potatoes (Yukon Gold, for example) or potatoes with properties specific for baking, barbecuing, or frying provided more convenience (value).

Today, potatoes and several other vegetables and fruits are offered not only by their brand and varietal names but with suggestions for the best culinary uses of that particular variety. The customer knows what variety to buy for the purpose in mind, or the customer can buy the variety, know how to best prepare it, and have the assurance of quality that the brand name confers. The drive to provide added value continued with prepeeling, dicing, and slicing the potatoes. This minimized preparation time for institutional users and concentrated waste in a central location where it might be more profitably used. The move now was to a new niche, a new market.

However, new problems were introduced. Added value requires application of advanced technology and labor to reduce losses from spoilage with a more fragile product and to prevent hazards of public health significance (see Pyke's statement opening this chapter). A new market meant new marketing and sales techniques. It also meant that a distribution network suitable for a fragile product had to be established.

Sophisticated further processing serves the needs of the food service sector with preprepared hash browns, baked potatoes, partially cooked french fry–style chips, stuffed potatoes, and so on. Each added value introduces the need for more advanced technologies to maintain safety and quality and provides greater marketing opportunities; these also require more complex distribution and sales systems.

Meat, poultry, and fish are also sold with more descriptive names, by brand names, and with preparation instructions or recipes describing how to cook the particular cut of meat or species of fish. Added value for both the consumer and the customer has been introduced.

C. CUSTOMERS AND CONSUMERS

The terms *customer* and *consumer* are often substituted one for the other: yet, they describe two different entities and should not be confused. A customer is one who buys in a marketplace. This is the person who is attracted by sales material, promotions, or tastings in the marketplace. The customer is the purchasing agent suffering the blandishments of technical sales representatives or hunting for suppliers able to offer ingredients and raw materials as well as reliability of delivery, quality, and price. Retailers are mostly interested in the customer since the customer buys.

Dr. Kurt Lewin (reported in Gibson, 1981) described the customer as a "gate-keeper." The gatekeepers are:

- Family members who decide what will be purchased for the household
- Purchasing departments of companies and stores or central commissaries for restaurant chains, institutions, hospitals, and the like who buy or send out tenders to suppliers with specifications for manufacturers, for the military, for caterers to penal institutions, and so on
- Chefs who plan menus and decide what raw produce, ingredients, or semifinished goods are purchased for the diners in a restaurant
- Owners who determine what pet foods will be purchased for their animals

In addition, the customer makes choices (i.e., purchases) according to their and others' likes and dislikes, allergies, disposable income, or commercial industrial requirements.

The consumer uses (consumes) what is purchased by the customer. The consumer can also be the customer. For example, a diner in a restaurant and a person walking down the street eating finger food are both customer and consumer. Obviously consumers influence what customers purchase or serve. Uneaten food returned to the restaurant's kitchen; prisoners rioting over the quality and variety of food prepared for them; ingredient suppliers losing contracts because of poor quality, inability to meet their clients' specifications, or late deliveries; or children refusing to eat their meals: all of these attest to the power of the consumer.

As noted by Fuller (2001), "There is the conflict between the consumer's hedonistic demand of 'I want' and the customer's practical barrier of 'I need or 'I can afford."

Therefore, marketing and sales personnel must clearly understand this distinction between customer and consumer. Sales personnel are concerned with retailing and retailers and only indirectly with customers unless their duties also include stocking store shelves.

D. Markets and Marketplaces

Market and *marketplace* are often used synonymously, but each has a unique meaning. A market is conceptual; it represents a need discovered in customers and consumers that marketing personnel hope to develop into a want, a potential to sell. That is, one can say that there is a market for organically grown vegetables and meats or that there is a market for low-calorie foods. This means there are customers in the many different marketplaces who will preferentially purchase organically grown products or low-calorie foods. Their needs must be developed into wants and satisfied with products.

Products are, however, sold in marketplaces, not in markets (despite the use in common parlance of *farmers' markets*). Marketplaces are real physical entities; they

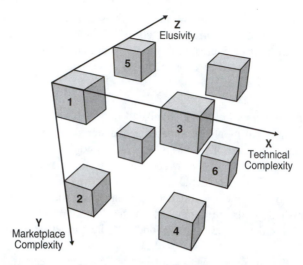

FIGURE 1.2 Product complexity, marketplace complexity, and consumer elusivity interactions characteristic of new products and their marketing.

are not conceptual. They range from farmers' roadside stands to giant food stores, beverage and snack bars in movie houses, and coin-operated food vending machines. Even electronic food marketplaces operating from Web sites are real marketplaces.

III. MARKETING CHARACTERISTICS OF NEW PRODUCTS

Figure 1.2 demonstrates three-dimensionally the difficulties accompanying the movement of known products in known markets into new and unknown markets or the movement of existing products to complex innovative products.

The *y*-axis is a measure of increasing marketplace complexity; the farther from the origin, the more complex the marketplace. This measure of complexity encompasses any one or all intangibles such as:

- The product is moved into marketplaces with increased activity by the competition.
- The product requires more sophisticated warehousing, distribution, and retailing activities.
- New food legislation and regulations or marketing customs (perhaps in foreign countries importing the product) need to be adhered to.
- The general economy of the country as well as specific economic factors in the area being serviced are altered drastically.
- The level, availability, or lack of marketing skills within the company to answer all the above challenges becomes apparent.
- The geography of the market area as well as the difficulty of selling novelty or educating the consumer present problems.

The x-axis is an index of increasing the technical complexity required for a product with innovation, creativity, and added value. The farther from the origin, the more creativity, innovation, or technical complexity there is in the product. Consequently more research, development time, and costs are involved. The product is also more fragile and sensitive to abuse; it requires care in handling and distribution channels to maintain its high quality throughout retailing and in the hands of the end user. More education of the customer and consumer is required to recognize the added-value features. Therefore, more creative marketing technology is needed.

The remaining axis, the z-axis (into the back of the page), is a measure of the volatility and elusiveness, and perhaps even the fickleness or incredulity, of consumers and customers. How else can one explain the growth in "healthy" foods when there has been "a dramatic increase in the per capita consumption of high calorie desserts, salted snack foods, and high calorie confections" noted as far back as 1986 (Gitelman, 1986) and continuing to this day? Or the horrendous outcry over obesity, which an October E-newsletter of the Institute of Food Technologists claims has quadrupled, with the contradictory belief of many marketers that consumers are on a wellness kick? (The average caloric intake by consumers is up over what it was 20 years ago. Today obesity is considered to be of epidemic proportions.) The z-axis represents the elusive consumer or customer who is targeted in the marketplace. Elusiveness can be likened to market segmentation, a new market niche. If a product designed for the general public (a rare event) is redesigned for the teenage market, that consumer has become more elusive and the market is segmented. If this hypothetical product is redesigned again for teenagers of single parents, more elusiveness is created. Marketing purists might cavil at this, claiming that elusiveness is really a variant of the y-axis or marketplace complexity. This is incorrect: it is not the volatility of the marketplace that is represented but the volatility of consumers and customers within these marketplaces.

Six numbered cubes in Figure 1.2 represent typical problems faced by developers as they attempt to bring new products to market. Cube 1, at the origin, depicts the situation of an established product in a known market (also encountered when a line extension is introduced into a known market). It is the *status quo*, the starting point.

Cube 2 represents taking the same established product into a more complex marketplace situation. This could be simply expanding into a new geographic area or perhaps into foreign marketplaces. The targeted customers and consumers have not changed. Development costs are not a factor in getting into the new marketplace. Only the costs associated with marketing (labels, promotional materials, and advertising), sales (brokers), and distribution increase.

Cube 3 is a product with added value (increased product complexity) introduced into a known marketplace and aimed at known consumers. Costs for development to add value as well as marketplace introduction and promotional material to educate the old customers and consumers now begin to escalate.

In the situation represented by Cube 4, a product with added value has been introduced into a more complex marketplace, one previously unknown to the developer. The company has simply expanded into a new marketplace. The targeted consumers remain the same, but there is a new playing field. Researching the new market area brings increased costs because the old marketing strategies may not be suitable.

Cube 5 represents another repositioning problem. An established product is targeted for an elusive consumer but in a familiar marketplace. Some examples are a popular antacid positioned as a calcium supplement for elderly women or a hand cream that proves to be an excellent insect repellent for campers. Costs can again increase significantly to reach these new targets that are more elusive. The situation is risky: A cosmetics company may not want its mystique associated with the sporting life and insect pests; or a manufacturer of a hot sauce with a high content of phytochemicals of the capscaicinoid family may not want to enter the quasi-medicinal arena based on its product's content of this nutraceutical.

Cube 6 represents the worst of all possible worlds. A technically complex product with added value is positioned for elusive consumers in a market foreign to the developer. The best example may be added-value products such as cheeses made with medium-chain-length fatty acids for people with digestive disorders in the health care market. This situation could result when a product is repositioned or a new market niche opens for it. The risk is high, development costs are high, and promotion can be difficult. An example of a successful product in this situation is ARM & HAMMER Baking Soda, for which marketers exploited a rediscovered property possessed by an old product (using baking soda as a deodorant, dentifrice, or bath powder).

Figures 1.1 and 1.2 can be compared using potatoes. Industrial products certainly drive technical complexity far to the right and at the same time push market complexity into new market areas, for example, the food ingredients and hotel and restaurant markets. Reformulated chips and other products demonstrate increased technical complexity (Cube 3) and represent a leisure snack market (Cube 4) but also a more elusive consumer, that is, the upscale, older consumer (Cube 6).

A. PRODUCT LIFE CYCLES

Every product has a life cycle, as depicted in Figure 1.3a. The horizontal axis is a measure of time. The vertical axis is an index of a product's acceptance, either volume of cases sold or sales dollars. Here case volume of product sold was chosen to follow life cycle. Five distinct phases of the life cycle can be discerned:

1. The introductory period is heavily supported by promotions, in-store demonstrations, and advertising to gain introduction. Sales volume is initially low as customers and consumers are educated about the product.
2. A strong growth period ensues when first-time customers begin repeat buying and new consumers are attracted. There is a positive acceleration of sales growth. Growth continues as new markets open, but continued promotion and expansion at the introductory pace are costly.
3. A decline in the growth of the sales volume begins. Growth accelerates negatively.
4. Next comes a no-growth period. Sales are constant, a sign of a stagnating market.

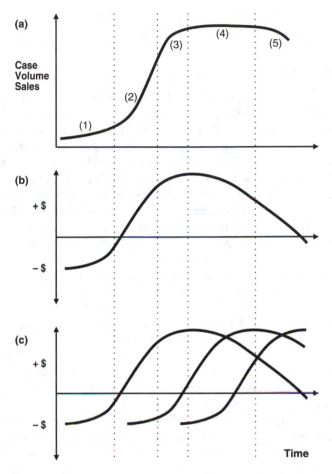

FIGURE 1.3 Characteristics of products, their life cycles, and profitability. (a) typical product life cycle; (b) the profit picture; (c) the contribution of new products to profitability.

5. A decline in volume begins. Newly introduced competitive products adversely affect sales; customers and consumers become indifferent to the old product. Promotions prove too costly to maintain sales volume.

Life cycle curves can be generated for product categories as well as specific products within a category. Instant coffee, as a product category, could be described as still in the growth phase. Nevertheless, the leading brands of instant coffee have changed places as their manufacturers go through different stages of the cycle at any given time. The sale of flour had for years been in a no-growth phase that was only slightly ruffled by the advent of cake mixes; now it is enjoying modest growth as many households are returning to the art of baking. During a period in the 1970s when meat prices soared because of a scarcity of beef, the sale of meat substitutes and extenders grew dramatically, and then plummeted drastically when meat became plentiful and prices fell. Meat substitutes never reached a no-growth phase (Phase

4); they just plummeted and their life cycle could be described as a spike. Life cycle curves are as varied as the products they represent.

B. THE PROFIT PICTURE

More revealing of the success of a product than sales is the profit brought in by those sales. The introductory phases will have minimal net profit (see Figure 1.3b, Phases 1 and 2) because these bear the costs of past research and development as well as the heavy costs for promotion, market penetration, and getting shelf space (slotting fees). There are improved profits during the latter part of the growth phase (Figure 1.3b, Phase 2). The improvement continues in Phase 3, but toward the end of this phase profits begin to drop off as costs for market expansion and costs to support the product against the competition begin to take their toll. During the no-growth phase (Phase 4), profits barely cover costs to support the product. The company eventually sees the product as an unprofitable item that costs too much to maintain. Manufacture of the product ceases.

To maintain the viability of the company, replacement products must be ready to launch to continue the flow of profits. In Figure 1.3c, two additional products have been launched by the company. These products maintain the company's net profit picture in a healthier state; the profits derived from each new product are additive, but so are the expenses of research and development and promotion. The smart company keeps a series of new products in various stages of development. Bogaty (1974) suggests that for every one product on the national market, two should be in test marketing. For each of those two, there should be four in the last stages of consumer testing. He works this idea backward, ultimately stating that 32 product ideas should be screened for development consideration. The cumulative profitability of these new food products promises, if they are successful, a good return on investment. To stay healthy, a company needs to constantly research markets and marketplaces for new product ideas and maintain in its portfolio many new product ideas and projects that complement its findings in market research.

IV. WHY UNDERTAKE NEW FOOD PRODUCT DEVELOPMENT?

The previous section highlighted two reasons to develop new products. First, they do not last forever. Second, they contribute enormously to a company's continuing profit picture.

Customers and consumers are deluged with new food products each year. Figure 1.4 presents statistics on new food product introductions (Friedman, 1990; Kantor, 1991; Harris, 2002). New product introductions are plotted on a logarithmic scale against time. The number of introductions, according to this data base, peaked at the astonishing number of 15,000 products in 1994 but has dropped precipitously since then. Harris (2002) ascribes the decline to several factors:

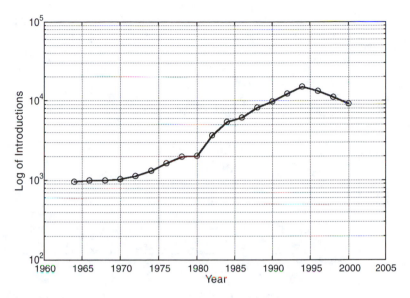

FIGURE 1.4 The rise and apparent decline of new food product introductions (logarithmic scale).

- There has been a consolidation of many food companies and hence a diminishing number of product lines. However, many spin-off brands have been sold and become the nucleus of new and smaller companies.
- Better market and consumer research have helped companies remove potential product failures much earlier in the development process. (Harris attaches great importance to efficient consumer response technology for this early culling of losers.)
- Retailers are giving more attention and space to their own private label products. This, in addition to slotting fees and promotional allowances, discourages many companies, especially small companies, from competing with new products.
- There may be product saturation in some product categories.

This graphic presentation of the data (Figure 1.4) and its interpretation are somewhat misleading. Figure 1.5 plots the percent change in the number of introductions in a 2-year period over those of the previous 2-year period. (This graphing procedure is discussed in Cleveland and McGill, 1984.) Here the rate of introductions of new products is seen to have been declining since 1982 and not just since 1994. This decline may be more readily observed from the dotted "best fit" curve (a fourth-order least squares fit).

Was the 80% increase of 1982 over 1980 simply an outlying spike? Such spikes are the bane of those familiar with control charts who have occasionally encountered these anomalies on process control charts. Such aberrations cause the quality control inspector to rush into the production manager's office screaming, "Stop production,

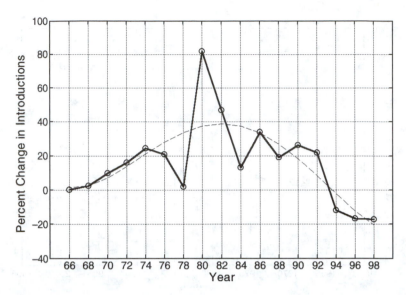

FIGURE 1.5 The percent biannual increase or decrease of new product introductions over several years.

the line is out of control!" Resampling on the line, of course, shows everything to be "in control." Were definitions of what constituted a new product introduction consistent throughout the data gathering?

The decline in new product introductions started much earlier than Harris suggests. Harris's reasons for the decline may very well be valid, but more reasons to mark the decline in new product introductions in this earlier period must be sought. Besides mergers, companies at this time (the 1980s) were getting down to their core businesses; that is they were downsizing, and often one of the first groups to be downsized was the research and development department. Hence, it might be suggested that the dismissal of technical staff and the break-up of technical departments may have presaged the decline in new product introductions. Whatever the reasons, they are not important to the discussions to follow.

However the decrease in introductions is viewed, whether beginning in 1994 or 1982, customers still see several thousand new food products entering the marketplace each year. Obviously, the sheer number of introductions shows that companies believe that new products are important to their economic futures.

Unfortunately few of these thousands of new product introductions will be rewarded with repeat purchases or even last a year from their introduction. They fail for a variety of reasons. The failures represent the ultimate screening for a new product, that is, no interest by the customer or by the consumer through the customer.

The odds against success are disheartening. Estimates of the failure rate range from 1 in 6 to 1 in 20. Skarra (1998) reports that 1 in 58 new product ideas develops into a successful new product. In my own experience, gained over a 4-year period with one company's product development program, for each product that went into test market, 13 others had received some development at the laboratory bench level or in the pilot plant before being rejected. Clausi (1974) estimated that in one 10-

year period General Foods Corporation conceptually tested, developed, and undertook home-use testing on more than seven products to get one considered suitable for test marketing. Less than half of those introduced to test markets were eventually successful. This leaves an astonishingly small number of products that will achieve their developers' goals of market placement.

Estimates for the number of new food product failures range as widely as estimates for new product introductions. Any estimate is beset with imprecision. At which stage was the estimate made? At the laboratory bench stage, in a mini-test market, in a regional test market, or a year after product launch? Was the estimate made after the product failed to reach a satisfactory market share? A "satisfactory market share" is a criterion established by the company's management and may not represent an unsuccessful product for customers but merely one that did not meet management's objective. Does loss of a product idea during in-house screening or test market count as a failure? Is a product that is successful in a regional market but fails nationally a failure? Or is one that just plods along growing slowly and steadily but fails to reach the profit targets in the time established by management (both of which may have been unreasonable targets) a failure? At what point during development and on what basis was the decision made respecting success or failure? Therein lies part of the imprecision for calculating failure rates.

If new food product development is fraught with so much difficulty, if it is so costly, and if it has a high rate of failure, why go into it? Would it not be simpler to save the expenses and coast along with the existing products? As Pyke (1971) stated at the beginning of this chapter, much science representing time and money has gone into this endeavor. The failure rate in new product development is, indeed, horrendous. The rewards, however, can mean the continued profitability of the company.

It would certainly be simpler but not profitable for very long to ignore new product development. Food companies have to be profitable to grow and to survive. "The engine which drives Enterprise is not Thrift, but Profit" (John Maynard Keynes in *A Treatise on Money*). New food products are one of the major avenues open to a food company to be profitable and survive. Some marketers would argue that new food product development is the only path the food company can follow for survival.

A. THE "WHY" OF "WHY UNDERTAKE NEW PRODUCT DEVELOPMENT?"

The need for new food product development is driven by five dominant forces:

- All products have life cycles. Eventually they die and must be replaced or reinvigorated by heavy marketing or see consumer rejuvenation if the brand or the manufacturer is to survive.
- New products offer the opportunity for aggressive growth to satisfy management's long-range business goals.
- New markets may be created, for example, organic or functional foods markets and companies are tempted to enter with their products. Marketplaces may change, for example, e-commerce has emerged, requiring new products more suited to respond to the changes.

- New knowledge and advancing technologies make technically feasible food product ideas that were once considered impossible. New knowledge in the health sciences suggests opportunities for new food products suited to the healthy lifestyles of today's consumers.
- Changes in government legislation, health programs, agricultural policy, or agricultural support programs could require the development of new food products.

1.　Corporate Reasons for New Product Development

Senior management, under the direction of the owners or the shareholders, creates a corporate business plan that sets out specific financial and growth objectives. Food companies can achieve growth in a limited number of ways, such as:

- Expanding into new geographic markets: This can be expensive and risky if the expansion is into a territory where a competitor has a strong foothold. For products with short shelf lives, the distribution system and its costs may limit such an expansion. Research and development will be necessary to increase shelf life. Export markets present their own unique hazards including the need to accommodate to local taste preferences.
- Achieving greater market penetration with a greater market share in existing markets: If the market is dominated by the competition, then this approach means slugging it out with competitors. Large sums of money are required for advertising and promotion.
- Developing new products: New markets can be opened up and contribute to growth and profitability (see Figure 1.4c). The umbrella of brands expands. Such products must be developed to bring in new profits. There are, however, the costs of their development and marketing to consider.
- Acquiring rivals (the competition) or smaller companies with similar or complementary products in widely separated regions of the country: A company can expand its brands into new markets and use those of its acquired company. (This possibility for growth is beyond the scope of this book.)

Each avenue above comes with expenses and problems, but each may be required to strengthen the company's profit position.

Management seeks ways to meet its financial objectives by reducing expenses and overhead costs. This is usually anathema to new product development. Expenses are reduced by:

- Reducing staff: This is explained by the company getting to its core business. The research and development staff are often the first to go.
- Cutting costs: This can involve implementation of an energy conservation program, altering benefits programs and cutting wages, restricting travel, outsourcing some functions, etc.

- Improving processing efficiency by adopting a waste management program.
- Adopting a sound process and quality control program to reduce line losses, reduce substandard quality, and control overfill, rework, and product returns.

These thrift measures may help the company's profitability but are of limited value for growth. Profit drives the enterprise, not thrift, to paraphrase Keynes.

When a plant operates seasonally, or has a slack season, its management has an incentive to even out production throughout the year. A plant operating year-round is more profitable than one idle most of the year. The slack season can be used to produce new products, putting the underutilized plant to work. This keeps trained workers employed throughout the year, reduces plant overhead, provides a more steady cash flow, and benefits the community.

2. Marketplace Reasons for New Product Development

Food purveying is taking many new forms. Warehouse stores, mail catalogue shopping, and teleshopping are changing customers' buying habits. The traditional supermarket is adopting a new concept in retailing: it is becoming a collection of food boutiques. Small mom and pop stores are becoming 24-hour convenience stores. The abundance of restaurants, diners, take-out food outlets, mobile canteens for work sites and the military, and deli counters in supermarkets makes ready-to-eat (quick serve or speed scratch) food abundantly available.

The diversity of customers and consumers in the many different marketplaces can be daunting; both are quixotic, impulsive, and fickle. The customer is the daily shopper at green grocers, butchers, and bakeries or the more conventional once-a-week shopper in the supermarket; the customer is also a restaurant patron, the commissary of a fast food chain buying its pickles and relishes in the food service arena, or the government buying in the institutional marketplace for military or penal institutions. Because the needs of the customer and the types of marketplaces change for a variety of reasons, food manufacturers serving those buying and selling sectors need to respond quickly to changes.

The profile of customers and consumers in any marketplace, but in particular in the retail marketplace, is constantly altering. The result is a changing pattern of buying habits. Many factors cause this:

- Population movements bring changes in the ethnic backgrounds of neighborhoods as immigrants bring their food habits with them. With these movements come consumers with different food needs.
- Populations age and children move away from home to seek their own livelihoods. Young couples just starting their careers, some with young families, move in, and the economic cross section of a community may change.
- Downtown neighborhoods, left empty because of movement of young people with children and businesses to the suburbs, become yuppified as developers move in. They are reborn as fashionable areas for young

professionals (often referred to as DINKs, dual income no kids) who have different food needs and habits.

- Companies change. Some fold or are absorbed by larger companies, and the result causes economic change in a neighborhood. Product mixes change and the companies (customers) change their suppliers.

No matter where the marketplaces are, they are in a constant state of flux with respect to their customers and consumers; the ethnic make-up, incomes, education, and lifestyles of customers and consumers in any community change. Local industries modernize, move, or fade away, and economic values change. All these changing factors determine what market niches for food products develop and what marketplaces are suitable for placing products in these evolving neighborhoods.

Another factor in the marketplace that influences its dynamics is the competition. The launch, by a competitor, of an improved product requires reaction from companies with similar products whose sales may suffer from this introduction. A new food product launch by a company requires retaliation on the part of other companies with similar, competitive products. Retaliatory action may involve new pricing strategies, promotional gimmicks, or the development of new products to combat the competitor's intrusion.

The food manufacturer must accept fluidity in the marketplace as a challenge. No one product is expected to answer all the demands of customers and consumers forever. Only a battery of new products will suffice to satisfy emerging market niches. The many changes in the marketplace can be great motivators for product development.

3. Technological Pressures Forcing New Product Development

Scientists are continually providing new knowledge about the physical and biological worlds. Technologists then translate this knowledge into new processing technologies and products with which to assist daily living.

With computers linked to information databases, any company can access vast quantities of business and technical information that previously was accessible only in specialized libraries in major urban centers. Expert technical information to assist in development programs is available to the smallest, most remote food processing company.

Advances in food packaging provide good examples of the impact of scientific discovery and its application. At one time the packaging industry relied primarily on tin-coated steel, glass, and aluminum, and the mainstays of packaging were the three-piece can with its lead-soldered side seam and the glass jar. This can gave way to the two-piece seamless can. Materials such as tin-free steel; new plastics; coated paperboard; composites of aluminum, plastic, and paper; and even edible food cartons have permitted the manufacture of a wide range of containers with unique properties to preserve and protect the high-quality shelf life of foods. There are now bio-based packaging materials with antibacterial properties (Shahidi et al., 1999; Petersen et al., 1999; Han, 2000). Protective packaging for containers is now microwavable, edible, biodegradable, and recyclable.

Greater knowledge of food spoilage mechanisms and preservative technologies has given rise to new products with better stability, quality, and nutrition. Improvements in retorting technology and equipment have elevated "canning," the workhorse of food preservation, to a technique that gives added value through improved texture, color, and flavor. New ingredient technologies can protect flavor (e.g., encapsulation), texture, and other quality attributes of products and maintain these through strenuous processing procedures.

Advances in nutritional knowledge have produced a growing awareness of the relationship of food to health and specifically of the function of certain foods in the amelioration of disease or its prevention. This has spawned a new class of preventive foods: foods to prevent cancer and heart disease, delay aging, and cure depression. Food products are now recognized as much for the absence of something (sodium, cholesterol, saturated fats, refined sugars, or high calorie counts) as for the presence of something (calcium, fiber, mono- or polyunsaturated fats, antioxidant vitamins, or nutraceuticals).

Technology has changed the various marketplaces and all the players in them in many ways:

- Customers and consumers have become more concerned about their health and the role of food and nutrition in their health and in the prevention of disease. (But even here there is a certain quixotic element.)
- A growing number of more discriminating shoppers are now aided by food nutrition labeling, in-store computers providing information about meal planning and recipes, and home computers on which they can comparison shop from the convenience of their homes and widen their horizons for more venues to shop in.
- Social scientists have developed better techniques to research and understand customer and consumer behavior and emotions. Such methodology provides marketing personnel with improved skills to conceive new products for customers and consumers and to develop better communication with customers and consumers.
- Food manufacturers are becoming adept at successfully developing new products based on these concepts for the various marketplaces.
- Retailers are using their knowledge of buyers' behaviors to attract customers (e.g., the use of food odors, in-store tastings, delis, and entertainment) and to serve their needs (advice on foods) to the retailers' advantage.

Food manufacturers cannot afford to be unaware of developments in science and technology and the influence these advances have on customers, consumers, and food retailers.

4. Governmental Influences Pushing New Product Development

The objectives of government respecting its enactment of food legislation are summed up by Wood (1985):

- To ensure that the food supply is safe and free from contamination within the limits of available knowledge and available at a cost affordable by the customer
- To develop in cooperation with food manufacturers, responsible consumer groups, and other interested groups standards of composition for foods and labeling standards that provide adequate information to customers in order for them make intelligent choices respecting the food they purchase
- To maintain fair trading practices and competition among retailers and manufacturers in such a way as to benefit customers

How these objectives influence the activities of food technologists, marketing personnel, and retailers is obvious.

Government at the federal level, state or provincial level, and municipal, local, or country level strongly influences the business activities of food companies. At the highest level, senior governments negotiate and establish international standards for products and trade practices among countries. Some of these trade agreements are among a few countries (the North American Free Trade Alliance) or among many countries, as in the European Union (EU). As recently as 1997, the EU adopted a Novel Foods Regulation (EU, 1997; see also Huggett and Conzelmann, 1997). Article 1 of the EU Regulation describes the regulation thusly: "1. This Regulation concerns the placing on the market within the Community of novel foods or novel food ingredients." Article 2 explains the scope of this regulation as including "foods and food ingredients which have not hitherto been used for human consumption to a significant degree within the Community."

Compounding the influence of these official levels of government are two more: international bodies and the quasi-governmental agencies. These also bring regulations to food manufacturing and international trade specifically and new food product development indirectly. Some examples of regulatory bodies at the international level are:

- International Standards Organization (ISO) and its recently published ISO 9000 and ISO 14000 series of directives on quality control and management and environmental awareness, respectively (Boudouropoulos and Arvanitoyannis, 1998)
- Codex Alimentarius Commission under the joint direction of the FAO/WHO Food Standards Programme

Quasi-governmental bodies do not have the legislative powers of the various tiers of government but do have the support of government and hence the effect of law. Classic examples are the various marketing boards that exist in many countries to regulate the local supply, importation, and price of many food commodities and ingredients derived from them. Other examples are professional and trade associations. These establish rules of conduct for their members, members' wage scales, and regulations that participating parties must adhere to. Table 1.4 summarizes a wide assortment of activities associated with government, international organizations, and quasi-governmental bodies.

TABLE 1.4
Various Food Business Activities over Which Governments in Different Forms and at Different Levels Exert Influence

Activity	Influence
Fiscal policy	Interest rates for development loans
	Grants-in-aid; research funding
	Taxation policy
Patents and copyrights	Copyright protection and licensing
	Research funding if guarantee of patent protection
Trade barriers	Tariffs and protectionism
	Standards of food product identity
	Availability and cost of ingredients
Environment protection	Waste disposal
	Recycling or reuse requirements for packaging materials
	Energy utilization and disposal
Marketing and trade practices	Product or advertising claims
	Billboard and advertising placements
	Zoning by-laws
	Store hours
	Container sizes
Employment practices	OSHA and worker safety
	Unemployment benefits
	Minimum wage levels
Health policy	Nutritional guidelines
	Nutritional labeling
Agricultural policy	Support programs for commodities
	Availability of commodities
Consumer protection	Product safety; safety of agricultural chemicals
	Labeling; product names; comparative advertising
	Inspection services

Adapted from Fuller, G.W., *Food, Consumers, and the Food Industry: Catastrophe or Opportunity?*, CRC Press, Boca Raton, FL, 2001.

A few of these require further discussion. Governments provide incentive programs for development to stimulate depressed regions and industries or to spur research and development for use of specific commodities. These opportunities allow eligible companies to undertake development programs, modernize equipment, or utilize new technologies for product development.

National health policies beget nutritional guidelines, and this invariably leads to standardized nutritional labeling legislation. Food companies wanting to adopt these guidelines as company policy for all their products may have to reformulate products to conform to the guidelines.

Standards of identity or nutrient guidelines for food products differ from one country to another. Food companies with a strong export trade will require reformulation of their products to meet foreign requirements.

TABLE 1.5
Overview of the Stages of New Product Development

Holmes (1968, 1977)	Crockett (1969)	Mattson (1970)
Company objectives	Search opportunities	Idea generation
Exploration	Translation of concepts into	Concept screening
Screening	products	Preliminary formulation
Business analysis	Marketing plan	Taste panels
Development	Implementation of marketing plan	Final formulation
Testing		Trial placement
Commercialization		Fine tuning
Product success		Package design
		Copackers
		Mini-market test
		Symbiotic distribution

Oickle (1990)	Graf and Saguy (1991)	Skarra (1998)
Exploration	Screening	Assessing management
Conception	Feasibility	commitment
Modeling (prototypes)	Development	Finding the right idea
Research and development	Commercialization	Developing the business case
Marketing plan	Maintenance	Development and
Market testing		commercialization
Major introduction		

Governments also regulate consumer protection and product safety. For example, several years ago, the U.S. government banned saccharin and cyclamates for safety reasons. Manufacturers of dietetic foods and low-calorie soft drinks reformulated their products. Some companies found suitable noncaloric alternatives to reformulate with. Other companies developed canned fruit packed in water or packed in their own fruit juices. That is, new product development was forced upon manufacturers through government regulatory changes.

Governments at all levels can have a tremendous influence on new product development. An excellent overview of the complex of food legislation and regulation in the U.S is presented in Looney et al. (2001).

V. PHASES IN NEW FOOD PRODUCT DEVELOPMENT

Most authors agree that new food product development can be divided into several distinct phases (Table 1.5). Few agree on the number, order, or names of the phases. For instance, Meyer (1984), not tabulated here, does agree with Mattson (1970) on 11 steps. Authors often then divide the stages into substages and even sub-substages. This is not a sign of confusion in analyzing and attempting to understand the process of new food product development. Rather it represents an evolving change in the thinking or philosophy of new product development. This change is well described by Earle (1997a), who describes four main stages in the development process:

- Development of a business strategy that describes the project
- The research and development phase of the project including the manufacturing design
- Development of programs for marketing, production, and quality assurance
- Organization of the production and distribution for the launch and the analysis of the launch sales data

Rather inappropriately, screening is depicted as a series of sequential stages in many descriptions of product development. Certainly there is some sequencing; a business plan must precede all else, and closure is a successful launch. However, when authors describe the phases of product development as a sequence, a one-after-the-other cascade from ideas to a final finished product, they both misstate and misinterpret the process. The phases do not start, proceed, and then finish as the next phase begins. The phases are not, strictly speaking, sequential: they often overlap and are concurrent. Projects might even return to the conceptual phase for a complete rethinking as new information arising from the development process becomes available (see, e.g., Bradbury et al., 1972).

Linnemann et al. (1998) have attempted to organize the development process into an integrated system. They recognize "seven *successive* steps" (italics added):

- Analyze socioeconomic developments in particular markets.
- Translate preferences and perceptions of consumers into consumer categories.
- Change consumer categories into "product assortments."
- Group "product assortments in product groups in different stages of the food supply chain."
- Identify processing technologies required for particular product groups.
- Analyze the state of the art in required processing technologies.
- Compare the state of the art of required processing technologies with future needs.

Linnemann et al. describe their model in some detail and conclude that "dedicated production systems that follow more closely market dynamics" are required; a structured, integrated approach is needed for the efficient use of both knowledge and labor required in product development, and integration of knowledge from many different areas of technology is required. They conclude that relationships among markets, consumer behavior, the variety of food product, and processing technologies are inadequate and require further refinement.

The starting point most generally agreed upon is to establish company objectives and to identify customer and consumer needs (Figure 1.6). Company objectives must be established in order for all personnel to know what is planned and why. This

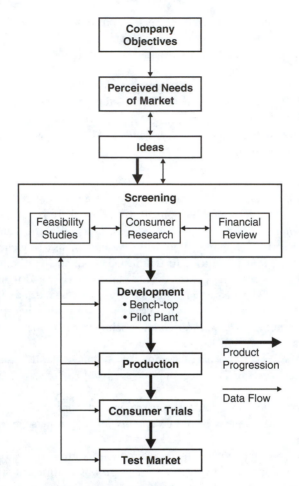

FIGURE 1.6 Phases in new food product development.

clarifies what senior management's goals are and underscores their determination and dedication to these objectives. This precedes everything else.

With this clarification of purpose, marketing personnel determine what new products would meet these company objectives. To do this, the perceived needs of both targeted customers and consumers must be determined through market and marketplace research. One does not create new products in a vacuum without a knowledge of the needs of customers and consumers. Unfortunately this lack of understanding is often the environment for product development in small companies where the boss creates new products in a vacuum. The thick arrows in Figure 1.6 show the advance of product ideas and later the product itself. Thin arrows indicate flow of data and information.

The next phases winnow all the ideas, reducing their number to a manageable few that are deemed the most worthy. Three parallel screening criteria are used:

- Is the idea feasible within the time frame required by the marketing department and dictated by senior management and within the skills level of all the responsible departments? This answer should be provided not only by the marketing department but also by the manufacturing, engineering, and research and development departments.
- Does the idea meet perceived consumer and customer needs? Does it resolve the conflict of "I want" vs. the customer's "I need"? Further market research will determine how this dichotomy is to be resolved.
- Will a financially sound business plan based on these new products stand up to critical analysis and meet objectives set by management?

The technical skills of the research and development department are brought into play as they proceed to create bench-top prototypes that match the product statement as closely as possible. The next few boxes in Figure 1.6 might suggest that research and development are dominant elements. This is an incorrect interpretation. As development proceeds, information is used to make more informed decisions. Such information aids in the development of standards for raw materials, ingredients, and packaging requirements and on process design and equipment requirements. All are needed for costing purposes. Decisions on the acceptability of prototypes through taste panel studies, for example, are constantly fed back to the technical development group.

A series of parallel events (not depicted in the figure) begins, based on the data obtained throughout the development process. An analysis of the business plan is now refined by the financial department with more complete information on ingredient, processing, and marketing costs. Sourcing of ingredients and packaging materials is carried out. Marketing people prepare draft labels and label statements, refine consumer analysis, plan marketing strategy, and develop promotional material for use in newspapers, flyers, radio, and television. The manufacturing department determines in-house production capabilities and manpower requirements.

As data produced at each phase are transformed into useful information, more positive decisions must be reached. Go or no-go decisions must be made. If the concept has to be changed, so does the product statement. It is quite common to see many changes in the direction of product development, and this results in numerous returns to the drawing board. All are necessary based on the accumulation of more accurate information. Development is a constantly evolving process keeping pace with a changing target.

By the time production samples have undergone consumer trials, management should be able to decide whether to go into a test market. Marketing, depending on the wishes of the company, may conduct mini-market tests, market tests in only one or two cities, or go directly into a regional launch (Figure 1.6).

The final phase of any new product development is an evaluation of that test. If the test market was unsuccessful, then the weaknesses must be determined and corrected before the next new product is developed. If the test market was successful, then the reason for its success must be determined. The strong points of the company's progress through the development process must be recognized. Such

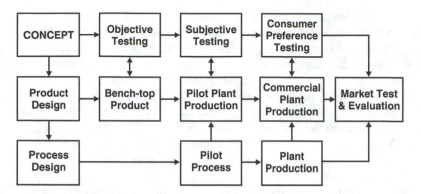

FIGURE 1.7 Idealized representation of activities flow in product development.

recognition (i.e., learning) allows the company to capitalize on what they did right or to avoid errors for application in future development programs.

Development is a progression from the intangible, the product idea embodied in a product statement or concept, to the tangible, the actual product with all the attributes stated in the concept and ready to be tested in the marketplace, the final screening. Food technologists begin by designing a product based on a concept statement and alter the product according to the results of sensory and consumer evaluations.

Figure 1.7 is a more realistic overview of the process shown in Figure 1.6. The upper flow in Figure 1.7 depicts efforts that are largely the responsibility of the marketing department.

Food technologists formulate a tangible bench-top, or prototype, product based on the product statement (middle flow in Figure 1.7). Marketing researchers use this to stimulate consumer research for further refinements of the concept or use it in mini-test markets for the same purpose. With a prototype product in hand, sensory evaluations can begin to guide food technologists in formulation refinements. Preliminary product safety and shelf stability testing may require further alterations both to the product and the concept in these early phases.

The bottom flow in Figure 1.7 is largely the domain of the engineering and production departments. Engineering and production personnel design processes for new products, incorporating changes determined by food technologists. At this stage, reliable sources for ingredients and raw materials should be found and product cost data confirmed. Potential copackers can be evaluated if it is thought that the product cannot be made in-house. As development progresses from left to right in Figure 1.7 there is a constant interplay and exchange of ideas between the other two flows. Each shapes and molds the other streams.

All development from idea generation to test market and even to a national launch must be viewed as one long, ongoing screening process. Screening is not a stage in development — it is synonymous with development. Each stage of development brings further data that, when translated into information, provide development teams with more refined tools with which to screen. As information is gathered from marketing personnel, technologists, or suppliers, the product becomes identifiable with the needs and demands of consumers. The result is the continuing interplay of market research, technology, and financial efforts of companies to produce the right products at the right prices for customers and consumers (Blanchfield, 1988).

The purpose of the development process is to move a desirable product to market with the least amount of uncertainty respecting its probability of success in the segment of the marketplace where it is to compete. Market, financial, and technical research (and their associated analyses) are nothing more or less than screening procedures for reducing, as much as possible, uncertainty in the development process.

2 The Generation of New Product Ideas

To as great a degree as sexuality, food is inseparable from the imagination.

Jean-François Revel (as cited in Robbins, 1987)

I. GETTING IDEAS

A. GENERAL GUIDELINES FOR IDEAS

Thinking up ideas for new food products is easy. There have been several excellent ideas both for ingredients and food products in the scientific and technical literature, e.g., leaf protein (Pirie, 1987), protein from exhausted bee bodies (Ryan et al., 1983), sugar and citric acid from corn cobs and husks (Hang and Woodams, 1999, 2000, 2001), and wine from cheese whey (Palmer, 1979). These are some of the more unusual ones. These products had a purpose; there was more than just idle scientific curiosity behind them. But was there a demand for the products derived from them? Who needed them? With the number of ingredients made from milk (see Chapter 10) and the plentiful availability of excellent wines made from grapes and other fruits, where was the need for a Château du Petit Lait du Fromage?

They were developed on a "it's a shame for these by-products to go to waste" reason. They were created to explore the utilitarian value of a waste product; this is itself valuable for environmental reasons where producers are faced with the problem of disposing of these by-products. The wine was ostensibly to demonstrate new and improved methodologies for pollution control since whey is a major waste product causing pollution. But all were product-oriented ideas and not consumer-oriented ideas; they led to producer- or technology-driven products, not to consumer-driven products. They are all technically feasible but derived in isolation from the real world of customers and consumers. They are examples of what I have described as "Little Jack Horner" research after the nursery rhyme character who "stuck in his thumb and pulled out a plumb and said 'what a good boy am I.'"

On the other hand, a Reuters (2000) report describes work in Brazil to create a diesel fuel from soy oil and sugar cane. Both sugar cane and soy beans are major crops in Brazil (Brazil's soy bean production is second only to production in the U.S.), and prices for this commodity are falling. The development of this diesel fuel was prompted by economic (and political) reasons but it is successful because there is a need. Opening up a new market for products that are falling rapidly in price and creating self-sufficiency in diesel fuel are the needs.

Thinking up ideas is *not* the problem. The problem, simply stated, is getting ideas that fit the following criteria:

- They satisfy the needs and desires of consumers and at the same time they attract customers, the gatekeepers to the consumers. The seed of an idea must come from those who will need and use the product.
- The ideas must also be within the skill level, technical capabilities, and managerial and financial resources of the producing company. These are often referred to as the *core competencies* of the company, that is, what the company does best. Ideas must be implementable.

There are no "perfect ideas" for new food products. The problem is not one of undertaking the further development of a perfect idea; it is the pulling together of the skills and resources to bring good ideas to fruition in the marketplace. Good ideas require that the company have the ability to take advantage of them with products that they ultimately introduce to their targeted marketplace. The company must be able to market and sell the product (being able to produce the product is not always necessary). The company finds itself in a dilemma trying to balance ideas based on market research of perceptions of customer and consumer needs against company skills and capabilities.

New product ideas are not likely to succeed if companies do not answer all consumer-oriented information generated from the marketplace and fully employ their strengths.

Earle (1997b) sees two prerequisites for successful innovation: a company that is innovation oriented and a positively responsive environment. The environment both within the company and external to it must be conducive to and accepting of innovation. Earle treats the subject from a general, conceptual, and philosophical point of view.

B. Sources of New Food Product Ideas

Food companies have three sources for ideas for new products (Table 2.1). These are only sources of ideas; the ideas must still satisfy the needs of customers and consumers. This detail of satisfying a need will be found only within the market-places where the customers and consumers are to be found with research that is done by marketing personnel.

Pressures (see Chapter 1) that drive a food company into new food product development are also sources of ideas. New technologies such as extrusion process-ing technology opened up a wide vista of ideas for snacks that compelled snack producers to develop new snack ideas. The development of a greater understanding of the principles of water activity and hurdle technology offered consumer product manufacturers new categories of semimoist foods ranging from dog food to fruit leathers to chilled foods; manufacturers looked closely at these new technologies for new product opportunities. Even the need to reduce the waste management costs by upgrading a useful by-product into an added-value product motivates a company into doing research to find new uses for the by-product *if* the needs of customer and consumer alike are satisfied by doing so.

TABLE 2.1
General Sources for New Food Product Ideas

General Source	Specific Impetus Providing Inspiration
The marketplace	Market research to identify customer and consumer needs; results of consumer profiling
	Retailers providing details of customers' buying habits and expressing their own requirement for products
	Distributors expressing their requirements for products and problems they encounter with handling
Within the company	Sales force's interaction with retail buyers and with individual customers in stores and their observations of competitive products and their placement within stores
	The need to recycle or reuse by-products more productively and efficiently to protect the environment and to reduce costs of waste removal
	Spontaneously generated ideas from employees
Outside the marketplace	Advances in science, technology, and nutritional knowledge
	Competitive intelligence gathering
	New market openings
	Review of food and cooking literature
	Review of technical, trade, and scientific literature
	Exposure to new sources of raw materials, of newly developed ingredients, of new equipment, and new consumer products at national and international food trade shows
	New government food legislation

New product development is neither an exercise to gussy up a waste product nor a means to display a company's technical skill to the consuming world. One does not start with either a waste by-product or a new widget maker and wonder what to do with it. If these tactics do satisfy a perceived market need, this is fine, but caution is advised. Technical innovation, *per se*, will never sell a product except to other technocrats. Nor is new food product development meant, necessarily, to be novel for the sake of novelty. New food products are meant to fill perceived needs, satisfy the desires of customers and consumers, and meet the expectations of those consumers for those products. The products must deliver what the company has promised in its promotional campaigns for these products.

1. The Many Marketplaces

The customer, frequently with the consumer in tow, is found in many diverse marketplaces (Table 2.2). This fact, logically, would make the marketplace the best site in which to find out what satisfies the perceived needs of at least one of these entities, the gatekeeper.

The observation that hits most is the great diversity of marketplaces that there are and with this the great diversity of consumers. Researchers must first decide in which marketplace the new product will be positioned and how best to conduct research in that marketplace. How one attempts to uncover the "perceived needs"

TABLE 2.2

General Classification of Marketplaces Selling Food Products and Ingredients

General Classification	Examples of Types
Artisan-based	Farm gate, farmers' markets, country fairs
	Butchers, greengrocers, specialty bakeries, cheese shops, fishmongers, wine stores
	Carriage trade and specialty gourmet stores such as coffee, tea, chocolate, and confectionery stores
Grocery (food) store	Family-owned-and-run small grocery stores and large supermarkets, often with artisan-based stores (butchers or greengrocers) within them
	Special food stores (organic foods, health food stores, stores catering to religious observances)
	Box stores (large sizes or case lots)
	Bulk stores (products are loose in bins)
Convenience store	Paper/magazine stores, tobacconists
	Drugstores* (chocolate bars, snacks, and small units of canned goods)
	Department stores,* often with a specialty food section
Food service	Full service restaurants; fast food restaurants with sit-down and take-out service; hotels; roadside diners; bars, especially bars with happy hours; coffee and tea houses
	Take-out restaurants, mobile street vendors, walk-up kiosks, vending machines, cafeterias (work sites, schools), mobile canteens
	Gourmet commissaries (Chinese dim sum, Indian curries, French croissants and patés)
	Transportation meals (in-flight and on-train meals)
	Institutionalized care feeding (nursing homes, hospitals, short-term care facilities), penal institutions
	Personal chefs, rent-a-chefs, home-care feeding for elderly
	Military feeding (messes, cafeterias, field kitchens, combat rations)
Industrial arena	Food trade exhibitions with demonstrations
	Supplier-sponsored short courses demonstrating special ingredients
	Sample distribution to industrial users
Electronic and postal services	Catalogue ordering services

* Local retailing legislation may restrict the sale of foods in these stores.

Adapted from Fuller, G.W., *Food, Consumers, and the Food Industry: Catastrophe or Opportunity?*, CRC Press, Boca Raton, FL, 2001.

of customers and consumers will be very different for industrial marketplaces, consumer product marketplaces, or hotel and restaurant marketplaces.

It is advisable first to define or, at best, to describe what is meant by "perceived needs" of either customers or consumers. Perceived needs are known neither by the customers, the consumers, nor the market researchers. Market research discovers a gap in the market, a need; the customers, or the consumers, when presented with the product or the concept embodying this need, have that flash of blinding light and say, "That's what we've been looking for." Neither the customers nor the consumers may ever have realized this need themselves. Marketing personnel did not create that need. Market research uncovered the need for the customers or consumers to discover for themselves.

Market research is actually, then, a means of screening new product ideas. Of all the ideas gleaned from all sources (Table 2.1), only those that uncover that hidden need will be most likely to succeed and offered to the technologists for development into a tangible product. The others will be rejected.

It is within these marketplaces (Table 2.2) that food manufacturers must attract customers and consumers with new products. Food retailing, wholesaling, and industrial sales have changed in response to the demands brought on by the changing lifestyles of customers and consumers in these marketplaces and by the very diversity of the marketplaces themselves. Each presents opportunities.

All information about the desires, buying habits, and any other characterizing traits of customers and consumers helps developers define products for the particular targeted consumer. As a result of the information, the needs and expectations of that targeted customer and consumer are better understood.

2.   Getting to Know Them

The simplest beginning to getting to know about customers and consumers is with census data. Demographic and psychographic information about the vast population of customers and consumers are essential to this understanding. This is readily available data but its inherent weaknesses for obtaining targeted information must be recognized.

a.   Census Data

In many countries demographic information (census data) provides statistical data about population size plus several other characteristics, usually at 10-year intervals. The information often includes the age distribution of the population, income distribution, the number of family units as well as nonfamily units, the number of children per family unit, the number of single-parent families, male to female ratios, the ethnic backgrounds of different geographic regions, whether families rent or own their homes, and so on. By itself, demographic information is not very inspiring for specific product ideas, but there are some (Foot with Stoffman, 2001) who believe that not enough effort has been made to use it for predictive purposes.

Taylor (2002) used Canadian census data to compare the top seven countries serving as sources for immigration into Canada for the years 1961 and 2001. Table 2.3 displays the countries from largest to smallest numbers of immigrants.

When the numbers of immigrants are examined, the extent of immigration and its impact on the food microcosm becomes more apparent. In 1961, the Italians, the largest group immigrating into Canada, amounted to fewer than 15,000 individuals. In 2001, the largest group immigrating into Canada, the Chinese, amounted to a little over 40,000 individuals. Two distinct observations are apparent from the data:

- The rate of immigration has obviously increased in this 40-year period. More people are on the move now than 40 years ago for a variety of reasons.
- The pattern of immigration has changed dramatically. Europe is no longer the origin of most immigrants; Asia has become the dominant source of immigrants.

TABLE 2.3
Top Country Sources for Immigration into Canada, 1961 and 2001

1961	2001
Italy	China
United Kingdom	India
United States	Pakistan
Germany	Philippines
Greece	South Korea
Portugal	United States
Poland	Iran

The exciting factor is that immigrants bring diversity with their cultural traditions and ethnic food customs wherever they settle. This diversity is exhibited by the arrival of restaurants and stores stocked with food products reflecting their food habits and cooking styles. Here there are ample opportunities for new food product ideas for any food manufacturer. The impact of immigrants on the food supply is such that their foods, once viewed as exotic, eventually become mainstream, as occurred with Italian and Chinese cuisine.

Another example of the value of demographic data to provide insight into population profiles and generate new product ideas can be seen in school registration records. The cultural diversity for five centrally located schools in Montreal is presented in Table 2.4. Any baker, butcher, grocer, restauranteur, delicatessen owner, or supermarket in these school districts who ignores the richness of the cultural diversity in their immediate vicinity by not providing food products, especially child- or teen-oriented snack food items reflecting the community profile, does so

TABLE 2.4
Cultural Diversity in Five Centrally Located Montreal Schools

District	Students from Other Cultures[a] (major mother tongue spoken in school)
SK [St. Kevin's]	95.2% (Tagalog 53%)
HB [Henri Beaulieu]	94.3% (Arabic 79%)
B [Barclay]	94.8% (French and English 57%)
CL [Camille Laurin]	94.3% (French and English 53%)
LV [La Voie]	93.6% (Tagalog 75%)

[a] Children born outside Canada or born to at least one immigrant parent.

From Lampert, A., *The Gazette, Montreal*, A1, Oct. 9, 2002.

at its risk. If there are children from a diverse ethnic background, there must also be parents who shop for foods required in their cuisine. This opens the door for consumer food product manufacturers and suppliers of ingredients and raw materials for these ethnic products.

These Canadian census figures are typical of population changes noted in other countries and reflect movements of people bringing their ethnic cooking traditions with them. They modify the available local products to suit their tastes. Whether this is called the localization of ethnic tastes or the "ethnicization" of local foods is immaterial. These people represent new product ideas and opportunities for the introduction of new products meeting their needs.

Per capita consumption figures are calculations based on what food has been produced and imported minus what was exported minus what, if any, was stockpiled and divided by the population. They are not consumption figures *per se*, but are "disappearance" figures; these foods disappeared or were used somehow and somewhere in the country. They provide a poor and indirect reflection of either customers' purchasing habits or consumers' usage habits and no information at all on the habits of specific segments or regions of the population. They purport to pertain to the average consumer, an entity that does not exist.

The term *average consumer* is often seen in the literature. Yet, it is so highly inaccurate that it is meaningless. An average is a statistical construct with significant meaning only if it is representative of a homogeneous group; that is, the members of the group have recognizably similar traits. Even a superficial reflection on the customers and consumers seen in any marketplace will reveal that they are *not* recognizably similar. There are children who can be subdivided into babies, toddlers (for whom purchases are made by parents, grandparents, and godparents), tweenies (a well branded group), and teenagers; adults are subclassified by so many niches as to be almost indefinable; by age, by ethnic grouping, by religious conviction, by income, by profession, and so on. Confound these with ethnic, religious, and cultural diversity and a world of market niches appears.

b. Other Sources of Data

One cannot speak about the average shopper or how much and when this person spends. To do so requires similarity of stores, geographic locations, and local residents. New product developers are not developing a product for an average consumer. They develop products for specific customers and their associated consumers (i.e., a targeted group), who represent niches in which these targeted individuals are to be found.

An astonishing amount of information can be obtained about customers and consumers using demographic data that has been broken down for cities and towns along with other databases. For example, much information can be gleaned by combining regional population statistics with:

- Magazine subscription lists or magazine sales in local areas. Sales of topical magazines reveal the interests (e.g., hobbies, cooking interests such as barbecuing, outdoor activities) of people in those particular areas

- The use of reward cards, which provide a wealth of information about purchases, the amounts purchased, and location of purchases
- Membership lists for trade and professional associations, which give evidence of educational levels and, indirectly, the income levels of the area

Consumer information and personal information about consumers are located in databases all over the country. When all sources are cross-referenced, surprisingly accurate pictures of customers and consumers in general, and individuals, can be had. An address provides information about income status when used in association with property tax records that identify expensive residential areas. This collation of data is called consumer profiling. The lack of privacy it represents in the computer age is a fact of life and a frightening concern for many people.

Indiscriminate use of demographic data can, however, be misleading without integration with other sources of information. Foot with Stoffman (2001), however, do not believe demographic data has been mined enough. If one were to use the following demographic data isolated from other databases, the result could be disastrous:

- The over-55s are the fastest growing segment of the North American population.
- They have the most wealth of any age category.
- Females make up the greater part of that older population because statistics show women outlive men. Consequently, many over-55s are either widowed or single.
- Women over 55 need extra calcium to prevent osteoporosis.
- Other physiological changes in aging for this population are general frailty, poor circulation, high blood pressure, a general loss of flavor and taste sensations, often a loss of teeth, and constipation.

Some product developer may incorrectly see an opportunity here to bring out a high-calcium, low-sodium, low-fat (simply because the percent of calories from fat in the diet is acknowledged to be too high), added-value product (older people can pay) that is a bland, soft food fortified with fiber and vitamins (for good measure) in an easy-open (over-55s lose their strength and dexterity), single serve (they live alone) container.

Certainly there are frail and incapacitated over-55s in nursing homes and other institutions. They are an elusive population. New products designed for their special needs will come from astute product developers in food service companies. But the over-55s are also people finishing schooling, which was interrupted by children. They join travel groups and travel to exotic places on conducted tours, for example with Elderhost™ or on tours sponsored by their university alma maters. They initiate their own Olympics for the elderly. They consult either professionally or voluntarily with small local businesses or with projects in third world countries.

The aging market, as one example, must be seen clearly as highly segmented. To capitalize on its vitality, the generation of new product ideas must come from knowing these consumer niches. More than demographic data is required for this understanding.

TABLE 2.5
Percentage of Sample Populations
Indicating Some Degree of Vegetarianism

Year	Vegetarians (%)
1984	2.1
1985	2.6
1986	2.7
1987	3.0
1988	3.0
1990	3.7
1993	4.3
1995	5.4
1997	5.0

From Vegetarian Resource Group, Web site, 2003; Vegetarian Society, Web site, 2003.

---

c.   Other Statistical Data: Vegetarianism

The practice of vegetarianism is also a highly segmented one, whose proponents may practice it for health or ethical reasons and may be rigorous, loose, or even part-time vegetarians. Some refuse all animal products. Others refuse red meats but will partake of poultry and seafood products. Others will eat animal products such as dairy products and eggs, for which there has not been a killing. The reasons for adopting a vegetarian diet are many, but do not involve developers.

The important issue is that there is a market for a myriad of products, as many customers and consumers believe they are adopting a healthier eating style; with the growing interest in nutraceutical factors found in many fruits and vegetables, they are buying into vegetarian products. Silver (2003b) states that the vegetarian market is worth about $1.25 billion, with the sale of meat alternatives having grown nearly 40% annually for the past 8 years. An interesting characteristic of this market is that many meat eaters eat vegetarian on occasion. One area of special activity for the food service sector is the many college and university cafeterias that now offer vegetarian menus. Fast food chains have also adapted their menus to include vegetarian food items.

Krizmanic (1992) reported on a survey that found that 7% of the American population considered themselves vegetarians. However, more recent statistics seem to agree that vegetarians represent approximately 3% of the population. Proportions of vegetarians in the population vary by area, age, ethnicity, and sex. But there is no agreement on whether this is a growing group or one that is relatively static. Data in Table 2.5 obtained from the Web sites of the Vegetarian Resource Group (2003) and the Vegetarian Society (2003) do suggest that there is modest growth in this area. The former group estimates the market potential as approximately 20 to 30% of the population!

A health reason for vegetarianism was headlined several years ago in a respected science magazine, "Surgeon General Says Get Healthy, Eat Less Meat" (Anon., 1979). Such reports have since been taken up by science editors of nearly every newspaper chain.

The vegetarian niche is attractive for developers in many ways:

1. It straddles the health food market or healthy food market, thus satisfying a need of customers and consumers.
2. It is common practice in many ethnic food styles. As such with an ethnic-inspired flavor addition, it fits many market niches.
3. There is already an established market for vegetarian foods, albeit a small one, so acceptance is already present.
4. It satisfies many market niches (main course, side dish, and ethnic cuisine), and thus many diverse consumers can be attracted to it.

A wider selection of many varieties of fresh vegetables and fruits has become available. These are becoming a more important and attractive part of meals, approaching, but not yet supplanting, meat as the main item. Fresh fruit is eaten more frequently. Undoubtedly, vegetarianism is growing.

The vegetarian sector also requires the use of new ingredients to replace meat and gelatin. This has been accomplished with soy-based products, textured wheat proteins, meaty vegetables, and rice starch.

d. Other Statistical Data: Obesity

CNN news flash viewed February 4, 2003: "London theater seats too narrow for U.S. behinds."

Governments are very concerned about obesity and its ramifications: it represents a major expense for the medical and health care systems. MacAulay (2003) describes obesity as an issue of national importance in the U.S. She mentions several programs and campaigns that the American government is implementing including some programs by individual states that might possibly tax high-fat, high-sugar foods. Car, airplane, and theater seats have to be redesigned to accommodate wider bottoms and bigger bellies. Obesity is a direct factor in many health and social problems (Lachance, 1994; Birmingham et al., 1999). According to P. James of the World Health Organization (WHO) obesity is on the increase, "doubling every five years" (Reuters, 1996) and it is an epidemic.

The National Center for Health Statistics (NCHS, 1999) reported on the basis of the 1999–2000 National Health and Nutrition Examination Survey (NHANES) that the proportion of overweight or obese (body mass index greater than 25.0) U.S. adults is estimated at 64%, up from the previous NHANES III report (1988–1994), which gave an estimate of 56%. Since the NHANES II survey (1976–1980) there has been a steady increase (age-adjusted) from 47 to 64% in the 20 to 74 year old category.

The data (NHANES, undated) for children whether boys or girls, male or female adolescents, or whether white, Hispanic, or Mexican American also shows an increase in obesity. The American Obesity Association has obesity data broken down by history, age, gender, educational level, and geographic distribution in the U.S. (A.O.A, 2002).

This increase in overweight and obesity results in health and social problems later in life. It is occurring despite a declared interest by the public for wanting good health, eating healthy foods, and living a healthy lifestyle. The meaning of these contradictory findings is unclear. It demonstrates the public's ignorance of nutrition, abetted by misleading or ambiguous advertising and overzealous promotion by food manufacturers. Gitelman (1986) commented on the same issue nearly 20 years ago. A colleague of mine, a biochemist no less, exclaimed to me when fat-reduced chips appeared, "Now I can eat all I want without worrying about calories!"

At the same time, skinless chicken breasts are disappearing from menus, and extra rich creamy foods and extra strong (i.e., cream) drinks are appearing. There seems to be market niches for a diversity of interpretations of both health foods and healthful foods and hearty foods. There is still a strong hedonistic market that can be catered to side by side with healthful foods.

Sloan (2003a) suggests this confused market demonstrates four product opportunities:

- Foods for both weight control or loss and foods for health (i.e., disease prevention)
- Foods that are low carbohydrate and high protein
- Foods that have minimal caloric density since the desire to lose weight is a high priority
- Repositioning existing "healthy" foods back to what they are: low-fat, low-carbohydrate foods

Political lobbying enters into this arena. Eilperin (2003) reports that the U.S. Sugar Association is angry at the WHO's recommendations to curb sugar intake to no more than 10% of daily calorie intake. Development teams should be alert to governmental and nongovernmental agencies' actions through their programs and perhaps to changes in taxation policies if "empty calories" should come under government scrutiny in this still confused arena.

e.   Environmentally, Ethically, Socially, and Religiously
Responsible Foods

There is a small but growing market for foods that are not classifiable by their nutrition or health-giving properties but on what may broadly be called ethical or social grounds. That is not to say that their adherents do not believe these foods to be healthier or that they necessarily are healthier. The marketing issue is that *consumers believe them to be so* and there is a market for them. The growing organic food market proves this to be the case. For developers this segment is full of paradoxes.

Environmental, ethical, and social concerns as elements affecting food products and their consumption are confusing. Some equate the concerns to the developed world's proprietary attitude to third world nations; others see a confusion of issues involving agricultural practices such as organic farming, factory farming, animal rights, and health concerns.

i.   Pollution and Environmental Concerns

Pollution and the environment are major concerns among people in the developed world and are an emerging concern among indigenous peoples in the less developed world. Since the Earth Summit in Rio de Janeiro, Brazil, in 1992, governments see that they must act, or be seen as acting, to reduce pollution. Dumping waste into oceans, lakes, and rivers or burying it in landfill sites is no longer acceptable. Medical wastes, dumped at sea, have washed up on the eastern seaboard beaches of the U.S. and has been clandestinely dumped in foreign countries (Patel, 1992). Cities and nations can no longer find landfill sites for garbage disposal. Disposal of waste is a significant cost for food companies who pay to have waste hauled away and then pay a tipping fee at a land disposal site.

To manage waste, less waste must be produced in the first place. Products or processes will be rejected if they produce excessive waste or if they arouse environmentalists to protest against the companies producing the waste. Therefore companies need to divert some of their technological skills to improving processes and products to make them less wasteful and to upgrading the waste into useful by-products. Products prepared and packaged to minimize waste in the home or in the area where the food is consumed or that centralize waste production where waste can be treated in an environmentally friendly fashion are becoming demands of both governments of all levels and customers.

Pollution and other environmental concerns caused by agriculture (especially large-scale factory farming operations that produce tremendous amounts of animal waste), fish farming, food manufacturing, and especially the food packaging industry are highly visible. Food containers, for example, are the most obvious feature of processed foods, especially if empty containers are seen strewn along highways, dumped into streams, or littering city streets. Yet the container is a necessary evil designed to give optimum protection to the product from harvest areas to the processing plant and thence throughout the stable life cycle of the finished product. Without the package, according to Akre (1991), nearly half the available foodstuff in the developing world would be lost to human consumption, whereas developed economies (which use packaging) lose no more than 2% of available food. Waste and its ever-present companion, pollution, are inevitable in the food microcosm; the mechanisms to deal with them effectively and efficiently are often lacking. Developers need to consider their actions respecting waste production and consider alternatives or devise systems to reduce waste and pollution. Environmental activists are a large enough political force that governments listen when they become concerned, active, and vocal (Akre, 1991).

ii.   Counterarguments to Recycling

Product developers are, therefore, compelled by public and governmental pressure to use either recyclable or reusable packaging or biodegradable material for packaging. Both alternatives present developers with a dilemma. Recycling or reusing requires a collection system, an added cost. Some experts argue that recycling is not energy efficient, requiring energy to return the recyclables, energy to pick up the recyclables and deliver them to a recycling center, and energy to process the recyclables. The news that biodegradable packaging material is more expensive and is not readily biodegradable in most landfill sites makes manufacturers reluctant to use these (Lingle, 1990). The use of degradable packaging materials raises several problems, not the least of which is the reaction of regulatory officials to contact of the degradable materials with foods. A similar challenge may be put to the edible films that are emerging.

Headford (1996) describes efforts for a database of the "global food industry" that will include environmental issues and a study of the reporting strategies respecting environmental issues of major food manufacturers and retailers. Her studies indicate that the major environmental issues for companies are energy, packaging, recycling, and pollution.

iii.   Meeting the Ethical Shopper's Demands

Ethics will have a greater role in shaping the purchasing habits of future consumers (Wilson, 1992). Guides are available for shoppers that provide information on the companies behind the food brands, for example, *The Ethical Shopper's Guide to Canadian Supermarket Products*. These guides describe a company's environmental record and its policy on women's issues, labor relations, and consumer issues in general. There have been similar developments in the investment community where stock investment plans are devoted entirely to holding stocks of companies, including food companies, whose policies are ethical and green. Kraft is reportedly planning to sell coffee certified by the Rainforest Alliance.

Social concerns are shaping food product development; old products are being redesigned to appear more ethically responsible. New products are being developed, for example, shade-grown coffee beans and products derived from them. Marketing and marketing policies are changing to accommodate these shifts.

It is still a confused area, and development teams are well advised to tread warily.

iv.   Alternative Formats for Selling

The influence of the so-called "alternative formats" for retailing food, the warehouse outlets, direct mail selling of food items, as well as direct sales (selling directly into the home) has not yet been fully assessed as to the changes these will bring to the food marketplace or even whether such alternatives will survive in the future. However, these alternatives will influence the development of new package formats and new products to satisfy the demands of these channels. Conventional retailers will have to combat the impact that alternatives have on consumers and their buying habits.

v.   Customer Fancies: Less Processed, More Natural, Organic Foods

These environmental concerns have indirectly led many consumers to want less processed, more natural foods. As a result, there is a general reluctance to accept highly processed foods — which the consumer understands as "ersatz" and "somehow not good." Lee (1989) refers to these people as "food neophobes," that is, people who consider new food technologies, food additives, and ingredients as untried, artificial, and hazardous. The interest of consumers in minimally processed chilled foods may reflect this desire for less processing and more naturalness. Busch

(1991), when discussing consumers' concerns about the growing application of biotechnology to the food microcosm, puts it this way:

> They desire foods that have been prepared in traditional ways, that contain few or no additives, that are "natural," and that are made neither from transgenic plant[s] or animals nor via new fermentation techniques.

This distrust of processing and high technology may be challenged and dismissed as irrational, but nevertheless, it is there and it is growing. It should be noted, however, that consumers espouse biotechnology in all its subdisciplines in matters of health and vanquishing disease.

Also, the demand for organically grown products, even organically grown added-value products, has increased. Jolly and his colleagues (1989) reviewed some startling historical statistics:

- A 1965 survey conducted in Pennsylvania showed that only 15% of respondents were greatly concerned about pesticides. A similar survey conducted in the same state in 1984 found that 78.7% of respondents were concerned about the use of pesticides.
- In 1965, 94% of respondents felt there had been adequate inspection of food purchased at retail, but this figure had dropped to 48.9% by 1984.

These figures indicate a growing lack of confidence in the safety of the food supply.

Jolly et al. (1989) found that 57% of all respondents in California judged organic foods to be better than nonorganic foods, and 35% considered them to be no different. Obviously there is a perception that organic produce has advantages. These advantages were listed in order of importance as:

- No pesticides or herbicides
- No artificial fertilizers
- No growth regulators
- Residue free

Campbell (1991) reviewed two Canadian surveys of consumers' concerns about the food supply. When asked about their most serious food concerns, most respondents (26%) answered pesticides; pollution (24%) was next and nutritional concerns followed (15%). When a multiple response was permitted, the answers were food poisoning (72%) and pollution and pesticides tied at 69%. Food additives rated low as a perceived hazard.

Whether scientists think pesticides are a hazard or not, other people — nonscientists, the consumers — do believe pesticide-free foods have an advantage for both health and safety. Many major retailers in many countries maintain an organic food section.

"Big Firms Get High on Organic Farming" read headlines in *The Wall Street Journal* (Nazario, 1989) in 1989. Many farmers and large companies with extensive farm holdings (Sunkist Growers, Inc. and Castle & Cooke, Inc.) have joined the movement to natural or organic farming. They find there is a demand for their organically grown products. Many realize that organic farming can be as profitable as so-called chemical farming. As Nazario wrote, "case studies have found that yields and profits can be just as high on an organic farm as on a non-organic farm."

It is worthwhile for any food technologist to visit a natural food store, examine the products, and talk with the people purchasing them. A complete line of fresh produce is available at elevated prices; people are buying despite the price differential, so deep is their distrust of nonorganically grown produce. Since large farming companies have moved into organic farming with their more efficient agricultural practices, prices for fresh organic produce are falling.

Irrational as it may be to food technologists, there is a strong demand for organically grown produce and the added-value products that are made from them. The organic food business is no longer "a counterculture business run by flaky hippies" (Nazario, 1989). McPhee (1992) reported that organic food introductions have increased 400% since 1986 and the organic beverage category has increased 1450%. Yet, the organic food industry is not without its problems. As McPhee noted, big companies are still testing the market cautiously, well aware that any movement into this new market could have repercussions to their regular lines.

The Institute of Food Technologists (IFT) published a *Scientific Perspective* (Newsome, 1990). In this paper an objective assessment of organically grown foods was presented. It pointed out that the claim that organically grown foods are healthier is scientifically baseless: IFT highlighted problems with organically grown foods citing the *Listeria* outbreak in cabbages caused by the use of sheep manure (natural fertilizer usage). Newsome concluded in this *Scientific Perspective* that:

- Organically grown foods were not superior to conventionally grown foods with respect to quality, safety, or nutrition.
- They are more expensive.
- They provide less product variety for consumers.
- Diets based only on organically produced foods may "present the risk of possible loss of balance and variety."

All the above may have been absolutely correct at the time the article was written; one might now rethink the latter three issues. But this is not the concern of developers. If consumers perceive there to be an advantage, if consumers are willing to alter shopping habits to go to specialty stores to purchase organically grown products, and if consumers are willing to pay more for organically grown foods, then one must admit there is a certain qualitative superiority since the consumer is right. Hauck (1992) described a ranching and meat operation in the U.S. that provides organic beef and lamb, with sales of $25 million in 1991 projected to $50 million in 1994. There is clearly a market for organic foods for consumers who believe these products to be superior.

The consumer is becoming better informed, better educated, and more vociferous. The result is a very strong consumerism movement. Food scientists and technologists in their ivory towers are being challenged. Their science and technology have apparently outstripped their communicative skills. Perhaps this poor communication has been a factor in a consumer mistrust, not only of science, but also of big business and, equally, big government (e.g., see O'Neill, 1992). Communication between consumers, on the one hand, and science and technology interests, big business, and big government is not well developed (Lee, 1989; Busch, 1991). Until communication has been improved and the vitriolic rhetoric that has developed is curbed, consumerism and all its manifestations may be a prominent consideration for new food product developers in the future.

Developers need to pay attention to the environmental, social, ethical, and religious aspects involved in product and process development as well as to the technological aspects of their development.

vi   Changing Food Habits and Lifestyles

Cooking, particularly gourmet cooking, has become a hobby for many consumers; they are becoming generally more interested in food and cooking. Proof of their interest is the burgeoning numbers of cookbooks describing a wide gamut of ethnic cuisines, the growing popularity of cooking schools, and the popularity of television cooking shows. Consumers are doing more cooking from scratch; that is, consumers have gone back to using traditional recipes for home cooking, especially for special occasions. Experimentation with new cuisines has led to the use of more exotic ingredients. This awakening to exciting foreign foods and cooking styles has provided food manufacturers with opportunities for new products and ingredients.

Families eat fewer meals together. This means that more meals are eaten away from home. In addition, grazing (snacking) is an established and preferred eating pattern for many people who accept several small meals a day as normal practice. Such a change in eating habits opens up opportunities for finger foods that are tasty and nutritious. For years food pushcarts have operated on the streets of large metropolitan areas to take advantage of the snackers and grazers. They serve the function of the fast food restaurant without the overhead of expensive real estate. Many of these pushcarts sell ethnic foods, which seem very suited to finger foods, for example, empañadas, mini-pizzas, calzones, and tacos. In addition, they offer fresh fruits and prepeeled raw vegetables; nutritious snacks have become a natural consequence of this change in eating habits.

There is a serious side to meal patterns and their influence in daily life. Chrononutrition, an offshoot of chronobiology, the study of biological rhythms, has emerged as a study of the "time-dependent features of nutrition" (Arendt, 1989). How does time or the timing of meals and the consequences of digestion affect food selection or other biological rhythms in the body? The practical application for these studies is the concern for the safety and efficiency of shift workers, airline crews, or, of course, the military, especially when on patrols, when the daily rhythms of these people are disrupted. The gastrointestinal problems of shift workers can lead to work loss, fatigue, and inattentiveness. Worker safety could be endangered.

Based on the findings of chrononutrition, foods, particularly snack foods, could be designed to provide proper nutrition to workers whose biological rhythms are disrupted by their work schedules.

f.   Surveys and Polling: Asking Questions

An assignment given to my students for a course on new food product development required that they keep a diary of all food purchased and eaten during a 1-week period. They were required to record portions purchased, cost per serving, type of packaging, country of origin, whether it was an added-value product, and whether special handling, storage, or preparation was required. In class the purchases were discussed with respect to quality, desirability, and satisfaction (as student consumers), ease of preparation (limited kitchen facilities were available in the dorms), packaging, ease of storage, ease of disposability or recyclability, cost vs. value, and so on. Then the students were to develop ideas for products they would have preferred to have eaten, that would have fit better with their lifestyle as students, and that would have made life, their "food life," more enjoyable.

This exercise produced a welter of ideas that would have delighted any product development manager. When confronted with a what-would-I-prefer situation the students began to look at what constituted their lifestyles: students are cash poor and have long working hours, limited food preparation facilities, limited time for preparation, and academic pressures. They came up with ideas that fit their needs. They looked more closely at themselves as consumers. They were no longer faceless consumers: they had an identity and were a definable consumer niche.

Psychographic data obtained through polls and surveys reflect behaviors and attitudes of customers and consumers. From an analysis of the data, developers can define the qualities that must be designed into products to meet the needs and expectations of targeted consumers. If people have more leisure time (based on demographic data) plus a desire for a healthier life style (based on psychographic data), this suggests ideas for nutritious snacks or low-calorie beverages such as fruit beers, wine coolers, or low-alcohol beers to accompany leisure activities.

In the 1960s, 1970s, and 1980s consumers were concerned about nutrition in general. Nutrition was, in some vague way, good for one's well-being without consumers being very sure why. Now consumers know why. Today, consumers focus on very specific health and nutrition issues that fit neatly together, such as:

- Products to fit a healthy diet, with foods that provide protection against diseases such as cancer, heart attacks, depression, and memory loss
- Products containing nutraceuticals (functional foods, probiotics and prebiotics, phytochemicals), for which a body of reputable literature is slowly appearing that suggests these may assist people with certain digestive problems, behavioral problems, and specific disease conditions

At the same time, consumers have not entirely abandoned such concerns as low-calorie, low-salt (sodium), high-fiber, and low-cholesterol foods. In addition, interest in natural or organic foods (by which consumers generally mean additive-free foods) is growing. All this information leads to new product ideas. Psychographic data plus demographic data provide the inspiration for food product ideas.

Consumers are slowly changing cooking and eating habits. Consumer magazines are enjoining their readers to fry less and broil more. Unfortunately most local governments are cutting back on educational funds and usually the first items to go are courses such as home economics (or household science); good food habits, meal planning, and general knowledge about food safety are the first items to go. Statistics show that most families use at least one (frequently more) factory-prepared food at their meals, so this advice to fry less and broil more falls often on deaf ears.

Meal patterns have changed. In Victorian days and earlier, four or five meals were the order of the day. The work habits of our modern world reduced that to three. However, my personal observations of food habits of most office workers and students today show that five or six meals are more common than the "three square" of more recent times. The following meal periods with typical fare are common with both students and office workers, particularly young female office workers:

1. Early morning: fruit juice, coffee or tea, cold cereal or toast, muffin, croissant or Danish. The beverages and the baked items were frequently eaten on the way to work or school, in class, or at work.
2. Mid-morning: coffee, milk, or juice and a baked item (Danish pastry or muffin), often with fruit, or fruit alone.
3. Noon: sandwich, yogurt, hamburger, or hot dog (the students' fare), cole slaw, and fried potatoes.
4. Mid-afternoon: similar to mid-morning with the possible addition of machine-dispensed foods such as soft drinks, peanuts, and chocolate bars.
5. Evening: Students relaxed over beer, chips, and other snacks in the campus buttery; office workers went home to a light deli snack (paté, small pizza, cole slaw, or bread — all items picked up on the way home) before going out for the evening.
6. Night: students ate supper in the cafeteria or at apartment residences.
7. Late night: both office workers and students usually ate a snack before retiring.

These are eating periods, snacking periods, or grazing periods. Today many people follow no traditional meal pattern regulated by the clock but eat when they are hungry.

Much useful information is obtained from data received firsthand from customers and consumers through surveys and polls, but there are some caveats to be observed.

By far the best and most interesting of several courses in statistics I have had was given by a professor who made a comment about surveys that has stuck with me as a cautionary principle: "Don't believe them." By way of explanation, he described what is often heard on the radio or TV news or read in newspapers: "This survey result is correct within ±x percentage points 95% of the time." The professor would get very excited saying, "No, no, no! That is an incomplete statement. It is wrong. They should say that this result is *either* right or wrong. *If* it is right then...."

Then he explained. The survey questionnaire could have been poorly prepared. The survey could have been improperly carried out or the results could have been misinterpreted. Without assurance of knowledge of the manner in which a survey has been conducted, no survey results should be accepted as fact but merely accepted as a news item filler.

A survey is a collation of many interviews. Interviews are simply one-on-one encounters between an interviewer and an interviewee during which the attitudes of the interviewee are sought. There are two types of interviews:

- Structured interviews in which the interviewer has a very specific list of questions to follow faithfully and poses these to the interviewee. The interviewer then records the answers by checking off the appropriate squares.
- Unstructured interviews in which the interviewer uses a prompt sheet rather than a list of questions with which to conduct the interview and record answers and opinions.

Cohen (1990a) discusses interview techniques as a tool in product development.

Whether a formal, structured questionnaire or a cue sheet of topics is used, surveys do permit interviewers to get personal opinions, comments, and reflections on a wide variety of topics of interest to the client. Unstructured interviews give great freedom to interviewers to probe very deeply into subjects. Herein is one of their weaknesses — interviewers may force "don't know" or "no opinion" responses into specious answers or opinions from interviewees.

Converse and Traugott (1986) discuss the errors that can arise in the analysis and interpretation of polls and surveys of the general public; these are equally valid for polls and surveys of customers and consumers. There are several sources of error:

- There is a volatility of opinion among respondents that can vary with time. This source of error is especially important to product developers since development does take time — time during which subjective opinions can change or be changed by external events.
- There is an inherent difficulty in measuring subjective states of respondents. How does one measure, for example, "is likely to buy," which can be found in questionnaires? It is a conditional question, i.e., the respondent "is likely to buy *if* such and such happens."
- Incomplete data, especially "no response" or "won't respond," compromise sample coverage.
- Measurement errors can be introduced by the interviewer, the respondent, or the questionnaire itself.
- Individual market research houses may have different techniques for weighting responses. These can generate bias in the interpretation of results.

Interviewers, using either structured questionnaires or prompt sheets, can sway respondents in many subtle ways, consciously or unconsciously. Voice, clothing, body language, physical appearance, age, or sex of the interviewer can trigger this influencing effect. To avoid any bias being introduced during an interview, an interviewer requires skill, training, and experience.

Unfortunately, many interviewers are not skilled, trained, or experienced. Often they are college students earning money on school breaks. An incident observed in the shopping mall of a small city provides an excellent example of inexperienced interviewers.

Two college-aged men were conducting a survey in the mall. I was curious as to what was being surveyed and frankly wondered why I wasn't being interviewed since males and females seemed to be the targets. I edged close enough to them to overhear one young man say to the other, "It's my turn to get the next [good looking girl] to interview."

The results of this survey from this particular portion of the interviewing team might be biased toward attractive young women. Properly trained interviewers add to the expense but their use is worth the investment. For companies employing market research houses, the caveat is "buyer beware" regarding how the survey or poll is conducted. If the survey is improperly conducted, the cost is twofold; the money spent on the survey is wasted and the information obtained is misleading and can lead to other mistakes.

The structure of questionnaires or their counterpart, prompt sheets, requires careful design to remove ambiguity. The language of the survey's preamble, and even the order of questions, must be designed to avoid any bias. The following hypothetical example illustrates how a change in a questionnaire could introduce a bias:

> Potential voters are being polled for their voting preferences for three political parties. Mr. Right leads the Right Wing Party. Ms. Left leads the Left Wing Party. Mr. Middle leads the Middle-of-the-Road Party. All the lead-in questions pertain to the platforms and public issue policies espoused by each of the parties. The last question, however, asks, "Would you vote for Ms. Left, Mr. Middle, or Mr. Right?" From delving into political philosophies, the survey has jumped to leadership personalities, introducing a very different slant to the survey.

Consciously or subconsciously, respondents express opinions or answers as directed or influenced by the wording in the questionnaire. Clients hiring market research houses need unbiased and objective data; they do not want data that support preconceived concepts about customers and consumers that either they may hold or that is held by the market research houses. If clients have preconceived, but erroneous, ideas of what their customers and consumers want, their desire for vindication of their view may influence the wording of the questionnaire. As Converse and Traugott (1986) state, different research houses may have their own reasons for weighting responses.

Data based on interviews must be cautiously interpreted. Interpretations based on data from a small sample of targeted consumers should not be extrapolated to broader populations without interposing strong reservations. The results, without the hyperbole of marketspeak introduced by the market research house or even by the client's own marketing department, should be seen as providing only a very general reflection of the targeted population.

g. Telephone and Mail Surveys

Telephone surveys reach a very disperse population in a large geographic area. They are not selective regarding who is contacted, and not all potential respondees have telephones. In telephone surveys, questionnaires must be brief because they are intrusive. Interviewers have interrupted respondents at home. Questions, nonetheless, can be somewhat more detailed (given the constraints mentioned above) since interviewers can explain difficult questions. When computers are handy, interviewers can key in respondents' answers, and the results are available in real time. There is no delay, and the development team can proceed rapidly. Interviewer bias in telephone surveys is much less since facelessness of both interviewee and interviewer removes any visual elements causing bias.

Mail surveys have no geographic bounds. Since most marketing research companies keep mailing lists of respondents characterized by income, religion, ethnic background, and so on, great selectivity of respondents is possible. These listings can be selected to represent the target population that a company desires information about, but they are far from randomly selected or representative of an elusive population. They are that market research house's private listings.

h. Advantages and Disadvantages of Mail and Telephone Surveys

Mail surveys are the least costly technique of gathering data. They have no interviewer bias, but that does not mean the survey questionnaire itself will be without bias. The explanatory text, the wording of questions themselves, and the objectives of the survey can influence responses.

Printed questions must be clear and self-explanatory on mail surveys. Unstructured surveys cannot be used since respondents have no contact for an explanation of ambiguous or complex questions. Any difficulty in understanding questions could discourage respondents from finishing the survey or may elicit false data based on the respondents' confusion. This need for simplicity in mail surveys may limit the kinds of information companies can obtain.

TABLE 2.6
Advantages and Disadvantages of Various Survey Techniques

Survey Characteristic	Type of Survey		
	One-on-One Exchanges	Mail	Telephone
Cost[a]	High	Low	Low to medium
Geographic scope	Limited in area	Limitless	Limitless (long distance costs)
Potential for bias	High	Low	Low to medium
Availability of results	Slow to medium	Slow	Rapid
Selectivity[b] of respondents	Nonselective to highly selective	Selective	Low
Style of questionnaire	Complex	Simple	Short (intrusive)
	Lengthy	Medium	

[a] Costs for any type of survey can be variable. Generally, personal interviews are more expensive, especially if specific interviewees are targeted. Long distance charges can add to telephone surveys unless done locally in several cities.

[b] If respondents in each survey technique are preselected from screened lists then the survey can be highly focused. Random selection of telephone numbers provides no selectivity. Accosting shoppers in a mall also provides little selectivity beyond selecting for obvious physical traits such as age, sex, skin color, weight, etc.

A disadvantage of mail surveys is a notoriously poor mail-back response rate. Response rates of 45 to 50% are considered excellent; lower rates of return are the norm. The slow return of completed questionnaires delays data analysis.

Nonresponses, whether in personal, telephone, or mail surveys of targeted populations, contribute significantly to sampling error (Converse and Traugott, 1986). Both telephone and personal surveys are intrusive; telephone surveys, especially during evening hours, can be especially irritating and invite negative survey results.

Table 2.6 compares the advantages and disadvantages of these surveys.

i.   A Modern Delphic Oracle

A novel method of surveying, especially to spot trends that might lead to new product development ideas, is the Delphi method of forecasting. It is based on questionnaires mailed to addressees whose opinions are valued (for example, senior company executives). These selected respondents are assumed to be acknowledged experts in the topics under consideration or are leaders in the various industries for which forecasts are sought. The process is much like conducting a think tank by mail at this initial phase of enquiry.

Opinions from the first survey are collated. A second questionnaire is formulated based on a consensus of the responses given in the first one. The same respondents are canvassed again, but this time the questionnaire is based on the expert opinion derived from the first survey. The respondents may or may not revise their opinions in the face of this new information. Again, the respondents' answers are collated. The researchers can decide to continue or not with yet another questionnaire digested and distilled from the additional, rethought information gained in the second survey. The survey may go for a third time (rarely are more than two performed), and again, the answers are collated. By this third round a semiquantitative analysis of future trends in particular fields of interest based on expert opinion has been obtained. The Delphi technique of surveying is a very powerful tool in evaluating and forecasting trends. A good example of a Delphi survey is the Food Update survey described by Katzenstein (1975).

j.   Elusive Populations

Reference has already been made to the difficulties of researching elusive populations (see Figure 1.2). Sampling these populations presents unusual problems for market researchers. Some examples of elusive populations are (Sudman et al., 1988):

- Racial, ethnic, or religious groups within larger populations
- Income levels of various groups (professionals, tradespeople, retirees, etc.)
- Persons with specific illnesses or disabilities

For the food product developer, some examples of elusive populations of customers and consumers might be:

- Users of organic foods and vegetarians (Nathan et al., 1994).
- Those following religiously or philosophically based diets or socially or culturally based dietary laws such as Halal, Kosher, Baháí faith (Solhjoo,

1994), or Zen macrobiotics (a diet based on a balance of principles of yin and yang) typical of Zen Buddhists. These diet principles regulate types and composition of foods to be consumed and their processing. Erhard (1971) and Glyer (1972) provide extensive reviews and discussions of the early literature on various cultural philosophies regarding diet.

- Those practicing ethical or socially responsible food selection (for example, buyers of coffee made from beans sold by workers' cooperatives rather than from multinational manufacturers).
- Those with food allergies or metabolic dysfunctions (Blades, 1996; Brown, 2002; Cullinane, 2002; Steinbock, 2002) resulting from genetic errors of metabolism or surgery or prophylaxis (e.g., immuno-suppressed individuals).
- Those who are overweight, underweight, or anemic and require special nutritionally designed diets.
- Those who use dietary supplements, such as athletes and the elderly.
- Teenagers, the elderly, and pensioners.

(The articles by Cullinane, Steinbock, and Brown discuss allergy problems and preventive action taken by a supermarket chain, an industrial caterer, and a company in the hotel, restaurant, and leisure activities markets, respectively.)

Each elusive population presents different levels of difficulty to survey. Obviously listings of elusive populations that would greatly assist sampling are not always readily available to researchers. Hospital or medical records, where accessible, or self-help or social associations to which members of these groups might belong may have listings of those with specific disorders. Some groups, such as teenagers and ethnic groups, can be geographically located within cities, for example, in high schools.

By any measure, sampling these populations is difficult and hence costly. Demographic data (census data) can assist market researchers by identifying where these groups are not present. Sudman et al. (1988) discuss sampling techniques for groups that are geographically clustered, more general techniques (network sampling) for nongeographically grouped populations, and sampling techniques for the most elusive of all populations, mobile populations, by capture–recapture methods.

3. The Seller/Retailer and Distributor as Sources of Ideas

a. Analysis of Purchases

Retailers gather much information about customer shopping habits through the universal product code printed on foodstuffs and recorded at the time of purchase. These include:

- What items are purchased and which items are not moving well. This simplifies stock keeping and opens the door to efficient consumer response.
- Which items are purchased as a function of the total dollar amount.
- Which stores have the highest average purchase per receipt.

- Which products are purchased together. Ideas for new product development (or for piggyback couponing) may come from frequent purchase combinations. If product A and product B are purchased together frequently, would combining the two in a single product or packaging them together as a single unit be a logical product idea?

Obviously some caution is needed in interpretation of data of this nature: products purchased together are not necessarily used together. Nevertheless, studies of purchase combinations do yield ideas for new products.

A similar caution applies to the interpretation of data describing case movements of goods between warehouses. These movements are what they are: product is being moved around, nothing more, nothing less. Case movement is only an indirect measure of customer interest, much like per capita consumption figures.

b. Marketplace Analysis

By studying all categories of products available in the marketplace, one can note gaps — products or categories of products that are not available. GAP analysis is another technique for generating ideas for product development. Simply put, marketing people (usually) select a particular product category and then examine the marketplace for empty spaces in that category.

In a typical GAP exercise, a grid is drawn. Each row of the grid describes some product attribute such as texture, flavor, color, particle size, function, application, etc. Columns might be labeled by form, that is, solid, liquid, or gas; or ready-to-serve, condensed, frozen, or dehydrated for soups and other items. When the grid is filled in with data from the marketplace, ideas for new products may be revealed by the empty spaces of the grid.

In Figure 2.1, for example, if hot pepper condiment sauces were the category selected, marketing staff would construct a grid as depicted. Horizontally, the form

Hot Sauce Flavor	Solid ──────────────────────▶ Gas				
	Sprinkle-on	Paste	Pour-on	Dispensed as Foam	Pressure-dispensed
Hot Pepper					
Ginger					
Garlic					
Mustard					
Horseradish					
Blends (e.g., Curries)					
Black Pepper					

FIGURE 2.1 Example of a GAP grid for hot condiments.

of the condiment sauce has been expanded from a liquid to a solid with two possible forms: a concentrated paste (squeeze tube dispensed) and a ground powder. The gaseous forms are also in two forms: dispensed as a foam or dispensed by gas as a liquid or paste, the latter pushing the product into more sophisticated packaging techniques.

Vertically, one can adapt various flavors based on a single variety of spice (e.g., black pepper). One can also base the sauce on various plants such as hot peppers, ginger root, or horseradish, with variations of heat level. Or combinations of spices and plants can be used. Still other adaptations to be used vertically in the grid are ethnic varieties of Indian curries or sauces based on Thai or Japanese cuisine.

As the grid is filled in, the following would be noted:

- There were numerous liquid pour-on sauces with wide variations in ethnic types of sauces, heat levels, and flavors.
- There was a large number of solid "sauces" (i.e., round sprinkle-on products), with an equally wide variation in heat levels, types, and flavor
- There was no "gas" product, that is, pressure dispensed — discounting, of course, pepper sprays used for incapacitating highly agitated, violent individuals or as a personal protection device.

Here is a product gap. Does this suggest that there is an opportunity for a foam- or spray-dispensed hot condiment sauce?

GAP analysis, a form of attribute analysis, looks at the marketplace for product vacuums. Where a space is noted on the grid, no such product exists — in some cases perhaps for very good reasons. No product may exist because both customers and consumers see no advantage for such a product or have no expectation of one. On the other hand, is there an undetected need for such a product, which has never been fulfilled, because such a product has never been presented? The main purpose of this exercise, however, is to stimulate thinking about unfulfilled market opportunities as the grid is stretched beyond its limits both horizontally and vertically.

GAP techniques can be applied broadly to financial and marketing matters as well. Again a grid is made. Columns could be labeled as years when products (in the rows) reach the ends of their effective life cycles. Or columns could be different sales regions and rows the products that marketing needs to introduce into those markets to keep them strong. Or columns could represent anticipated income from different geographic areas projected for the next 5 years. Rows represent products both existing and new to fill the gaps opened up on the grid.

4. Other Sources of Ideas

a. Ideas from within the Company

Ideas for products for new market opportunities can spring from many areas and activities within large and small food companies. Internally generated ideas must still lead to products that satisfy the needs and desires of customers and consumers; ideas for products are not intended to merely satisfy the personal whims of individuals within the company.

First and foremost as sources of ideas for new products are the company's retail and industrial sales personnel and technical sales representatives. Salespersons are sensors in the marketplaces they serve; they are the eyes and ears of the company. Where the retail sales force restocks shelves in grocery stores and supermarkets, they can review the shelves for the competitors' products or check in-store displays. They have closest contact with the customer or with the retail store manager as they discuss orders. These people meet consumers; they see the placing, price, and condition of their competitors' products and are aware of promotional displays. They know what the competition is doing at the store level. Their discussions with store management can reveal weaknesses in products, their packaging, or deliveries. Store returns may point to problems with formulations, and complaints by customers to store management may uncover faulty preparation instructions.

The sales force is the strongest resource a company has for determining what is happening in the marketplaces they service. They provide competitive intelligence by reporting the earliest signal of competitive activities in the marketplace that needs to be heeded by the company. By listening to retailers, customers, and industrial users, sales representatives present their company with ideas of what is wanted in a new product and they provide information about how much customers are willing to pay for this innovation. Some of the product's design work has been done for the company that listens.

A novel approach is used by one company to give this consumer and marketplace awareness to its engineering, research and development, and distribution departments. Members of these departments are required at regular intervals to accompany a sales representative on customer contact visits. Such visits serve several purposes:

- Customers (i.e., retailers and the client's research and development personnel) are impressed by the concern displayed by the company.
- Staff of these sometimes insular departments are exposed to the environment of customers and the marketplaces in which their products compete.
- These individuals see opportunities for improving their products and packaging or for new product ideas.

b.　Ideas from Customer and Consumer Contacts

When consumers or customers take time to e-mail, write letters to, or telephone food manufacturers, they are expressing a need to be heard. They may want to:

- Vent anger about a failed product. This provides clues regarding faulty preparation instructions or the need to reformulate to improve a product.
- Express pleasure about a favorite product. This suggests ways to improve a product by enhancing those pleasurable features.
- Enquire about a product's suitability for a special diet. Is this perhaps an unexpected opportunity for product extension?
- Seek clarification about cooking instructions. This suggests cooking instructions are not clear.

- Offer useful information or new uses they found when using the product, suggesting possible line extensions for the product. These could be product maintenance opportunities.

Customers and consumers need to be respectfully listened to, and their letters should be responded to in a constructive manner.

Hotline telephone numbers (1-800 numbers) are used by many food companies on their labels and advertising brochures; these let customers and consumers directly call the consumer relations department, where consumers' needs are addressed immediately. In addition, companies with staff skilled in consumer relations can elicit background information on the callers to get user profiles. Valuable psychographic information so obtained assists food companies in their marketing strategies and new product development plans. Better still are Web sites, for example Kraft Kitchens, where customers who sign on are "personally" corresponded with.

Product complaints can be likened to the tip of an iceberg: much is hidden beneath the surface. Estimates to evaluate the significance of complaints vary widely. Ross (1980) gave me an estimate that for each consumer who complains there are eight who did not and will not try that product again. Other estimates provided by colleagues suggest that there are 20 or more consumers who do not complain for each consumer who does. Graham (1990) reports that for each consumer who complains there are 50 who do not.

All complaints should, first, be acknowledged. Then the complaints should be classified according to their nature, the product involved and its code identification, identification and location of the complainant, place of purchase, and the time when the purchase was made and the defect noticed.

From these records, the astute, consumer-oriented company looks for common threads. For example:

- A hidden product defect (if it is a common recurring complaint) may indicate a need to improve a food product by reformulation or new process technology. Similar complaints from across a broad marketing region might suggest that something serious happened or is happening in manufacturing or in the distribution channels.
- A package weakness (poorly positioned labels, scuffed labels, overgluing and glue stains, dented cans indicative of rough handling, leaking lids, etc.) may indicate poor package design or a problem in transportation and distribution.
- An idea for a new product may be suggested by complaints. If products consistently fail to meet consumers' needs or expectations as evidenced by complaints, then an analysis of complaints may find ideas for new products more closely designed to meet those needs (Cooper, 1990).

Daniel (1984) describes an elaborate computerized system for organizing consumer complaints of a hypothetical food company with a 1-800 telephone number. It is designed for a quality control function but could be adapted for an approach more oriented to consumer relations. Cooper (1990) describes the planning and implementation of a program to organize consumer complaints to provide:

- Weekly summaries of all complaints
- Complaints broken down by product or brand
- Complaints broken down by specific factory

Such information in the hands of plant managers, quality control managers, as well as senior management alerts all to a potential breakdown in the processing system and can also motivate process or product improvement or new product ideas.

The Internet has brought a new means for airing complaints: Web sites on which people can post complaints anonymously. These sites are commonly called "gripe sites." Such sites require monitoring by a company, as they may contain malicious, misleading information about a company or its products that needs to be countered early. Damaging information may arise from disgruntled employees, cranks, or competitors.

Complaint files should not be handled facilely by marketing personnel with a form letter and coupons, thereafter discarding the correspondence. All consumer complaints, indeed any consumer communication, should be catalogued and cross-referenced under the complainant's name, street address, city and postal code, and the nature of the complaint or the comments made. These files may contain information about consumers and how they use products; they may identify product or package defects that require correction; they may pinpoint the sources of crank complaints by those seeking free coupons; or they may identify competitive activity. In new product introductions such interpretative information can be useful.

c.   In-House Product and Process Research and Development

All food companies conduct experimental trials in their laboratories or pilot plants or on their processing lines to test the use of new equipment, new ingredients with improved properties, and new supplier sources of raw material. These studies and experimental trials require careful documentation, cataloguing, and cross-referencing and then storage in a central repository where they can be accessed. However, so simple a task is rarely undertaken by many companies. In my experience with both small and large food companies, a company's experimental plant trials or other research projects are either not written up at all or if they are written up are hidden away in privately held files scattered throughout the plant. This unavailability of information represents a significant loss of research and development dollars.

These records have more than historical value; they may contain clues to new products that will be found by future technologists. At the very least, they are records of past product ideas and formulations that were rejected. Reasons for rejection of products or processes 10, 5, or even 2 years ago may not be valid today. Times and technology change, as do customer and consumer needs. Projects impossible a short time ago could be within present processing skills through advances in process, package, and ingredient technology.

A later research team will be doomed to waste time and money "reinventing the wheel" when they could have been building on the work of past groups. If a test is worth doing, it is also worth recording and filing the results competently so that others can find the data and understand the information therein.

Information retrieval can be confounded by bureaucratic red tape:

> In one large multinational company, all projects did require write-up and were deposited with management. Indeed, even laboratory work notebooks were confiscated. I had reported on several projects, mainly on the rheological properties of molten chocolate, fondants, and syrups. I needed one of my reports on previous work for reference on a current project. To my amazement I was refused access. I did not have security clearance for access to my own research work!

All reports of process and product experimentation that have been conducted need to be collated, organized, and catalogued with computer access within an information management system that permits ready accessibility.

d.   Collective Memory

In-house reports for computerized information management retrieval systems as presented in the previous section can be developed further to create a collective memory of what happened when problems were encountered and how those problems were solved. In the history of any company, people have encountered and solved problems in manufacturing; they have overcome short supplies of raw materials by either a combination of reformulation or novel processing steps; people have resolved failures in products or packaging. Ideas and experiences that company employees have had are important to the company as a resource. If the knowledge and experience of the cadre of old-timers and pensioners can be organized into an accessible body of information, then companies have an asset valuable for the clues to problem solving and suggestions for new product ideas or process improvement contained within them.

This collective memory is a tool that no company should allow to be lost. Whitney (1989) describes the development of expert systems using the skills and experience of company personnel. A general discussion of expert systems is provided by McLellan (1989).

5.   The External Environment as a Source of Ideas

Ideas generated exclusively from within companies carry risks. They may bring a too introspective range of ideas for development. Exploration of opportunities derived from ideas outside the company will balance this.

a.   The Competition

It is a marketing axiom that a wise company knows what its competitors are doing. To state the blatantly obvious, the competition also competes for the customer's attention and dollars and for the tastes and preferences of the consumer. Any activity by competitors in the marketplace may require retaliatory marketplace action. This action can be the impetus for an accelerated product development program requiring the generation of new ideas for products with which to counter the competition's new introductions.

There are several ways (see, e.g., Table 4.1) in which the activities of the competition can be followed. Each one provides little pieces of information that when integrated together reveal a broad picture of the competition's direction. This activity is called competitive intelligence gathering and is discussed more fully in Chapter 4.

Many food companies omit even simple intelligence gathering activities; they are unaware of what is going on in the marketplaces they service. When management do not visit stores to learn what is happening on store shelves in the marketplace, they are probably equally unaware of who their competition is or how their products look side by side with their competitors' products and how these are judged by customers and consumers.

If competitive products are more successful than yours, then why? Are there ideas here for improvement of existing products? Are there ideas for demonstrating a perceived edge or difference over the competition that is discernible by the consumer?

Inspections of competitive products, including sensory and compositional analyses, provide data that allow:

- Comparison of ingredients to provide an approximation of ingredient costs and therefore an estimate of the competition's cost margins (i.e., competitive intelligence gathering)
- An assessment of quality characteristics of the company's self-manufactured products against competitive products and a determination of the appeal to the competitor's customers
- An evaluation of flavor preferences with taste panels
- An evaluation of package and label appearances to gauge what the customer sees on the store shelves before a package is picked up in a supermarket

Maintaining a strong competitive program of product development requires such activity.

In addition, the information obtained as outlined above can provide ideas for new products. Companies can indulge in free association of ideas to generate new product concepts. An attribute of one leading competitor's product put together with features of another market leader combined with something from a third product may lead to a new product concept. Such free thinking can be very exciting for the developer. Competitive products on store shelves are probably the source of new product ideas for many companies. This activity has now been graced with a name, *benchmarking*.

b. Food Conferences, Exhibits, Trade Shows, and Research Symposia

Attendance at domestic and international food and equipment trade fairs is essential for members of the development team; these are places where companies often showcase their new developments and advanced technologies. Attendees have access to vast arrays of food products, ingredients, and the latest developments in processing equipment. They see and sample a variety of consumer products, ingredients, and food equipment from around the world. They network with other international business people to discuss export market opportunities and discuss their processing problems with technical sales representatives. In other words, they get maximum exposure to a broad range of ideas, products, and contacts from many countries. Such awareness of the wide variability and availability of food products internationally plus creative thinking can lead to ideas for new products.

Similarly, domestic and international technical meetings on food topics are sources of new knowledge and ideas. Conferences where technical papers are presented and discussed reveal novel research and development activity that may suggest new possibilities for new products and could also reveal the research activities of competitors. For example:

- Technical papers provide access to research and development findings a year or two before these results are published in journals and several years before they appear in books.
- Authors of the papers can be quizzed about details of their research methodologies and about directions that more recent, unreported developments have taken. Such discussions are particularly valuable in poster sessions, where more probing, directed, and self-interested questions can be asked. Useful contacts (networking) with experts active in a company's interest areas and with academics can be made.

Who should go to shows and conferences? The obvious answer is: those who will be the most interested in personal development and who will return the most benefit from the experience to their companies.

c. Public Libraries

Public libraries have sections on food and cooking with extensive collections of cookbooks filled with recipes. Recipes based on local, national, and international cuisines provide ideas for new food products or serve as starting points for benchtop test products. More importantly, public libraries have librarians, who are professionally trained in finding sources of specialized information.

Computer-assisted information retrieval systems, common in all libraries, give subscribers access to databases to fit any need for information. These services are reasonably priced and can complement the resources of small library collections.

d. Specialized Libraries and Library Services

Without access to information, businesses would be helpless. Information about markets, food legislation, trade statistics, financial analyses of companies, consumerism, and consumer trends are necessary tools for strategic planning. Equally valuable for food technologists is access to technical information. However, most food manufacturers do not have a staffed on-site library (Goldman, 1983). One way around the problem of lack of library facilities is to subscribe to a database. Databases are available from publishers on CD-ROM or DVD or on-line.

Williams (1985) provides a general discussion of the basic types of databases available: bibliographic databases, full-text databases, and numeric databases. She also describes and discusses aids to on-line retrieval such as user-friendly front ends, intermediary systems that help the searcher with questions, and gateway systems that help the searcher reach other databases within the system. Access to technical literature through these databases reduces the time spent in literature reviews, provides a wide access to literature beyond the budget means of most food companies, and provides scientists in development with the technical literature support they need.

Buxton (1991) describes some general databases available in the U.K. and explains the terminology used (in user-friendly language). Hill (1991) describes the International Food Information Service, a database producer whose main product is the well known *Food Science and Technology Abstracts* (*FSTA*), which is available in printed form, on-line, on magnetic tape, and on CD-ROM. The database contains original abstracts, authors' summaries, and, in approximately 10% of the cases, title-only entries. Classification in *FSTA* is tabulated and on-line hosts for accessing the database are described.

A different database compiled for retrospective searching was described by Mundy (1991). The Institute for Scientific Information (ISI), based in the U.S., develops citation indexes. Well known to most food technologists is the *Science Citation Index©* (*SCI*). This index comprises four interrelated indexes. *SCI* can be used to find recent papers on a particular topic that use earlier papers in that field. The subject index uses key words or phrases to provide a list of authors using these reference phrases in their titles. The corporate index provides information on what has been published by researchers at a particular company, and this leads to an index of authors from that location. *SCI* is available on CD and on-line and was previously available on magnetic tape. ISI also produces the popular *Current Contents©,* which is available on-line and was available on magnetic tape.

The Information Group of the Leatherhead Food Research Association (U.K.) has produced Foodline©, which comprises scientific and technical, marketing, and legislation databases (Kernon, 1991). These can be accessed on-line. The scientific and technology database FROSTI© (Food Research on Scientific and Technical Information) dates from 1974. All entries contain a concise abstract of the article with key word cross-referencing. The marketing database FOMAD© (Food Market Data) is supported by two databases, FLAIRS UK© (Food Launch Awareness in the Retail Sector) and FLAIRS NOVEL©. The latter contains information about products described as novel with respect to attributes and that are introduced anywhere in the world. FOREGE© (Food Regulation Enquiries), a legislation database, provides information concerning permitted food additives worldwide (Kernon, 1991).

An interesting demonstration of the value of computers for storing and sorting data for application in predictive techniques in microbiology is discussed in Gibbs and Williams, 1990; Cole, 1991; Walker and Jones, 1992; Williams et al., 1992; and Buchanan, 1993.

O'Brien (1991) provides an overview of available food and food-related databases. For readers interested in the topic or perhaps interested in subscribing to a service, some difficulties and problems associated with food databases are discussed by Pennington and Butrum (1991), especially the virtually inescapable translation problems associated with food nomenclature, food descriptions, as well as taxonomy of foods as these vary from country to country. These are major problems in international trade. For example, fish and fish products have a rich variety of local, regional, and taxonomic names that present difficulties in translation. Klensin (1991) describes the use of multimedia (and hypermedia) techniques to answer some of these difficulties described by Pennington and Butrum (1991).

Stored technical information is useful only if it can be accessed. If that information could be organized and structured so as to be available to assist the user just as the apprentice in real life has the expert to lean on, it would be a valuable tool. Anderson and coworkers (1985) describe developments in intelligent computer-assisted instruction that are "programs that simulate understanding of the domain they teach" and can interact with the student according to a strategy. This is, in essence, an expert system.

The use of expert systems in the food industry is comparatively new. An oft cited example is the Campbell Soup Company's retrieval of the expert knowledge of their hydrostatic cooker operator before his retirement and the development of this information into an expert system (Whitney, 1989). The chief value of expert systems is in improved training programs for personnel, in having an "expert" on site in times of crises to prevent the introduction of hazards into the food, in the design of new processes, in developing better control systems for safer food, and in the preservation of the accumulated wisdom of experts. Bush (1989), McLellan (1989), and Herrod (1989) describe in detail what is involved in the development of expert systems and the immense amount of effort required in their production.

Specialized libraries, for example, business libraries, reference libraries, technical (medical or engineering) libraries, and patent libraries, provide information and statistics on a wide variety of subjects valuable to developers. Many magazines, trade journals, and technical journals can be accessed via the Internet. The patent literature can be an interesting source of data for investigating what is patented and what patents are pending. From the pending applications, a company can find information revealing directions in research and development that competitors are pursuing.

Technical libraries are equally valuable in providing ideas for new products or new developments in the science and technologies of foods, food processing and preservation, and nutrition (see Table 2.7). Specific ideas such as these, ranging from snack foods to ingredients, may not have direct interest for a developer, but they can provide guidelines for formulations or processing and experimental protocols or suggest products that can be adapted from these ideas.

The value of reviewing scientific and technical literature is threefold:

1. The subject matter of articles has value for what it describes about a food or process. Typical examples are Babic and coworkers (1992), who describe stabilization systems for chilled ready-to-use vegetables (with carrots as the example) and Slade and Levine (1991) and Best (1992), who describe water relationships in foods, their implications for the quality and safety of many foods, and their control to stabilize foods.

TABLE 2.7
Examples of New Processes and Products in the Technical Literature

Type of Product	Product or Process Described	Reference
Jellies	Complete process for extraction of juices from wild native fruits and their manufacture into jellies	Mazza, 1979
Vegetable pie filling	Carrot pie filling preparation	Saldana et al., 1980
Beverage	Nutritious, chocolate-flavored shake-type drink prepared from chick peas	Fernandez de Tonella et al., 1981
Snack	Nutritious West African snack food, akara, prepared from cowpeas	McWatters et al., 1990, 1992
Fruit leather	Fruit sheet made from prickly pear with formulation, composition, and sensory data included	Ewaidah and Hassan, 1992
Fruit leather	Detailed manufacture including packaging and storage of fruit leathers from mango, banana, guava, and mixed fruits	Amoriggi, 1992
Snack and bakery item	Process for explosion puffing of bananas described	Saca and Lozano, 1992
Ingredient	Synthesis of valuable ingredient, vanillin, described using methodology allowing vanillin to be called natural	Thibault et al., 1998
Extruded snack item	Formulation of corn flour, green gram dhal, and gum snack food. Primarily a demonstration of response surface technique for formulation optimization	Thakur and Saxena, 2000
Gruel-based foods	Processing conditions described for preparation of ethnic gruels such as porridge, atole, kishk, and trahana	Tamime et al., 2000

2. Information on the authors, and where the work was performed, provides contacts with whom the product developer may wish to communicate in future.

3. There are acknowledgments at the end of the article listing the supporters of that particular research. This information plus the authors' addresses may identify who sponsored the work and thus reveal competitive activity in a particular field (all part of competitive intelligence gathering).

e. Trade Literature

Ingredient and equipment suppliers provide trade literature, newsletters, and bulletins describing their new products and their applications. This literature can be surprisingly fertile ground for new product ideas or information about new ingredients that will inspire new ideas for product development. For example, Jones (1992) surveyed food labeling directives for European Economic Community flavoring in a very timely review published in *Dragoco Reports* published by Dragoco,

Inc., which would be invaluable for developers who may be contemplating entering the European Community with flavored foods.

Ingredient suppliers often provide sample recipes employing their products with their promotional and technical bulletins. For example, the *California Raisin Report* Winter 1993 issue describes kosher foods and provides recipes for various baked goods suitable for Jewish traditional holidays. The American Spice Trade Association supports a column, "Flavor Secrets," published in *Prepared Foods*. The March 1993 issue entitled *"Comfort" Pies* describes meat pies from various countries, provides a recipe for one, and advertises the availability of pilot recipes for others.

f.   Government Publications

There is a wealth of new product ideas in the deluge of literature available from governments and their various departments and agencies. Governments regularly promote the use of agricultural commodities or underutilized crops. They provide recipes using the foodstuffs, with manufacturing directions and occasionally market test data. Where these fit both the manufacturing capabilities and resources of a food company and the demands of consumers as indicated by market research, this readily available source for increasing a company's ability to generate new food product ideas should not be overlooked.

Government publications are also valuable sources of much demographic data such as population movements, age composition of the population, incomes, food and nutrient consumption per day (USDA, 1980), meal patterns, and so on. Disclosure on companies who have received research monies, grants-in-aid, or development loans provides information on activities in food research and plant construction among competitors.

Descriptions of developments in regional food research laboratories and agricultural research stations are published regularly. For example, the Food Research and Development Centre at St. Hyacinthe, Quebec, Agriculture Canada, publishes an information bulletin in one issue of which Gélinas (1991) described work on frozen bread doughs. The National Academy of Sciences (NAS, 1975) published a book describing the properties of underexploited food plants with promising economic value. Included in this are descriptions of the vegetable chaya, now readily available in supermarkets; winged beans; the cereal quinua (quinoa) and grain amaranths, both of which have become more commonly used products; and the oilseeds of jojoba and buffalo gourd. Specialty stores in many large cities currently stock all of these. Another example from the Food Development Division (FDD, 1990), Agriculture Canada, is an extensive study on modified atmosphere packaging.

II. KNOWING, UNDERSTANDING, AND
ENCOURAGING CREATIVITY

Edwin Land (1963), then president and director of the Polaroid Corporation, on the occasion of the 50th anniversary of the Mellon Institute, said:

> In our laboratories we have again and again deliberately taken people without scientific training, taken people from the production line, put them into research situations in

association with competent research people, and just let them be apprentices. What we find is an amazing thing.... In about two years we find that these people, unless they are sick or somehow unhealthy, have become an almost Pygmalion problem; they have become creative. If there is anything unpleasant to an unprepared administrator it is to find himself surrounded by creative people, and when the creative people are not trained it is even worse. They have two unpleasant characteristics: first, they want to do something by themselves and they have some pretty good ideas that do not fit in with policy; secondly, they have the most naive, uncharming and unbecoming direct insight into what is fallacious in what you are doing, and that, of course, is a blow to policy. I do not want to romanticize these people. I am simply reporting on what we seem to find is a fact ... and you have to find out what to do with these awakened people.

James (1890) describes genius as "little more than the faculty of perceiving in an unhabitual way." Another writer (Anon., 1988a) commented as follows about creative people:

Highly creative people are eccentric in the literal sense of the word. They have less respect for precedent and more willingness to take risks than others. They are less likely to be motivated by money or career advancement than by the inner satisfaction of hatching and carrying out ideas. In conventional corporate circles, such traits can look quite eccentric indeed.

Stuller (1982) recognizes several types of creativity and provides examples of each:

- Theoretical (Albert Einstein, Sir Isaac Newton)
- Applied (the Curies, Henry Ford, Alexander G. Bell, and Thomas A. Edison)
- Inspired (artists and composers)
- Imaginative (writers and poets)
- Prescriptive (thinkers such as Plato, Machiavelli, and Martin Luther)
- Natural (dancers, musicians, singers, and sports figures)

However he classifies creativeness, he also recognizes a common theme throughout: creative people have the ability to make associations of dissimilar things. That is, their creative ability is not compartmentalized.

A. Generation of New Product Ideas: Reality

The means to develop ideas for new product development are well within the capabilities of all food companies that have a desire to pursue them. Of course, the rub is do they?

Goldman (1983) surveyed 47 food companies in southern Ontario ranging in size from those with fewer than 100 employees to those with more than 500 employees for their management practices regarding food product development. The companies represented canning, freezing, bottling, drying, and chilling, as well as fresh-pack operations. The products processed were meats, fish, cereals, fruits and vegetables, beverages, dairy, confectionery, and pet foods.

The method most used for idea generation was imitation of products of competitors already in the marketplace. This leads to products of the me-too variety with little innovation or originality and no thought for either the customer's or the consumer's needs and expectations. Techniques employing focus group discussions (Marlow, 1987; Cohen, 1990a) and brainstorming sessions were next in frequency of use. (Focus groups will be discussed in greater depth in Chapter 4.) Surprisingly, lowest for frequency of use were those techniques most easily performed or most readily available: attribute testing, recipe books, company personnel suggestion boxes, and the patent literature.

Goldman (1983) noted that as the level of formalized organization of new product development increased, there was a greater tendency to make use of a wider variety of techniques for idea generation. This might also suggest that where companies employed a more disciplined approach to new product development, formal techniques for idea generation were considered a more valuable tool.

1. A Caution about Copycat Products

According to Goldman as well as from my personal experience, the disparity between what idea generation could and should be and what it is in reality is enormous and somewhat terrifying. Companies waste money and effort on new food products by following competitors onto the store shelves with me-too products. This is risky business at best because:

- The originator of the new product has researched the market to obtain a clear picture of the needs and expectations of customers and consumers. The copycat manufacturer does not have this picture.
- The originator is into the market first with a carefully planned marketing program. It will cost the copycat producer more marketing dollars to get market penetration into the competitor's already established market.
- The originator has the processing know-how and has established distribution channels. The manufacturer of the copycat product must learn the technology, and time is not on this company's side.

Copycat products bear a strong element of risk. The copycat does not have the history of development backing up its efforts.

Looking at competing products on the shelves and analyzing a competitor's product should be used not for copycat products but for ideas that will compete with the next generation of products aimed at the constantly changing consumer. That is, today's products on today's shelves are the precursors of tomorrow's products. This is where the developer should be looking. No company can spend too much time generating ideas based on all the information that can be gleaned from all available sources. Shrewd screening will weed out bad and unprofitable ideas later in the process of new product development.

3 Organizing for New Product Development

The best of all rulers is but a shadowy presence to his subjects.

Lao Tzu, Book 1, *Tao Te Ching*

I. THE STRUCTURE OF ORGANIZATIONS

Governments, companies, associations, or social clubs — all these groups require organizations of some kind to function in an orderly manner. Large companies demand very elaborate organizations (often more than they need); small companies usually work with very little formal organization. Organization facilitates, or is supposed to, the practice of management and management techniques and establishes lines of communication, control, and authority.

Organization is considered by many to be important for proper and effective management of new product development; others see organization, especially an overly bureaucratic one, as hampering creativity. However, as will become clear shortly, the practice of management and management techniques (and indirectly the internal structure of the organization) will vary widely with who is managing whom and for what purpose. Managing food scientists is very different from managing marketing or production personnel.

A. Types of Organizations

In the Godkin Lecture, *Science and Government*, Snow (1961) describes three kinds of organizations, which he refers to as *closed politics*. In the management of such organizations, there is "no appeal [hence their closed nature] to a larger assembly" such as a group of opinion, an electorate, or various social forces (p. 56). These systems of closed politics are:

- Committee politics
- Hierarchical politics
- Court politics

Snow's closed politics are characteristic of governance systems in all organizations from committees formed to arrange the company Christmas party to the management of the affairs of tennis clubs to the management of small and large companies; they even describe the internal workings of governments and particularly the most sensitive workings of government, cabinets. Cabinets are organizations that govern without any direct appeal to any electorate. Individually, parliamentarians may be elected, but the internal committees of governments and cabinets are usually appointed.

1. Committee Politics

In structures described as committee politics, all members of the committee have an equal voice and vote. In many new product development teams this may be described as the team approach to organizing for development. In practice this equal voice and equal vote philosophy is rarely attained. Those who have worked extensively on social, church, or school committees or in new product development teams will recognize this at once. There are always those who have "more equality" because of their demeanor, personality, training, years of experience ("I remember when we tried to do that years ago and it didn't work then."), rank within the group, or relationship by blood or marriage to senior management. In an entirely different context, that of sensory evaluation, Peters (1987) discusses the effect of position power, interpersonal power, and personal power on group dynamics. So it is in committee politics.

> I worked with one company in which meetings were held around a circular table in the belief, quite mistaken, that all seated around it were equals — King Arthur and the Knights of the Round Table style. When the president sat at the meetings we all knew where the power and authority was.

In small- and medium-sized companies committee politics is often the form of governance that prevails, but it seldom works successfully because of its inevitable weakness; it is easily influenced by position power. Committee politics can be found in large companies at all levels and especially within subdivisions of large companies. New product development often must try to flourish within it. This structure is not conducive to either good managerial relationships or team work.

2. Hierarchical Politics

Hierarchical politics describes the highly articulated organization typical of governance in multiplant, multicompany organizations. It is the big company structure. Power, it is assumed, is "up the line," resting in some person with ultimate, absolute power somewhere at the top of the pyramid. This assumption is seldom accurate. Snow (1961, p. 60) stated the problem succinctly regarding hierarchical organizations: "To get anything done ... you have got to carry people at all sorts of levels. It is their decisions, their acquiescence or enthusiasm (above all, the absence of their passive resistance)" that decides what gets done.

This structure is characteristic of systems used for new product development in large food companies that try to hide their bigness by breaking up their new product development teams into groups led by brand or product managers. Authority, despite efforts to make the structure appear as a democratic, cooperative team approach, is usually well defined and linked by solid and dotted lines to some member of senior management through a brand manager.

3. Court Politics

Snow's third organizational structure, court politics, is more complex, but at the same time, is one very familiar to most people in large or small organizations; it permeates many organizations. There is power — the boss or president has it by position. There is, however, another kind of power; this power is "under the table" power, unofficial managerial authority, or an undefined ability or knack for "getting things done" that is exerted through some person who possesses a concentration of power, influence, or contacts. This is not necessarily the boss or a person up the line of authority. This individual can be likened to the *l'eminence grise,* the unobtrusive facilitator, a reference to Père Joseph, a Capuchin monk who was the private secretary to and was very influential with France's Cardinal Richelieu — the power behind the throne. Within all organizations there usually exists a person who subtly wields and exercises more power and influence than either title or position would warrant. This person manages, somehow, to get things done, often done his or her way.

Such are the organizational closed politics to be encountered within managerial structures including new product development teams. These are the organizational structures that exist and the best must be made of them.

B. Organizing for Product Development

Although most companies try to organize their staff in some meaningful way for product development purposes, I have consulted with some companies in which informality of organization was the rule:

> In one company, the quality control manager and the research and development manager were one and the same person. He had hastily prepared a sample of a new product on the spur of the moment at the request of a senior manager. The product had been circulated to key personnel and displayed at an international trade fair where it proved popular. Orders came for a product whose formulation had not been finalized or approved, whose raw materials and ingredients had not been sourced for pricing or characterized for purchase standards, and whose shelf stability had not been established. Much postdevelopment work was needed when the product proved to be unstable in the marketplace.

Such informal systems nearly always lead to disaster, frustration, and misunderstanding.

Two questions emerge in a product development organization: the organization is meant for what? and meant for whom? There are two goals in organizing for new product development. The first is to create a system for communication between responsible parties for delegating responsibility and command (or leadership to keep the peace, to keep direction focused, to have authority to get things done, etc.). This system would allow orderly progression from concept to a finished product that has been "signed off on" to the team's satisfaction. This can be looked upon as the physical organization that answers the "what" question posed above.

The second goal is much simpler: to create an environment for creativity, to facilitate innovation and discovery for those involved in the development process. "Meaningful for whom" has now been answered.

1. Organizing for "the What": The Physical Structure

One must separate the organization (that is, the structure) and the modus operandi of management; structure should facilitate management. Unfortunately it does not always facilitate but often hampers management. There is no shortage of management information replete with charts for the physical organization of research and development. With their boxes and solid and dotted lines describing lines of either authority or communication they can be very impressive. A reader wanting these can refer to several excellent papers that are still pertinent despite their age:

- Mardon et al. (1970) discuss at some length the problems of administrating technical departments of multiplant companies.
- The role of the technical manager is discussed from a very human perspective by Head (1971). Head sees this manager with only two resources: people and their skills; and material resources (equipment and laboratory facilities). The manager "inherits a situation, good, bad, or indifferent." How to make the best of the good, bad, or indifferent situation is discussed.
- Aram (1973) describes informal networks (the "undergrounds") that evolve in companies for research and development and supports their encouragement rather than making any attempt to control them with formal management techniques. There is a striking similarity between Aram's undergrounds and Snow's court politics.

Head (1971) shows his disdain toward organizational charts with the following delightful comment, if one queries all the boxes and lines:

You will probably be submerged in a torrent of peculiar terminology about "line control finance wise" and "inter- and intra-functional communication channels." Initially, one thing only will be clear — the appalling debasement of the English language.

One must get beyond organizational charts for effective product development. Yet many companies live by these charts and dote on the dotted and solid lines linking the various boxes. Head would most surely have been a devotee of Aram's unstructured underground and Snow's court politics.

2. Organizing for Whom: The Human Side

The purpose of organization for new product development is to develop a cohesive team of diverse talents, to motivate this team, and to direct their talents toward the creation of a specified product required for the company's business plans. One is not attempting to organize rivalries.

> The future manager will become steadily more active in catalysing the participation process among his subordinates: equally, he will expect, in increasing measure, to participate with his own masters (Head, 1971).

Organization is not solely, then, either for lines of command or for lines of communication, but for lines of participation and facilitation. A strong element in Aram's underground research groups is the assemblage of those who can contribute and participate.

The difficulty for organization is twofold: there must be management of scientists and technologists on the one hand, and on the other there must be management of marketing personnel. These parties march to different drum beats. Organization is necessary to facilitate communication between the multidisciplinary segments that make up the new product development team, that is, lateral communication between the pyramidal structures that develop in large corporations but also communication up and down within these pyramidal structures. Communication implies participation, and this in turn implies that those who participate also contribute.

Another demand on organization is that it must satisfy the personal needs of scientists and technologists for credit recognition and esteem and it must foster creativity, that is, thinking outside the conventional wisdom while keeping them part of the team and not apart from it.

Hierarchical politics for product development should stop at the product development manager (or by whatever name the function might be given). The product development manager is best situated to motivate the new product development team in directions the management desires and to provide lateral communication between the members.

Large companies must control the resources they possess through a formalized hierarchy of inter- or multidisciplinary management teams (portfolios, as one company terms them). Small companies have more informally structured organizations in which, as in large companies, personalities can dominate (i.e., hierarchical systems controlled by the boss or some member of the boss's family). The new product development manager (if indeed there is one in the smaller companies) must be able to ease the project through the various departments involved and smooth the way with a strong cohesive team spirit.

Managers of product development have two resources: physical plant and skilled people (Head, 1971). Managers must harness physical as well as human resources. The plant is inert and immutable; it can only be used to complement the human resources. Therefore people are the most promising resource for developing creativity and innovation. Innovative people will use physical plant facilities effectively and efficiently whether these be test equipment in-house, at equipment suppliers, or equipment with copackers. The touch of the manager must be deft: too much control or too much pressure can stifle creativity and innovation. On the other hand, too little control provides no certainty that innovative product development will ever result. Organization is necessary to a degree; otherwise chaos would rule in the company.

3. Organization and Management

Organization is not the equivalent of management: organization cannot manage innovation and creativity. It is here that the subtle difference between organization as most know it and Snow's governance systems of closed politics becomes apparent. Management can foster innovation and creativity; management involves people. Organization implies planned systems of predictable activities, all of which are coordinated in a controlled fashion. Ultimately, these systems should be interfaced with the other systems that, together, make companies function effectively. If the unexpected happens in this network of systems, the organization of the company should be so structured that remedial activities swing into action to control the unexpected event.

A company's organization can be likened to the human body. When the unpredictable happens to the human body, for example, an invasion by a virus, defense mechanisms as represented by the body's immune system come into play. An elaborate system of activities are coordinated to combat the viral invader. That is the function of organization.

Innovation and creativity are generally recognized as involving activities that are unpredictable, uncoordinated, uncontrolled, and uncontrollable and are certainly to some degree unplanned. Managers of, for example, established brand products cannot be expected to manage product development of those products within their brand domain; these established business managers work within a system. Those engaged in new product development generally take extreme risks and often work within minimal systems.

This being so, does not a paradoxical situation arise whereby if organization is imposed, creativity and innovativeness are stifled? Actually less coordination, less control, less planning, and less bureaucracy do not lead to more creativity or innovation but to chaos and randomness (Aram, 1973). If, for example, a development group cannot be productive, innovative, creative, or inventive within time and budget constraints imposed by senior management who are responsible for the business goals of the company, there is no expectation that it will be more effective in these endeavors if no limits are provided by senior management.

4. Creativity: Thinking Differently

Establishing an environment in which people can be creative and generate new product ideas is a top priority for any company. There are some key phrases to attempting to establish this environment: "perceiving in an unhabitual way"; "have pretty good ideas that do not fit in with policy"; "are eccentric"; and "have less respect for precedent" or "make associations of dissimilar things." Land (1963) latched on to the idea by putting production people into unfamiliar situations where they might make associations of dissimilar things (their production background and a research situation) to develop an atmosphere for creativity and innovation. Too often, one's own disciplines fetter one's mind with unwritten or even written strictures on what is proper, accepted, or the correct way to do something. A set of laws, regulations, orthodoxy, or peer pressure govern one's thinking and funnel it along acceptable (proper, orthodox, correct) lines. This direction of thought can stifle creative and innovative approaches to problems. Land took people out of the humdrum and allowed them to develop and become creative.

Children are unencumbered by this rigidity of thinking. It is only acquired as they grow up, become educated and learn the accepted orthodoxies, encounter peer pressure, and begin to fear looking the fool or being on the outside. Young children have a wonderful capacity to put together unrelated ideas in implausible and improbable ways, as can be seen in their stories or drawings. There is no embarrassment in freely associating seemingly bizarre ideas. The undisciplined and childlike mind has not yet been confined to the path of "correct" thinking. This spirit of free association of ideas was very apparent when I worked as a YMCA instructor and youth leader:

> I played a story game with my young charges. I started the story, reached an impossible situation with the characters, and then passed the story on to one of the children. The rules were simple: the next story teller had to start after a count of five; no magic was permitted and no violence; new characters could be introduced at any (and usually most appropriate) times. Children were eliminated when they could not either extricate the characters from whatever misadventure they were in or start the story on time. Few children were ever eliminated, but they loved to eliminate, it grieves me to say, their leader. The inventiveness of these children was mind-boggling.

Creative people — writers, artists, and inventors — seem gifted and inspired to the ordinary people of this world. What is not fully appreciated by the rest of us is that creative people are also very hard workers who figuratively fill many a waste paper receptacle with discarded ideas, outlines, and schematics. Ideas come only after much study, thought, research, and experimentation, plus plain hard work.

A shout of "Eureka!" followed by the appearance of the scantily clad body of Archimedes heralded the discovery of a basic physical principle that has found uses in many branches of science. The idea, so we are told, came as he entered his bath and noted the displacement of the bath water. It makes an interesting apocryphal story. If truth be told, the basic concept was probably the result of a vast amount of thought and experimentation that only clicked together at the last moment.

At the beginning, there must, then, be an atmosphere for idea generation where strictures imposed by discipline, training, peer pressure, and peer ridicule are removed. In this atmosphere, the purpose is to glean ideas by appealing to the childlike qualities in people. This appeal requires that *all* ideas deserve a respectful hearing regardless of whether they emanate from the janitor, the technical director, the sales person restocking shelves, or the boss's wife (where many of them do come from in both small and large food companies and a source of which I can speak from painful experience).

Comments such as "we did that 20 years ago" or "what good or use is that?" (stock phrases of one vice president of research and development I worked with) are clearly not going to promote an atmosphere in which creativity or innovation will flourish. People will hold back their ideas fearing a rebuke — peer pressure — from their associates. The NIH (not invented here) syndrome must not be allowed to prevail in the creative atmosphere when a company wants to generate new ideas.

All ideas eventually should be screened, evaluated, and accepted for further development or rejected for valid, documented reasons.

Slavish attention to facts, to logic, or to reason (the refuge of technologists) will stifle, at this early stage, any ideas leading to creativity and innovation. According to Sinki (1986), such technical snobbism rates high in creating technical myopia. This myopia he defines variously as the inability to make crucial connections between ideas and applications, the difficulty in "making the translation from abstract to concrete terms," or "why ideas get aborted in their early stages." This is the blindfold that one's training, education, and experience can put on the free association of ideas from other disciplines to create something greater. Information is important; of course, it can assist the generation of ideas, but it can also limit the capacity for bouncing ideas off people. Too much information can intimidate and funnel thinking into conventional channels.

There must be good communication between people from the various disciplines within a company. As a factor in technical myopia whereby there is the separation of theory and practice, Sinki (1986) cites a lack of communication between scientists and entrepreneurs, that is, those who make the idea work. Lack of correlation between unrelated disciplines ("creativeness … ability to make associations of dissimilar things"), another indication of broken communication, contributes heavily to technical myopia.

The final contributors to Sinki's depiction of technical myopia are lack of perseverance and a failure phobia, that is, not "having the guts" to take ideas on to innovative products. How much of creativity or innovation is plain, old-fashioned hard work?

a. Summation

Organizing for new product development seems, then, almost to be a contradiction in terms. In reality, there is no contradiction here but rather a statement of necessity. Innovation and creativity cannot flourish in a bureaucracy with all the pejorative connotations of this word. However, some organization is required to keep the activities of technologists, engineers, market researchers, and production personnel under fiscal control, communicating, and provided with support resources to be creative.

New product development managers must create within the systems of closed politics an atmosphere that fosters creativity. At the same time, these managers must create an atmosphere that harnesses the skills of creative people while encouraging them to work cooperatively as a team with a common goal, that is, fulfilling the company's objectives.

An examination of the elements of creativity (Stuller, 1982) seems to contradict all the precepts of good organizational structure and regard organization as an anathema to creativity, very much to be avoided (Table 3.1). Challenging assumptions, for example, especially those held dear by management, can cause management to feel its leadership is being questioned. A willingness to take risks can be frightening for conservative elements within a company such as the financial department. A summation of Table 3.1, to use today's trite jargon, is "thinking outside the box." The last element, networking, is strongly reminiscent of Aram's undergrounds and Snow's court politics.

TABLE 3.1
Elements of Creativity in People

Creative people challenge assumptions. They ask "why?"

Creative people recognize similarities in patterns, events, occurrences, and concepts.

Creative people connect arrays of events and note new ways to see the strange as familiar and the familiar as strange.

Creative people are willing to take risks.

Creative people are opportunistic. They use chance to advantage.

Being wrong neither concerns nor frightens creative people.

Creative people network. They make contacts with other creative people.

Based on Stuller, J., *Sky*, 11, 37, 1982.

In Chapter 1 creativity and innovation were described by their dictionary definitions, and explanations of their role in new product development were detailed. A fuller discussion is necessary here to introduce some of the challenges of managing for new product development.

Interpretations and definitions of what constitutes creativity and innovation abound in the literature. For example, H. J. Thamhain (as cited in Dziezak, 1990) describes innovation as

a process of applying technology in a new way to a specific product, service or process for the purpose of improving the item or developing something new.

Bradbury and coauthors (1972) put forward a more precise definition of innovation, consequently narrowing its meaning:

the recognition that an opportunity or a threat exists and which is concluded when a practicable solution to the problem posed by the threat has been adopted or a practical means of grasping the opportunity has been realized.

Akio Morita, chair of Sony and inventor of the Walkman® (trademark of the Sony Corporation), was reported as saying that companies place too much emphasis on basic research to their own detriment (Geake and Coghlan, 1992). Reliance on basic research prevents companies from being competitive. The Walkman did not contain any new technology; its secret was new packaging and marketing. Morita saw certain qualities in various new developments and put these together with the perception of a consumer need. This is innovation.

Bradbury and colleagues continue with definitions for discovery: "finding or uncovering new knowledge" and, leading directly from this definition, invention is: "discovery which is perceived to possess utility." Bronowski (1987; but see Stent 1987 for comparison) used the terms *discovery*, *invention*, and *creation* in well-defined ways but added the element of "personalness." As examples, Bronowski cites Columbus, the first European to "discover" the West Indies; however, these islands were already there. If not Columbus, then someone else would have bumped into them. Likewise, Alexander Bell invented the telephone, but the basic elements of the telephone were there; if not Bell, then someone else. Neither event can be called a creation since Bronowski contends they were not personal enough. On the other hand, Shakespeare's *Othello* is a creation despite Shakespeare's reliance on sources written by other authors because this work is personal.

Whatever definition or interpretation one wishes to apply, creativity, innovation, discovery, and invention require harnessing (to direct the skills into desired channels) and encouragement. Companies interested in new products must organize their staff in a meaningful way to direct their staff's activities to the companies' goals and to manage their physical resources economically and effectively and at the same time to foster creativity and innovation within individuals. To foster growth, managers of new product development must be successful communicators, facilitators, and humanists and be a "shadowy presence."

C. Defining Research

Research means different things to different people. For instance, to the person in the street, the word may conjure up images of complex laboratory setups with white-lab-coated scientists who are far removed from normal mundane life and activities.

> My neighbor once remarked upon the seeming coolness of a couple to neighborhood activities. I dropped the comment that both were research scientists and probably kept erratic schedules. My neighbor instantly understood; they were not normal and that explained their remoteness.

Many lay persons would deny that they ever conduct research. Yet these same persons will examine brochures about cars, visit several car showrooms, hold discussions with and question numerous car salespersons, and bargain with financial institutions for the most advantageous payment terms, as well as perform an Internet search on cars and suppliers and join Web chat sites for comments from owners of particular vehicles. They would never dream of calling any of these activities research.

1. Classification of Research

There are two broad classifications of research: basic, or fundamental, research, which may also be called pure research, and applied research. Fundamental research is very loosely described as research for the sake of knowledge without thought of commercial exploitation. The descriptive terms for the two categories are by no means clear-cut. For example, *basic research* has also been used to describe research for which there is a possibility of exploitation (Gibbons et al., 1970). Applied research was referred to by Gibbons and colleagues (1970) as mission-oriented research, that is, research directed to some specific goal.

The following classification may remove, or perhaps add to, the confusion:

- Interest-for-interest's-sake research, which can best be described as research with no foreseeable application. It is the dilettante's research, just to satisfy curiosity (Gibbons et al., 1970). Muller (1980), in a very interesting paper decrying bureaucracy's stifling of innovation by pettifogging funding policies, might term this *seed* research, that is, research time and effort to follow up ideas. There are no time constraints in this research.

- Basic (pure) research, which (no matter how much the academics might argue contrariwise) is always undertaken with some expectation of an application in the future — if for no other reason than to get a research grant. Nevertheless, Muller (1980) quotes Wernher von Braun as having said, "Basic research is what I'm doing when I don't know what I am doing." If time constraints exist, and they usually do, they are measured in years rather than months.
- Goal-directed research, which is research directed to a specific mission, the application of which is quite apparent. The application need is at most a few months. This would be described by Gibbons et al. (1970) as mission-oriented research.

Readers must accept for themselves the meanings that apply within their new product development environment and how they apply these terms to what they do.

All new product development falls in the goal-directed category. That goal is to increase profitability, to gain market share, to exploit a perceived market need, or to counter the activities of competitors in the marketplace with a new product. Small companies cannot afford any other type of research save that directed to protect and expand their profitability. Large companies may separate their research and their development departments with research confined generally to basic research, that is, with some expectation of future exploitation. This research is usually directed to objectives to be accomplished in 3 to 5 years, that is, longer-term research. This longer-term research is often contracted out to universities or other research institutes. Developmental programs are focused on goal-directed research aimed at very specific, short-term, food product objectives.

An interesting recent development is the pooling of research resources. Large corporations pool their resources with government or university consortia whereby cooperative ventures dedicated to longer-ranged projects of research can be undertaken.

Interest-for-interest's-sake research is rarely knowingly undertaken by food companies. It has never, to my knowledge, been engaged in for new food product development. Some explanation of that statement is required. Research with no foreseeable application may be undertaken by food companies to accumulate knowledge or to gain experience in some area of science. While there is no foreseeable application, nonetheless companies may have ulterior motives (not necessarily based on the science or its outcome) in doing the work. For example, support of graduate students in some esoteric field of food science qualifies as interest-for-interest's-sake research. If there is a good likelihood that those graduate students may be hired by the sponsoring companies, then is this research with no foreseeable value? Sponsoring companies have ample opportunity to evaluate their candidates.

Research with no foreseeable application may be undertaken within a large company in the manner of "underground research and development" without management knowing it is being undertaken (Aram, 1973). Aram, in a study of a company involved in research and coincidentally in innovation, noted that informal networks developed within the organization. It was through these informal systems that innovative research occurred. Aram found

> the cross-departmental informal organization ... had the connotation of an activity that was disguised, if not almost illicit. Part of its attraction and its effectiveness seemed not to be managed.

Two individuals from one particular underground group are cited, one from new product sales and the other from product engineering, whose group was responsible for ten patent applications.

2. Fluidity as an Organizational Tool in Research

Seeing how the other half lives, walking in another person's shoes, and bearing another's burden are all clichéd adages used to develop understanding between people. They are all colorfully illustrative of creating cohesiveness within a group, and many managements have applied the concept to the new product development team. It is a direct break with the rigid organizational structure.

Many large companies create this understanding by fluidity within their organizations, whereby technical people must be prepared to be transferred with a product as it matures from the laboratory bench through engineering and production to marketing. In this manner, language differences soon disappear. All members of the team begin to talk the same language or at least are understood by the other members of the team. Since the members of the new product development team come from different disciplines within the company, they have different sets of work values or interpretations of company objectives.

Communication laterally between team members and vertically within participating departments is encouraged with fluidity. This is important. Managers must have communication skills to sell technology as well as the innovative skills of the members of the group to others vertically and laterally within the system. The manager must make this a cohesive group. By moving technologists with products that they developed, the professionalism of many technical people is meshed with the professionalism of other members of the team and also with the commercial and business interests of the company. A greater understanding of the contributions made by all results. It is a form of technology transfer.

The marrying of research and development personnel with marketing personnel has been described as the food industry's rewriting of the television series *The Odd Couple* (Hegenbart, 1990). Hegenbart discusses the turmoil that often arises between these two elements of the team. Research and development personnel are quite used to resolving problems — it's their job. What is not so second nature to them is the resolution of relationships between differently trained people, with a different language and understanding. An early paper by Denton (1989) discusses steps to resolving conflicts between two groups with dissimilar work habits and goals. Managers must not take sides in disputes but should define the contretemps, limit its spread among the team members, and get both sides talking about resolving the conflict together.

Fluidity of movement is essential within any new product development group since the skills in one group may complement the skills in another group in unexpected ways.

D. The New Product Development Team

The new product development team should embody skills necessary to evaluate ideas that justify further exploratory studies and to screen out all ideas not suitable for further consideration or development at that particular time. The team's resources should include the following:

- Management skills: There must be an authority who can manage disputes, provide facilities, and be an enabler.
- Engineering skills: In small companies the engineering skill may be entirely within the production department; they determine processing feasibility.
- Advice and guidance from the production department: Again an assessment of processing capability and scheduling of processing trials.
- An ability to analyze financial data fairly and critically to monitor expenses.
- A source of legal advice respecting food legislation, patents and copyrights, protection of intellectual property and the company's interests in contract negotiations with research institutes and outside laboratories.
- A research and development department with skilled food technologists.
- Marketing and sales data and the skills to obtain customer and consumer information.
- An ability to research reliable sources of necessary materials, that is, a competent purchasing department.
- A skilled traffic department for dictating the proper warehousing and distribution needs.
- A knowledgeable quality control department to provide guidance and advice respecting safety and quality design for the product and advice on analytical methodologies to determine hazard levels.

Food companies have some or all of these skills and resources available to them within the departments housed in their organizational structure. Skills that are lacking in-house can be readily obtained through private outside companies or individuals.

Development teams in small food companies are smaller and often supported by outside skills and resources (lawyers, accountants, and market research firms). In smaller companies plant managers double as plant engineers, quality control managers frequently are responsible for research and development, and even presidents serve other roles, perhaps as financial officers or purchasing agents. This resort to outside resources is often the situation in larger companies, but development teams are seldom smaller.

Communication among a small company's team members is usually good. The greatest danger in such a closely knit setting of the small company or in the small product development team of the large company is that one strong-willed individual — the company president-owner or the technologist dedicated to a pet project — will dominate (an example of position power on group dynamics; Peters, 1987). All attempts at unbiased screening are futile.

In large companies several new product teams may be working independently of each other on a number of different projects. Each has a product manager who acts as the recording secretary of the team. These managers report up the line of communication, which may involve several levels of more and more senior management. The various teams rarely communicate. As a result, communication can be more difficult among individuals of different teams both horizontally and vertically. Duplication of effort can occur as a result.

As work proceeds, dominant roles within the team vary, as major activities in the development process vary. At some phase, chef-food technologists dominate as they develop recipes and prototype products. At another stage, marketing may be conducting sensory evaluation sessions on small consumer panels, and other team members await their results. However, none of the individuals ever stop having an input into the team.

Development teams usually remain in place until the project has been released to production as a part of the company's product line. This movement of the individual team members with the project gives the members a greater appreciation of all aspects of product development. They see the whole picture and understand product development's complexities.

E.   Criteria for Screening

Each person on the development team contributes criteria for screening that are unique to the skills they bring. The knowledge, experience, and training of the members can be applied to product concepts, prototypes, and at all stages of development to narrow a broad range of product ideas down to a few that seem capable of success given the reality of the customer and consumer research of the times.

Development is a complex system of screening steps that require some criteria to ascertain whether ideas should be developed further. Does the project go forward or not? And if it does proceed, how? Criteria for screening can be summed up by one word: *capability*. This is a facile summing up since capability can mean different things. However, defined criteria can emerge from probing deeper into what is meant by capability (Table 3.2).

Unforeseen problems inevitably arise over how and by whom the criteria are to be applied. There is often confrontation; challenging people's skills and opinions can frustrate and demoralize the team. In large companies with their multiplicity of divisions and departments, leaders of new product development teams face challenges as friction may build between people or departments. Managers must control

TABLE 3.2
General Criteria for Screening New Food Product Development Ideas

Criterion	Comments
Marketability	Is the product easily (considering promotional costs) marketable within the company's umbrella of products or brands?
	Does the company have the marketing skills in-house to market the product?
Technical feasibility	Can a quality product be developed within reasonable cost and time constraints?
	Does the company have the technical skills in-house to develop a quality product within time and cost constraints?
	Should outside development resources be employed to complement the in-house development process?
Manufacturing capability	Does the plant have the manufacturing capability to make the product at a cost and quality desired by management?
	If the plant is unable to manufacture product would use of a copacker be justified since increased costs and decreased profitability would result?
	Is there justification for entering into a partnership with another company?
Financial capability	New product development costs money. Is the company financially healthy enough to assume the burden? The financial department must monitor costs, project profits, and keep management aware of the financial risks of the project.

group dynamics to apply criteria evenly, justly, and without personal prejudices of the individual resource persons influencing the screening process.

The most frequent source of friction invariably arises between members of the research and development and marketing groups. Marketing people live in a world of optimism, chutzpah, hyperbole, and persuasion where sooner rather than later is more appropriate. Technical people, by contrast, prefer a world of logical methodology where organized skepticism is the rule. Technical people live in a world of verifiable facts. They keep perfecting and testing; they are inveterate tinkerers seeking protection and solace behind irrefutable data, much to the annoyance of marketing personnel. Technologists are devil's advocates, doubting Thomases. They never want to let their pet projects go; they become very attached to them.

A common complaint of marketing personnel is that research and development people are intractable and inflexible and do not, or cannot, respond to the rapidly changing environment in the marketplace. Marketing personnel see the marketplace as highly volatile and requiring rapid about-faces. Marketing personnel complain that technologists cannot keep abreast of changing market conditions and react negatively to changing ideas. Research and development people, they say, are too absorbed in their science and not absorbed enough in customers and consumers. This latter charge deservedly earns the criticism from marketing people that technical personnel are against everything and lack imagination. Technologists do not understand that the introduction of a new product must be timed precisely and that speed is essential. So say the marketers.

Technical people complain about time. Marketing does not give them time: time to research and develop the project, time to test all the variables, and time to retest and retest. Technical people often view the development process through blinkers, seeing only a narrow field dominated by the particular scientific disciplines they serve. They see ideas as opportunities for technological challenge. They fail to appreciate that ideas must lead to products that satisfy the demands of the customers and consumers on whom the company depends for its survival. They are frequently reluctant to accept the ideas of others, calling these impractical or not in agreement with their own. In particular, ideas from sales and marketing personnel are considered not only impractical but — sin of sins! — unscientific.

Marketing and technical people speak different languages and use different measuring tools in their trades. Each is skeptical of the merits of the others' tools of the trade. The vagueness ("airy-fairyness" as I heard it put) of terms used in quantitative scaling techniques in consumer research and concept testing disturbs the technical person used to logical methodology and verified data with statistically significant results. As French physiologist Claude Bernard has said, "Science repulses the indefinite."

This language issue can be a very real one. For example:

I used the words *rheological properties* in a product development review meeting to describe the flow properties of a sauce I was developing. This engendered hoots of laughter from the marketing personnel amid pleas to speak English. Yet they felt no discomfort about dropping such terms as *perceptual mapping* or *nonmetric multidimensional scaling analysis* during the same meeting.

Confrontations are not confined to research and development and marketing personnel. Production personnel and marketing personnel also gripe about whether new product development is involved. Production personnel live in an ordered world ruled by scheduling of labor, supplies, and produce. Any disruption to this order affects their bottom line. New product development, especially the sudden scheduling of plant runs, disrupts their production schedules and this in turn may affect their year-end bonuses. The importance of new product development, if it is not properly communicated, can be strongly resented.

Tensions arise and are quite normal during the push for new product development. They arise between all segments of the new product development team: between marketing, production, and technical personnel. What one does not expect, and does not want, are problems at the interface between each of these groups that result in the breakdown of communications. As James Humes aptly put it, "The art of communication is the language of leadership."

This then, is the environment of problems and constraints in which new product development must be managed productively and efficiently. Whether the company is large or small, the same problems exist. They differ only in size.

1. Applying the Criteria

Applying these criteria will involve applying subjective evaluations to the ideas collected; that is, people will have to evaluate other people's ideas as fairly as possible. No matter how objective people try to be, the final decision is essentially a subjective one. Difficult judgmental decisions are required. No matter how great individuals think their ideas are, certain bitter facts must be faced; not all ideas will succeed. Ideas submitted must:

- Satisfy the goals for new product success set by senior management
- Lead to profitable products according to criteria established by management
- Satisfy needs and wants of customers and consumers
- Respect certain financial impositions set by management (i.e., be developed or developable within certain budgetary constraints)
- Be within the marketability and sales skills of the company

Ideas for new food products have now been gathered from every source imaginable (including market research). Some ideas may appear very logical on the basis of initial marketplace data. At first glance, others may appear to be quite illogical or bizarre. However, the wildest ideas, if shaped to customer and consumer needs, have an element of brilliance if shaped by skilled personnel dedicated to the growth of the company. Screening promotes the advancement of those product ideas most likely to meet the needs and expectations of customers and consumers and to satisfy the goals of the company for growth with successful launches of new products.

The sole purpose of screening ideas is to improve the odds of the success of a concept to be developed into a new product in the marketplace. Screening does not eliminate ideas. No idea should ever be thrown out. Ideas may be inappropriate for further development today for any number of reasons. However, in the future, the reasons for having abandoned those ideas may be no longer valid. Markets change and develop with time. Problems encountered today become surmountable in the future. All rejected ideas should be recorded with the reasons for nonpursuance stated, cross-referenced, and then filed.

Several elements are encompassed in screening new product ideas. First, the new product development team must have clearly stated company objectives establishing the goals to be reached within a specific time frame and within stated costs. These guide the evaluation process through subsequent phases of screening and development by assessing how closely products in development meet these objectives. The team cannot operate divorced from the financial, strategic, and tactical planning of the company.

Second, there must be an organization with a leader to coordinate the tasks involved with development and to manage a team to carry out these tasks. This leader will have the responsibility of deciding, based on the collective advice of the team, whether to advance an idea for further exploration.

Finally, the new product development team requires physical facilities, ideally a laboratory, test kitchen, and pilot plant facilities or access to these. As well, there must be marketing skills to explore ideas for products that meet the needs and expectations of consumers. If a small food company has no such in-house capabilities, new product development companies and market research companies are available who will research markets and develop products for a fee.

Screening is not a one-time incident at the start of the development process. Development takes time. Six months, a year, five years or more could elapse depending on the nature of the product under development. During this period the marketplace is changing. If nothing else, the originally targeted consumers are getting older and being exposed to other market stimuli. Screening proceeds throughout the entire new product development process.

F. CONSTRAINTS TO INNOVATION

1. The Corporate Entity

The reluctance of food companies, especially large multinational corporations, to engage heavily in basic research is easily understood. It is expensive. Any corporate entity would welcome a major technical breakthrough that could enhance its competitive edge. This desire for a breakthrough must be tempered by consideration of long-range corporate financial goals, of shareholders' desire to make an annual profit, and of the need to stay within the financial constraints of an annual budget.

Management, including corporate management in large companies, has a time horizon rarely fixed more than a year or two ahead. Graduates of most business schools have had it drilled into them that profit is the name of the game, and profit is rarely viewed as anything more than short-term profit. The pressure for short-term gain forces all within the development team to look to quarterly or semiannual profits. This directs new product goals toward me-too products or products with only incremental improvements or advances. Short-term profit has no interest in long-term research. It is regrettable because as Dean (1974) put it succinctly; looking to short-term profit "is like looking for the leak in the bottom of my canoe as I drift toward the unseen waterfall."

Marketing and production departments have time horizons set closely within the annual budgetary plans of the corporation. As a consequence, these departments are usually smiled upon by senior management, who are often less tolerant of the longer time lines of research and development groups. It is easier for senior management to plan with short time horizons than with long ones. Senior management looking at bottom lines on a quarterly, semiyearly or at most yearly basis are much more apt to cut their long-term research projects when it appears that budgets may be exceeded or predicted return is not what was projected.

Risk capital is expensive. Simply put, why take risks with large expenditures on innovative and creative research if the rewards cannot be assured to justify the expenses incurred? Only a very small proportion of research ideas ever lead to the development of an innovative product. Patents are also unlikely to be financially rewarding. Indeed, roughly 2% of all patents survive their full lives.

Management's attitude about its own competitive edge and supremacy of the technology the company possesses can act as a constraint. Technologically successful organizations do not have a strong incentive to embark on a heavy program of new product development if their management believes that they have a proven superiority in technology over their competitors to provide products with a seemingly never-ending global acceptance. There is little impetus to engage in a heavy program of innovative product development. The philosophy "if it ain't broke, don't fix it" prevails.

The history of many technological developments has shown that there is an inordinately long lapse of time between the actual discovery or invention and an innovative application in the form of a profitable new product. One estimate puts this time interval between invention and a profitable product at roughly 11 years (Bradbury et al., 1972). Port and Carey (1997) report on a survey that found 15 to 25 years is common for what is called radical innovation. To add insult to injury, the company reaping the benefits of the new technology frequently was not the company that made the original discovery. One need only follow the changes in market leaders of instant coffees. Such observations do not encourage companies to engage in long-term research.

2. Communication Problems

Communication problems between people, between departments in a company, or between regional manufacturing plants in a company are difficult to handle at any time. In new food product development, they can be particularly disruptive. Unfortunately, communication problems frequently represent conflicts of personalities. Which came first is a moot point: the people problem or the communication problem between the people, their departments, or their manufacturing plants.

Communication problems are exacerbated in multiplant companies, in which communication between plants (lateral communication) or even between the technical staff within these plants can be poor or nonexistent.

> I had an assignment to evaluate the in-plant quality control systems of several regional plants of a large pickle manufacturing company. At one plant a manufacturing problem affecting quality had been successfully resolved several months previously. This same problem was unresolved at a sister plant (making the same product!) not 600 miles distant where I visited a week later.

Lateral communication between the several plants of this corporate giant was nonexistent in passing information to, or even sharing information with, other plants. The reason: the plant where the problem was solved met and even surpassed their production quotas, but the other did not; ergo the former was better run. The result was duplication of research effort at the plant still trying to resolve the problem.

Large multinationals have tended to centralize their research and development resources. The reasoning is fairly easy to follow. By incorporating all the expensive research and development equipment in one facility, together with all the pilot plant equipment and libraries to support the technical staff, there should be great economies of money, no duplication of facilities, and better communications, yes? No, not always. This has frequently produced corporate ivory towers of research and development divorced from the manufacturing, technical, and developmental problems of regional plants. Centralization has often exacerbated communication problems.

I have experienced or been informed of the following communication problems in large multinational, multiplant companies with centralized research departments:

1. I worked in the research and development laboratory of a regional plant of a large multinational, multibrand company. When the vice president of research of the international parent company visited, we were routinely warned not to discuss any projects that we were engaged in with him should we be asked. We cleared our laboratory work benches of all working apparatus. The fear was that our projects would be confiscated by central research and development headquartered in Europe.
2. In another company where the head office research and development laboratories and technical library rivaled that of many small universities, enquiries to this center by laboratory personnel situated at regional plants for journals, books, research reports, or information were discouraged by the local plant managers for fear that such contacts would result in "them" meddling in, or taking over, research projects in the branch operations. In short, central research and development would want to know why somebody wanted this information.
3. A research scientist with a multinational company told me how her company had conducted a new product development program at its corporate research headquarters in Europe on a confectionery product and test marketed the product there. The product was destined for the North American market. The product failed when introduced in North America. No North American input had been asked for.
4. In a similar situation a fruit juice developed and test marketed in Europe at the international corporate research and development headquarters of another company was packaged in material neither approved for use nor available in the North American market. It, too, was meant for manufacture by the company's subsidiaries in the North American market.

These are unconscionable and inexcusable breakdowns in communication. They depict the worst examples of the polarization of effort and lack of communication between regional plants and large far away corporate ivory tower research centers that can occur in the management of the research and development.

a.   Management and Technology Transfer

Technical communication or technology transfer, that is, the dissemination of useful technical developments within the company, requires different managerial skills than does the management of technical people; they are quite distinct activities requiring very different skills. To manage technology transfer within a company requires someone with communication skills, not necessarily someone who can prepare a soundly designed experiment or write a good technical report. To manage technologists and scientists requires the ability to encourage and inspire people, to protect them from bureaucratic intervention, and to challenge them. Managing research is the function of managers of research establishments. Management of the transfer of technology, a quite different matter, should not be relegated to laboratory managers (or supervisors or directors of laboratories or project leaders) but to individuals with skills to communicate technology. This is a vastly different skill.

Not understanding this distinction leads directly to another form of communications breakdown: the transfer of technology from the research and development resource to centers within the company that are able to utilize the information often fails because of the NIH syndrome (variously interpreted as not invented here or not interested here). When people have not had an opportunity to be part of the development and have not been encouraged to see how this development might assist the objectives assigned by management, a sudden attack of NIH may, and usually does, occur. This syndrome often appears when research is farmed out to external research and development companies.

Properly communicated to the nontechnical others in a company, technical information may spark an idea in the nontechnical community and lead to promising ideas for new products.

3. Personnel Issues

Job security concerns all personnel. One's personal security within a company influences one's productivity. If the innovativeness and creativity of technologists and engineers are paramount to their security, then their contribution to product development will suffer or the developer will become a nervous wreck. A failed product introduction is felt by the technical staff very personally. It reflects on them and causes them to wonder about their future.

By the odds, any new product development venture is likely on the basis of the statistics of new product successes to fail. This is the environment that all on the product development team live in. Production and engineering personnel are the least vulnerable members of the team when there is a new product failure, but marketing and, especially, technical personnel stand on the front line facing the odds of failure. Brand and product managers are apart from the team and, despite product failures, are often unscathed by the failure and continue to move up the corporate ladder. The technologists must have great sympathy for a remark attributed to Churchill, the gist of which is, when England wins the nation shouts, "God save the queen," but when England loses they shout, "Down with the prime minister." In the event of a new product failure, corporate management must be a just and forgiving management. Management must allow for failure in such a high risk enterprise.

Product failure must be carefully analyzed to determine what factors were incorrectly carried out or poorly interpreted. This analysis must be conducted constructively: it is not a witch hunt but a learning experience for all to benefit by. On the other hand, errors must be rooted out and corrected. Weaknesses in the development process must be strengthened. Reassignment or retraining of staff may be necessary, and management must handle this positively to encourage their staff's development.

In the same manner, in the event of a successful product launch, just as much is to be learned by an in-depth analysis of why and how success was attained. A keen understanding and knowledge of the strengths of the total development process will be invaluable for future projects. A secondary benefit derives from this analysis when management can suitably reward the achievement of the team members. This secondary benefit can do much for fostering innovation by providing a sense of security and appreciation of one's effort.

Close on the heels of the preceding is the need for management in their organizing efforts for innovation to develop and encourage young scientists and engineers for the future of the company's growth requirements. The rewarding of achievement as well as the learning by analysis of new product development successes and failures will promote the growth of new skills within the company.

Management must accept some blame if innovation has been constrained. If management is unable to defuse the conflicts between such disparate groups as marketing, production, and technical personnel and cannot oil the frictions discussed above, then management is at fault. Personnel will be unable to work productively. Innovation will die. Morale will suffer. Management must be able to unify the new product development team and get them to work as a cohesive force.

Another human problem must be introduced here. This is more a feature of the multiplant, multinational organization. Much product development work, especially that involved with leading-edge technology, is multidisciplinary work. Specialists employed by the parent company work at separate locations. Often, consulting academics at distant universities or research institutes will be employed for a new product development project to complement the team's efforts. Members of a team may never meet except electronically through the exchange of e-mailed reports, through written reports, or by video conferencing. This remoteness can be a dampener to the team spirit of the product development group.

Companies with such development projects must make sure all the distant members contributing to the project are assembled together frequently enough to exchange ideas in person rather than impersonally. Communication among the team members is improved. An added benefit can be the increased productivity created as members of the team interact by bouncing ideas off one another.

4 The Strategists: Their Impact on Screening for Product Development

You can never plan the future from the past.

Edmund Burke, "Letter to a Member of the National Assembly" (1791)

I. THE STRATEGISTS

Three groups within the company develop the strategy that provides the direction for new product development and hence the nature of the screening tools. These groups are:

- Senior management: the boss, the owner in small companies, corporate management in large companies, or the CEO (chief executive officer) and COO (chief operating officer) either alone or together with the board. Senior management provides direction and defines what the company is.
- The head of finance: the chief financial officer (also part of corporate management), the accountant, cynically the "bean counter" or the book-keeper who may be either an internal resource or an external resource such as a chartered accountant. These monitor the progress of development programs and advise management on the financial health of the company.
- The head of marketing (known, in my experience as a food technologist, as the "walkers-on-air"): the head of marketing in small companies is often the sales manager working in consultation with outside marketing research and public relations companies although sales and marketing are worlds apart in philosophy. Their function is to research the possible needs or niches in whatever marketplaces are aimed at and provide information on targeted customers and consumers in these marketplaces.

The latter two also serve as information gathering elements that contribute to policy by feeding information to senior management that is the basis for policy making; senior management always has the final decision-making responsibility. All that follows from idea gathering to the finished product can reasonably be expected to satisfy the goals and expectations established by senior management.

II. SENIOR MANAGEMENT AND ITS ROLE IN DEVELOPMENT

A. DEFINING THE COMPANY

The role of senior management is to define the company, to protect and build on this identity with sound business programs, and to attend to the broad aims and policies of the company. This definition of what the company is and its communication and understanding throughout the organization are essential for effective new product development. It is only by understanding what the company is that true direction can be given to product development.

For example, many food companies have spent considerable effort exploiting the use of nutraceuticals (i.e., phytochemicals, prebiotics, probiotics, and functional foods) in their food products. Some have added mood altering phytochemicals to foods. Are these companies fogging up their identities? Some companies have had to wrestle with the dilemma of whether they are a food company, for example, a snack food company, or a pharmaceutical company dispensing medications for real or presumed health problems of self-medicating customers. Getting into what has been called the "wellness market" presents some very definite strategic changes in direction.

> I worked with a company whose products naturally contained phytochemicals that had been determined to have significant health benefits. I suggested to the president that we use this feature in promotions. He and his management were adamant that they were a leisure food company and neither a pharmaceutical nor a health food company. Consumers used their product because it added pleasure and sophistication to their food; they did not use their product because it aided in warding off disease. They also had no desire to wrestle with the legal problems involving health claims.

It is not enough that the definition of the company be in terms of either products or processing capability (see Levitt, 1975). It is shortsightedness on the part of management to classify the company as a food canning company, a frozen food plant, or an ingredient manufacturing company. The company must be defined in terms of what values, services, or assistance it provides its customers and consumers. It is this knowledge that gives direction to the new product development program. It also allows marketing to find its way. In short, senior management must clearly provide an answer to "what business are we in?" Levitt (1975) provides the now classic example of the railroad industry, which declined because of a lack of definition although the need for passenger traffic and freight transportation did not decline. This need was filled by other means of transportation and not by the railroads. The reason for the decline of the railroads according to Levitt was that the railroads defined themselves as being in the railroad business exclusively: they should have looked upon themselves as being in the transportation business. In short, the railroads became product oriented when they should have been customer oriented.

B. AN INVOLVED MANAGEMENT

Management is part of the product development team if for no other reason than to see and be seen as having a deep interest in, and providing support for, new product development. In new product development, management performs several functions:

- Management establishes early and clearly its interest in and commitment to the new development project. Both the interest and commitment play an important role in encouraging team spirit.
- Management ascertains that the company's business objectives are strictly adhered to and that divergent paths of endeavor are not dissipating the energies of the team.
- Management ensures that ideas selected for development fit the corporate (or brand) image of the company (i.e., they answer to what business the company is in).
- Management's presence can hold in check the rivalries and abrasiveness that may arise among the disparate members of the development team as pressures and deadlines take their toll on even the most integrated venture team.
- Management has an opportunity to assess the strengths and weaknesses of individuals possessing different skills as they work under pressure with other members of the team. This allows justly rewarding good work — in itself a morale booster — and lets management earmark a cadre of future leaders.
- Management removes obstacles from the development process, thereby confirming their commitment to the objectives of the development team.

In small companies, senior management (often owner-presidents) are understandably very intimately involved with their companies' growth and especially with new product development plans. They are often micromanagers.

In one food plant I consulted with, the owner often worked on the line going from one work station to another assisting wherever he saw a need or an occasion to instruct workers in how he wanted things done at that particular moment. There were no written job descriptions, and confusion usually followed in his wake. He alone handled all aspects of new product development in a seat-of-the-pants manner, informally and haphazardly. There were no formulations approved and signed off on by either management or the quality control department. There were no consumer research or market surveys: if the competition did it, the company followed. I had been called in to correct an instability occurring in a newly introduced smoked salmon paté. No stability testing had been done on this product before introduction. Both the formulation and manufacturing procedures were secure in the head of the owner and were related to me on the plant floor. I made suggestions on reformulation based on compositional and microbiological analyses done by an outside laboratory and data provided by the owner. My suggested changes were conducted on the line. I also indicated sanitary violations and violations respecting good manufacturing practice (GMP) that would certainly contribute to lowering microbial loads and for which I strongly urged written sanitation and GMP protocols as well as approved and written formulations.

Lack of a clear definition of what the product concept is can cause terrible friction. This lack of clarity was brought glaringly home to me on one assignment.

I was called in because the president of a small company was not pleased with progress on the development of a vegetarian spread. When I arrived, the project had gone from a product spreadable at room temperature to one spreadable at refrigerator temperature, from one with a smooth buttery texture to one with a coarse egg salad–like texture, and eventually to one that could be grilled on toast pieces and finally to a REPFED (refrigerated processed foods of extended durability). There was no written product description. As test samples were developed, the president-owner would sample them and usually comment negatively on color, flavor, or texture. The concept was always changing direction — all to the frustration of the technical director. My final, and terminal, report to the president placed blame solely at his door for the lack of clarity and progress in product development and suggested strongly he put in writing a description of what he wanted the product to be.

I found myself in a somewhat similar situation on my first new product assignment when I was just a greenhorn fresh out of university. This time it was not a small company where the lack of communication occurred but a large multinational, multiproduct company. The company headquarters were in New York City, where the product manager was also stationed, and I worked at a branch plant in upper New York state.

My assignment was to make a stable icing product — that was it, simply a shelf stable icing. The product manager visited regularly on Fridays to review what had been done and taste samples. (I never knew who the other members of the team were, if indeed there was a team.) After weeks of frustration — my samples were either too sweet, too sticky, too dry, too runny, too chocolaty, too smooth, were not freeze stable, and so on — I complained to our plant's research director, a capable veteran with many years of experience, that I was going nowhere. The next visit I met with the director and the product manager. The director said only, "Give us the complete product concept and its desired quality attributes." This was done rather reluctantly and the product was quickly accomplished after I was told it was to be identical to a competitor's product.

Goldman (1983), in her survey of the product development management habits of food companies in southern Ontario, found that in just under half of companies surveyed with 99 or fewer personnel, the president claimed the main responsibility for product development. In confirmation of Goldman's findings, from my own experience, in all the small plants under 200 employees, the president-owner has had a major impact on the direction, even the specific product, in new product development. Unfortunately, this was often a negative impact.

As companies increase in size, the presidents' involvement in product development becomes, of necessity, more remote. More junior staff such as product managers take over. Only 12.5% of presidents of companies with more than 500 employees affirmed they had the main responsibility for product development (Goldman,

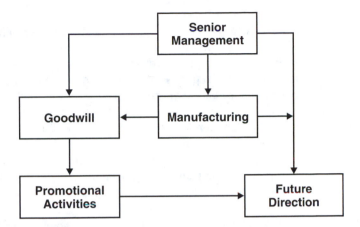

FIGURE 4.1 Understanding the company's corporate identity in new product development.

1983). In large companies, product managers serve this function and maintain a close liaison with senior management, usually at the vice presidential level.

In these larger companies a more organized approach to product development was noted by Goldman (1983). In these companies, 37.5% of survey respondents indicated their job function as product development. On the other hand, no person from small companies who responded to Goldman's survey described their job function solely as product development. In smaller companies, staff wear several hats.

The intensity of the involvement of senior management obviously varies widely with the size of the company. Usually their greatest involvement is in the initial screening phases when they decide whether the ideas fit the goals of the company and fit the brand or the company's image. Their involvement tapers off until the time when go or no-go decisions are to be made.

C. SHAPING THE COMPANY'S OBJECTIVES

Objectives should precede direction in new product development. That is, the direction of new product development requires that companies must first have company objectives. This, in turn, requires that companies know who they are, what business they are in, and where they want to go. When senior management are clear on these issues, then a company's product development objectives can be stated clearly, be understood by all, and be seen to focus on attaining these company objectives.

Senior management in many small companies, owners or presidents or both, often keep this information to themselves. Confusion arises within the marketing and technical development team members because of this privacy. They do not know what they are developing or for whom they are developing it.

The importance of understanding who or what the company is and how this awareness can affect the company are discussed in two excellent papers by Levitt (1960, 1975). An understanding of what the company is provides direction to the ultimate course of new product development. Figure 4.1 depicts a broad picture of the typical company's identity as composed of several elements:

- The owners have objectives; they have dreams and a desire to provide direction to their company. They also have altruistic and humanitarian feelings that lead them to indulge in their favorite charities, to support artistic endeavors, to become patrons of the arts, or to participate in or promote sports activities at the professional, amateur, or community level. These indulgences may be entirely personal or may be well orchestrated financial moves for the company's benefit or product promotion. These activities provide excellent opportunities for networking for senior management and marketing personnel.
- The quality products or services produced in an environmentally friendly and socially progressive company plus support of the arts, sports, or charities lead directly to a goodwill factor that enhances the value of the company and provides promotional opportunities.
- Both the above have a direct and indirect impact on the company's future: its new product or new service development, its acquisition program, or its geographic expansion plans. A simple example might explain this better: participation by a community-spirited confectionery company in local sporting, artistic, or social events might suggest to developers to work toward a nutritional bar or adult candies with a sophisticated taste and for marketers to use such events for promotional purposes.

This, then, is the company, with its tangible and intangible assets that influence its direction within the milieu it and its senior management are themselves influenced and move.

1. Company Objectives that Shape Product Development

Companies are rarely so single- and bloody-minded as to want only to make more money. Of course, management decides whether short-term profits are required. This objective directs the team to product development ideas that will return quick profits, usually products requiring little development time and effort such as line extensions. Where management does not feel constrained respecting its financial situation or where they are looking much further into the future, they opt for pursuing ideas that require a less restricted time horizon, more effort, and a greater amount of money to accomplish.

But the objectives driving the development team are usually only partially financial. Some other objectives are:

- Management's desire to grow; growth depends heavily on new product development.
- Management's need to expand into new geographic marketplaces to lessen dependence on the vagaries of the economic climate of local or regional markets that may be dominated by a major industry or a major competitor.
- Management's need to reduce dependence on commodity-type products and increase profit margins and competitiveness with more added-value products; management wants a broader range of added-value products.

- Management's desire to expand its product base and thereby reduce the company's dependency on one or two bell-ringer products for profitability.
- Management's desire for greater local or regional market penetration with a broader range of products to maintain and protect a competitive position.

These objectives can be compressed into three broad ambitions:

- Financial objectives such as obtaining a greater return on investment for the owners, their partners, or their investors.
- Strategic objectives, often defensive in nature, to protect the company's market position against inroads by the competition, or offensive to gain market share from the competition.
- Tactical objectives by which the company charts its path toward its goals. These are derived from its strategic objectives.

When these aims have been clearly stated and communicated to the product development team, they are used to develop criteria for screening and selecting one product idea over another. For example, if the plan of action is to go toe-to-toe with a competitor either to gain market share or to defend an already established market being threatened by a competitor, then this objective influences the type of product screened for, the criteria used in the screening process, and the time available for development. In such circumstances, the company wants new products as soon as possible.

2.   Sanctioned Espionage or Competitive Intelligence?

Senior management cannot lead or provide direction in a vacuum. They need vast amounts of information and careful assimilation of this information to provide the intelligence required to move advantageously in their marketplaces.

A recent cartoon depicted a candidate for a job in his interview with a personnel manager. The manager says, "Your several convictions for computer hacking will help you fit into our corporate plans." Simple humor or cynical sarcasm?

There is a saying, anonymous as far as I know, "Know your enemy." Knowing what competitors are doing is important for planning counterstrategies that help a company survive. For example, answers to such questions as what promotions are the competition running in the marketplace, what new products are being introduced, and what market launches into new territories are taking place provide knowledge by which a probing company can protect its market position. It is best, however, to know what the competitor is doing *before* the competitor does it. Knowing that they *have* done it is far too late for defensive action. That is the key in competitive intelligence gathering.

Competitive intelligence is generally the domain of senior management — there are vice presidents of competitive intelligence — but the function has also been considered a marketing duty. Companies have used many techniques to spy on their competitors. Espionage is a dirty word that suggests the unethical collection of information, a James Bond sort of activity. Competitive intelligence, however, can be collected in ethical and respectable ways.

Intelligence gathering masquerades under many innocuous names in corporate directories: corporate intelligence, competitive intelligence, market research, strategic business development, or business planning. Such activities, if carried out by trade officials assigned to embassies in foreign countries, would be cause for the officials to be accused of spying and declared *persona non grata*. Yet it is nothing more than the systematic gathering of information about the competition from as many sources as possible and the intelligent interpretation and use of this information to plan strategy against the competition or to counter the competition's strategy.

Corporate intelligence gathering has become very professional. Globally the collecting and analysis of information is expected to become an industry exceeding U.S. $110 billion (Lewandowski, 1999). Those interested in intelligence gathering can join the Society of Competitive Intelligence Professionals in the U.S.A.

The first line of developing a competitive intelligence network involves being in the marketplace to see what is going on. Traditionally, this information gathering has been carried on innocuously by sales people, but information found its way back to management in a haphazard fashion. A company's sales force has the greatest advantage for obtaining feedback from the marketplaces the company services. This includes information such as the following:

- What product facings the competition has, how extensive these are, and where these products are located in the store.
- What new product introductions the competition is launching, in what geographical areas these new products are being launched, and who they are targeting. Additionally, sales people can get a general idea of the success of the products.
- The extent of the competition's advertising and promotional materials in the marketplaces and, again, who these are aimed at.
- How the pricing of similar competitive products compares to the company's own products.
- What deals the competition may be making with the outlets carrying their products.

In short, sales people on site in the marketplaces are able to gather a lot of information about their competition in these marketplaces.

This information — indeed all information about the competition gleaned from any source by the company's staff — is fed back to a central repository, the war room. The term *war room* is found in a glossary used by one competitive intelligence firm (Crone, 1999). The war room is defined as a central location in a company where competitive intelligence obtained from all sources is sifted and analyzed to reach strategic and tactical decisions.

An ongoing activity in any food company is the examination and analysis of competitive products by grading, tasting, and chemical and physical analysis. The ingredients found and identified are costed, and based on the amounts found a crude formulation and costing could be established to determine profit margins.

Competitive information gathering requires that this information find its way back in a very focused manner to those who are able to analyze it and get information to management in a timely fashion. Many large companies require that their staff be aware of any actions by their competitors and report such information back to the war room.

There are other ways of gathering information in an ethical manner (Table 4.1). Some of these activities are broadly classified as networking, i.e., building up a broad association of contacts and using these as conduits for information. As such they are part and parcel of every business person's activities. Networking can be a source of critical information that the analytical wizards of the war room can utilize.

All organizations from chambers of commerce to professional associations meet regularly for business and for pleasure. At these meetings invited speakers may reveal some of their company's business activities. During informal sessions, cocktail hours, or receptions they may discuss or let fall some unguarded words describing their company's programs. Companies with active competitive information programs use such settings to ferret out snippets of information.

The presence of new names or the absence of old names of executives in company literature hint strongly at a corporate policy change. The management style and policies of an executive in previous positions indicate how that executive will operate in a new role. For example, a company's rival hires a new executive CEO and this is widely circulated in the news media. A check of the electronic databases gets information to profile the new executive. Since executives have established patterns of behavior (their previous record was, after all, why they were chosen for their new roles) it can be assumed the new CEO will act as profiled. The company can then revise its plans and prepare itself to counteract the anticipated actions of the rival company.

The technical and patent literature that is publicly available reveals what research the competition has supported, in what institutes it has been performed, and what patents have been granted or are pending. Noting the time lag in such publications, the war room analysts can prepare time lines of its competitor's developmental activities.

More surreptitious activity is arranging to get a seat at a conference dinner beside a competitor's director of research or one of its senior technologists. I personally have always found the conversation around the tables at, for example, the Institute for Food Technologists' (IFT) annual and quarterly meetings, to be very useful opportunities for glimpses of other company's technical activities. The information came either from academics to whom the work was contracted, the competitors themselves in careless, cross-table talk, or suppliers looking for customers ("Didn't you know that So and So company were using our ..."). One company I worked with would spread out its technical staff at the luncheons and dinners at conventions such that at each table where a competitor was recognized one of our staff attempted to be seated there also. We never sat together as a group.

Companies plan social outings for their suppliers, clients, and potential clients at trade shows or expositions; these are carefully orchestrated affairs. Seating arrangements at dinner tables, foursomes for golf, and outings for spouses are arranged so that maximum benefits to obtaining information are had. My wife has

TABLE 4.1
Commonly Used Sources of Competitive Information Gathering

Source of Information	Specific Activity or Information
Print media	Executive moves and removals (reported in business newspapers and trade journals) and mergers suggest policy changes within companies.
	Press releases issued by companies describing activities.
	Help wanted advertisements, especially those specifying the need for particular skills or training.
	Articles in scientific and technical journals in which researchers indicate source of funding by companies and hence reveal direction of research interests of these companies.
	Major equipment purchases announced by companies or major equipment sales contracts made by companies.
	Listings of land purchases by companies.
Corporate publications	Annual and quarterly reports record personnel changes, policy directions, and financial status (purchases and sales of properties and assets) of companies.
	Corporate Web sites with product information.
Conferences and trade shows	Exhibition booths at trade shows demonstrate latest equipment and products.
	Suppliers often inadvertently or purposely reveal other companies' buying activities as a sales ploy.
	Speeches or panel presentations made by senior management, especially in question and answer periods.
	Research papers delivered by technical staff or presented by research groups supported by companies.
Pro bono activities	Speeches by senior executives at charity events, chambers of commerce, fraternal organizations, Young Presidents Associations, etc.
	Volunteer participation in activities of professional associations, e.g., Institute of Food Technologists.
	Participation in alumni association activities.
Access to government information	Information and forms filed with various government agencies.
	Grant applications requesting cooperative research ventures.
	Searches of patent notices and patents granted and pending, which provide information of a company's research direction.
Company-sponsored social events	Receptions or dinners.
	Sports outings such as golf tournaments.

Adapted from Fuller, G.W., *Food, Consumers, and the Food Industry: Catastrophe or Opportunity?*, CRC Press, Boca Raton, FL, 2001. With permission.

provided me with many interesting bits of information as she attended spousal programs at IFT conferences.

Through access to information regulations researchers can explore grant applications, land rezoning applications, and temporary import permits or any of a multitude of forms their competitors may have filed with their governments. Each piece of information has significance and may be pieced together to form a bigger picture.

3. Benchmarking

Much of the foregoing may be recognized as a form of benchmarking, a comparatively new management technique. Benchmarking is a derivative of intelligence gathering and as such is a tool wielded by management to attain its goals for the company. A company practicing benchmarking closely examines its own operations in all aspects. Then it compares itself with the leaders in the same field, often successful competitors. The company then applies the techniques it has determined its competitors to have to its own operations.

The principle of benchmarking is a process of constant improvement of services, of key quality characteristics of products, and of all areas of operations by comparison with the successful leaders in the industry or product category; it is a form of imitative strategy to become more like, that is, as successful as, one's competitor. Companies must know who their competition is and what the strengths and weaknesses of their rivals are. Such knowledge involves researching their own customers as well as those of their competition. With this knowledge, management strategies are developed with which companies move quickly in the marketplace to gain superiority with products and services.

There is a side issue, competitive cost benchmarking. Knowing the competitor's costs for a product leads directly to being able to estimate profit margins. Such knowledge can be used in the marketplace to wreak havoc on the competitor's product launch.

III. FINANCE DEPARTMENT: THE CAUTIONARY HAND IN DEVELOPMENT

The Golden Rule: Whoever has the gold, makes the rules.

Anonymous

A. FINANCE'S NOT SO PASSIVE ROLE IN DEVELOPMENT

A very blunt fact that must be learned by all in any company is that the company must ultimately be successful. Being successful requires making money for its survival. A company is certainly not in business to lose money or to keep its employees happy and busy. Only if the company is making money can it enjoy the luxury of new product development, and it requires new products to make money. The financial department's duties include:

- Monitoring the day-to-day financial affairs of the company
- Alerting management to the financial health of the company and the financial consequences of any action, precipitate or otherwise, that the company undertakes
- Advising management of the company's financial ability to undertake, or its need for, new product development for its continued health

- Tracking the costs of development and comparing these against predictions of expected benefits, that is, cost/benefit ratio

Another fact of life to be learned by the novice product developer is this: the finance department often has its own plans for how the company's objectives can be met. It views with disdain all ideas that require spending money on uncertain high-risk projects, which new product development is. Financial people usually are, by nature, conservative people. They have their own ideas of how to make money for the company in the financial markets they are familiar with.

As a senior food scientist I attended a food product development meeting with senior management, marketing, and production personnel to discuss a new product development program. That meeting was demolished when the senior vice president (corporate) of finance bluntly pointed out to the assembled staff that by transferring company funds into various foreign currencies and by investing in bonds or stocks his department would make more profit more surely for the company than our food division would!

In many ways, the financial department shapes many of the activities for new product development. It has the advantage of knowing intimately what the financial health of the company is and hence how much financial exposure the company can withstand. It knows how much money is readily available for development, and hence how much development can be afforded. Its criteria for screening hinge on many intangibles such as expected profitability, probability of success of any ventures undertaken (i.e., risk assessment); financial stability of the company is the only tangible. Two out of these three criteria are intangibles, that is, outright guesses.

The financial member of the team monitors the costs of the project. These require careful tracking to ensure that the project is within budgetary limits, is in keeping with expected probability of profit generation, and has minimal financial risk exposure. Marketing personnel, with this cost information, can compare the projected introduction costs with sales predictions. The result, it is hoped, is a reasonably accurate estimate of the net profits that the new product is expected to earn. As development progresses and more data are accumulated for more accurate estimations, financial accounting is clarified. Management must assess all inputs to determine whether company objectives are striven toward in a prudent manner.

B. THE FINANCIAL REALITIES OF PRODUCT DEVELOPMENT

The relationship between developmental progress and costs can be seen in Figure 4.2. In the upper graph, the number of ideas under consideration or in development is charted against time. The number of ideas decreases during preliminary evaluation and recipe development as screening takes its toll. Sensory testing, further consumer research, and production scale-up weed out the unpromising ones until few are left, from which finally one is selected for introduction. The upper graph continues, but now sales volume of the newly launched product is plotted against time. The lower graph plots money flow (costs are –$; profit is +$) against the same timescale as the upper graph.

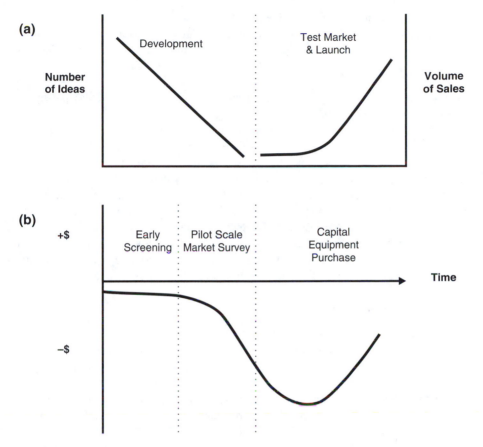

FIGURE 4.2 The relationship between (a) the course of development and (b) the costs of development.

Costs are minimal in the early phases of screening. It costs little to evaluate ideas with readily available marketing, production, technical, and financial data. When preliminary work begins on evaluating and tasting formulations and when outside marketing, consumer research, and development companies are hired for more sophisticated studies, then expenses can rapidly mount. Sensory testing eventually goes beyond small in-house panels to a larger scale that requires large quantities of product. These are used for home-use tests, mini-tastings at trade fairs or county fairs, in supermarkets, or free sample distribution with follow-up questionnaires or interviews.

Plant trials are costly in two ways to a company:

- They require labor, and perhaps extra labor, and they use up raw materials, all for a product that brings in no returns. That is, they produce an overhead for which there is no recompense except the gathering of consumer information.

- They are disruptive of routine plant production from which the company makes its money. Disruption of routine plant production can be a great annoyance to production personnel. Production targets are compromised and perhaps the financial health of the company will suffer.

An intangible cost, therefore, can cause animosity between production personnel and the development team.

Costs increase as development proceeds and research techniques become more sophisticated. New equipment may have to be purchased, specialized equipment designed *de novo*, or pilot plant facilities leased from equipment manufacturers. If new equipment has to be designed and fabricated, a steep increase in costs can be expected. Costs of having products packaged by a specialty packer or copacked by another manufacturer will impact on the profit picture. Costs for mini-test markets and their attendant market research also increase the amount of red ink.

A successful test market justifies a wider initial launch area, but the costs of a wider introduction can be astronomical. Eventually, sales volume, it is hoped, overcomes the red ink. Figure 4.2 should be compared to Figures 1.3a and 1.3b.

1. Slotting Fees

Not all new product expenses can be laid at the doorsteps of food companies. Retailers demand expenses that are major deterrents to product introductions for both large and small food companies. These expenses are slotting fees or slotting allowances. Slotting fees are monies that retailers require to be paid by manufacturers in order that the retailers allow them shelf space, particularly important for new products.

Slotting fees are an impediment to product development for all companies, especially medium and small manufacturers. Retailers justify these fees as necessary to compensate them for the costs and risks they face by putting new products on their shelves. If a new product goes on the shelves, the retailers argue, some established product must be removed or be given reduced facings. This represents a loss of income for retailers. If the new products should fail, retailers lose in two ways. They lose good returns because of the newly introduced product's poor return and they lose the income that would have been generated by the established products that were displaced. Manufacturers object to these fees as too expensive, unfair, and punitive. "Criticism for slotting fees is rooted less in the need to recover the cost of new product introductions and more in potential abuse" (Hollingsworth, 2000). In-depth discussions of slotting fees are presented by Hollingsworth (2000) and Stanton (undated).

Large food companies do have some clout that is not available to small companies, yet even the giants must face giant retailers. Big companies can pay more for space and squeeze the small manufacturers off the shelves. This is the complaint of small companies. As a result, smaller companies are forced into smaller markets (niches) and marketplaces for introductions of their new products. Another tactic of big companies that I was informed of but have never personally encountered is the following: Large companies would deliberately short a bell-ringer product from a retailer who did not allow shelf space for the introduction of a new product. Competing retailers who cooperated with the large companies got the shorted item. When noncooperating retailers allotted shelf space on the large company's terms, then the shortages disappeared and the shelves were filled with the bell-ringer product.

2. Financial Criteria

Financial criteria for screening depend very heavily on what the company's objectives are and how strongly senior management backs these objectives. Objectives vary widely with the economic environment a company finds itself in. Here are two situations with financial criteria that are very different:

1. A food manufacturer is dependent on seasonal processing of locally grown produce. Management chooses to broaden their product base with the development of added-value products and thereby reduce dependence on seasonally grown commodities. They believe this would allow them to keep a trained work force year-round with some community goodwill generated, reduce their overheads, and increase their return on investment. Financial success is measured by:

- Less reliance on seasonal manufacturing and its correlative, developing a year-round manufacturing operation
- Gaining a foothold in a new business in new marketplaces with added-value products

Their investment is for the longer term and it is expensive. They can be more patient in their expectations of a return on their investment.

2. A company is fighting intrusive action by competitors into its marketplace. It needs to protect its market share or to regain lost market share. This company's timeframe is shorter. Its investment will concentrate on aggressive marketing programs as well as new products such as line extensions or the closely related new, improved product with reduced risk, less development time, and less costly research. They require a more aggressive product maintenance program. The company's resources will be directed toward projects consistent with the restraint it faces.

The finance departments of both companies need, nevertheless, to monitor expenses (and rewards) and keep management aware of the financial health of the company.

The new, improved product is not without some risk. If it is one that is "new, with reduced fat" or "new, with improved nutrition" there can be a backlash. Customers and consumers might wonder why it was not improved before or why it was not originally put out "improved." This is a point marketers should consider.

As development progresses, the direct costs of development and the indirect costs associated with the new product's impact on the company's existing infrastructure become more apparent. Both are factors many companies figure into their costs. For example, the costs of development are treated as a loan; the current cost of money, the loan, is added to the development costs. The finance department argues that this money could have been earning interest for the company. The financial department argues that this lost interest should be an expense of development. More reliable forecasting of sales volumes based on consumer research should give clearer information of when expenses will be recovered. If these projections do not fit the company's objectives for both time of financial success and expected rate of returns, then screening begins. Ideas that do not fit are terminated, and the data and research work to date are catalogued and filed for future reference.

3. A Cautionary Note

Financial criteria must be applied fairly. Company controllers can, by using accepted accounting techniques, assess certain expenses as assets (investments) and can regulate the rate of depreciation. How the accounting is carried out can reflect badly or well on product development according to the attitude of the financial department toward the project. Management must recognize any bias that financial people introduce into their financial statements of the progress of development. Where short-term financial gain is favored, policies are usually directed to cutting development budgets, and thus companies risk losing technical skills that could be their salvation in the future.

Projects lost for financial reasons might be the cause of younger staff being discouraged to see challenging work terminated. Projects can be a training ground for younger staff and allow senior management opportunities to evaluate personnel for those who will be the company's leaders tomorrow.

C. Financial Tools

A number of rough rules of thumb have been generated to estimate the potential profitability of projects. They are all crude tools based on the most reliable data the strategists can get; in short, they are guesses used to estimate a guess. As arbitrary estimates of the economic viability of a project, they can be useful to counterbalance the intuition and "gut feeling" that is frequently behind the unwarranted, continued support of pet projects of questionable probability of success. Nevertheless, these tools for project cost estimations are themselves based on assumptions (that may not be true), on estimates (that may be biased by those deeply involved in the projects), and on faulty interpretation of customer research.

1. Comparing Costs with Anticipated Revenues

The simplest and crudest technique is to compare total projected costs against the projected gross sales for the period within which the company wants its payback. Table 4.2 describes projected costs. It rapidly becomes apparent that this index is based on uncertain costs in combination with equally uncertain income derived from projections of hypothetical sales figures. The potentially misleading nature of this criterion highlights its danger if it is wielded to screen out products that are rightfully worthy of development. (Admittedly, as development progresses some of the cost figures emerge more clearly.)

The very crudeness of this technique underlines its shortcomings. This measure focuses only on direct costs associated with new products. The impact new products have in other business areas is ignored. For example:

TABLE 4.2
An Analysis of Projected Costs of New Product Development and Introduction

Breakdown of Projected Costs

Production costs (?)

Overhead costs: These can be assumed to be the same as those for similar in-house product, but the novelty of the new product may confound this assumption.

Copacker fees and costs if product cannot be produced in-house.

Capital expenditures for new equipment or other plant facilities (special handling equipment, storage or warehousing facilities).

Packaging and labeling materials (new containers, packaging and container design, label design).

Raw material and ingredients: Seasonality, quality standard demands, and availability may challenge financial assumptions.

Process and product quality control (additional impact on facilities, analyses for nutritional labeling).

Development costs (?)

Market research (focus groups, taste testings [mini-market tests, in-home placements, etc.], surveys.

Technical research (laboratory research, pilot plant trials, and test kitchen development).

Consulting fees for outside market or technical research.

Marketing and sales costs

Test market and follow-up.

Advertising materials (brochures, handbills, coupons).

Advertising preparation, promotional costs, and media costs.

Slotting fees and other retail trade promotion practices (Stanton, undated)

Note: (?) indicates costs that cannot be projected accurately.

- Were the production department's additional costs of added downtime for line changeovers, extra labor costs, and costs due to regular production disruptions included?
- What extra costs occurred due to warehousing?
- Were the extra costs for sales calls included, whether calls made by the company's own sales force or an agent's, or the costs of extra sales people? A broader product line will be more difficult to sell in the few minutes sales personnel are allotted with the retailer's purchasing agent.

Comparison of total projected development costs and total projected sales alone is not a reliable index. Introductory slotting and promotional costs are variable but are highest when attempting to get market penetration for new products. These costs may weaken when penetration has been made and retailers welcome the new addition to their shelves.

In addition, there is no accounting for management's gratification period. In short, what is management's time horizon? Many products fail because management may have unrealistic expectations of a satisfactory payback period. Most companies want rewards as soon as possible. No new product is ever introduced to the market with the implied intention of withdrawing the product within a foreseeable time-frame. Therefore, management's time horizon is unreal and introduces a real conflict within the development team and with retailers.

2. Probability Index

Attempts have been made to improve the predictability of success. One such attempt, the profitability index (Holmes, 1968), compares the expected return to the total cost:

$$(\text{return/total cost}) \times (\text{probability of success}) = \text{profitability index}$$

This ignores all indirect costs such as slotting fees. This is hardly a great improvement since a guess (expected return) is divided by another guess (total cost) and the answer multiplied by a guess.

The shortcomings of all arbitrary indices to predict success, profitability, or market penetration are due to the imprecise nature upon which the indices are based. Predictions of sales by marketing for new products that have no proven sales record are imprecise. The indices make no allowance for the retaliatory action of competitors such as advertising, promotions, or pricing wars. Estimates of the probability of success, time for completion, and costs are exactly that: estimates. They are only as good as the information that went into them. Garbage in, garbage out, as the saying goes. The indices remain tools to assist decision making, not tools to replace decision making.

3. Other Tools

Malpas (1977) discussed the use of Boston experience (or learning) curves for what could be another criterion for determining whether research and development dollars should be spent. In general, when volume units of a product double, costs usually fall by approximately 20 to 25%. If this generalization does not hold, then it is time to seek a new process. Argote and Epple (1990) discuss the value of learning curves in nonfood manufacturing and cite their value in pricing, marketing, and predicting competitors' costs.

The competition's retaliatory action in the marketplace will confound all these indices and will have an enormous impact on cost calculations. All predictors fail to give due concern to what the competition may do or be planning to do. They, too, are indulging in competitive intelligence gathering.

IV. DEVELOPMENT OF STRATEGY FROM MARKETING'S PERSPECTIVE

Failing to see relevance when it is present is a form of ignorance inadvertently encouraged by traditional practice in science.

Gerald Zaltman, Joseph C. Wilson Professor of Business Administration at the Harvard Business School and codirector of the Mind of the Market Laboratory (Zaltman, 2000b)

... one of the more pernicious scientific fallacies: assuming that the absence of evidence amounted to evidence of absence.

Pearce (1996)

Marketing personnel must tread a fine line in what is as yet an imperfect science marked with imprecision and inadequate tools that provide more subjective than objective evidence. Researching people and their likes and dislikes and what motivates them to buy and understanding what influences them is slowly gaining in improved technology and understanding, but as a science it is, as yet, looked down upon by true scientists.

Science earns its reputation for objectivity by treating the perils of subjectivity with the greatest respect.

Cole (1985)

A. Marketing's Functions

Marketing people have three primary functions:

- To develop marketing strategies based on their research to understand customer and consumer needs and to be knowledgeable about the marketplaces in which the customers and consumers they have targeted for research are to be found.
- To develop sales and promotional materials suitable for use in the marketplaces where they have found their customers and consumers. "Marketing [focuses] on the needs of the buyer" (Levitt, 1975). Levitt does not distinguish between customer and consumer as used throughout this book. He comments further that marketing " [is preoccupied] with the idea of satisfying the needs of the customer by means of the product ... and finally consuming it." These sales and promotional materials must also reflect the wants of the retailer in the particular marketplace.
- To monitor the customer and the consumer to ensure that both are pleased with the product at the time of its purchase and with its first use and remain satisfied and happy about repeat purchases. This is essential for product maintenance.

The first two functions require extensive customer and consumer research; the latter requires an active system of product inspection in the marketplace, a vigorous product integrity program within the food plant, and good relationships with retailers in the marketplace.

Attendance to these functions results in a marketing plan that is unique to the nature of the company and the type of products under consideration. The plan is partly historical, wherein the business situation past, present, and future is described. Also included are the opportunities and difficulties facing the company, which is by way of justifying specific, and it is hoped realistic, marketing objectives (with a selection of product ideas) based on management policies. Marketing strategies and action programs to accomplish these objectives are detailed and responsibilities for them outlined. Then the mechanics (time tables, controls, sales projections, budgets, etc.) of carrying out the strategies are described.

All successful product development starts with the consumer just as all successful selling starts with the customer. Coincidentally, selling must include the needs of the retailer in the environment in which the customer and consumer are to be found. Senior management in small companies often make a fatal mistake: they start with preconceived ideas of what they or the company's owners think they intuitively know consumers will want. This is flawed thinking. The misconception (not unknown in large companies) of what consumers want frequently has its genesis in a feeling ("a gut feeling") a member of senior management has for some pet product idea. "It will sell. I know it. I have this gut feeling." They may very well be correct but they also have a very good chance of being wrong.

Marketing strategists have a misconception that is somewhat more prevalent in technology-dominated food companies: that technologically advanced products will sell. Technologists have convinced the rest of the development team that gimmickry in packaging or product disguised as novelty or innovation will draw customers. Wrong! Customers determine what will sell if and when they and their consumers' needs are satisfied.

B. Market Research

> The difficulty is to find the common denominator that governs the actions of men.
>
> W. Somerset Maugham, *A Writer's Notebook*

Market research is an organized and unbiased investigation to:

- Measure qualitatively and quantitatively all factors influencing the marketplace
- Provide information from the data gathered to guide marketers in decision making

The subtle oxymoron "organized and unbiased" stands out like a sore thumb; herein lies the potential weakness in the interpretation of market research data. If any investigation is organized, *a priori*, some bias is introduced by the researcher. It is the market researcher who:

- Selects the populations to be interviewed
- Selects the research techniques used in studying customers and consumers
- Prepares the wording of questionnaires and other survey materials
- Analyzes and decides how the data are to be weighted and interpreted

In the preconscious process of converting primary data of our experience step by step into structures, information is necessarily lost, because the creation of structures, or the recognition of patterns, is nothing else than the selective destruction of information (Stent, 1987).

The decisions made by the market research company necessarily introduce a bias. Market research companies do not want to antagonize and lose their clients. When clients show great affection for pet projects, the research company might feel obliged to shape research methods or to interpret findings to encourage their clients to continue a liaison with them. This is not a condemnation of market research firms or of techniques for market research. These latter are becoming very sophisticated and extremely high tech. However, an objective assessment of the reality of the market-place is obtained only with a careful and close liaison between the company's own marketing and sales personnel and the other members of the development team and the outside market research company. Interpretation of consumer data is subjective.

A further danger in market research arises when expansion into foreign markets is considered. Foreign markets are not extensions of the domestic market and cannot be considered as anything but a new market with new customers and consumers. The buying habits of customers are different no matter how close the countries may be geographically, culturally, and linguistically. Countless times my American clients have plaintively asked me, "But why are Canadian food laws and regulations different from ours?" or "Why is that flavor popular in Canada? It isn't popular in the States." Or more commonly, "Why do Canadians celebrate Thanksgiving at a different time than we do?" (this said in response to discussions on timing of promotions or arranging visits to their clients). Developers must recognize that food laws and customer and consumer habits and tastes differ from country to country. A classic example of these differences was demonstrated when Snapple™ first appeared in glass in Japan, where most such drinks are dispensed from vending machines for which cans are more customary for the Japanese. Again, in questionnaires used in customer research, some cultures are reluctant to offend and may distort answers and convey the wrong message to survey takers. Telephone surveys may be skewed when it is found that in some countries there is not a heavy concentration of telephones and hence fewer telephone users.

C. THE TIME ELEMENT: A FACTOR IN DEVELOPMENT

The days, months, or even years necessary for development may confound developers in two ways:

- Volatility of those originally surveyed for their preferences: They change their opinions subject to local or regional happenings.
- Aging of the targeted population: At different ages, people have different needs, preferences, and opinions.

Suppose an analysis of demographic data predicts that the population in a given market area will drop by a million people over a period of 4 to 5 years; the consequences are that a million meals or meal items will be lost each day of each year in that market area. Such losses are not uncommon as the economies of communities change. A company has less than 4 years to find new products or develop new markets to fill the void caused by the gradual loss of those meal items as people move away. Four or five years is not unusually long for the development of some types of products nor is it a desirable gestation period for the development of new markets for which a reengineering of the company may be necessary.

When a new product is introduced, the originally surveyed targeted consumers are several years, months, or weeks older than when the developer originally conceived of the product and tested the concept with them. The new replacement consumers, who have been subjected to a totally different set of stimuli, were never surveyed.

Time is not on the developer's side. Developers are not aiming for today's customers and consumers. They are always developing for some targeted customer or consumer in the future. For these reasons, when products for development are conceived, they must not be based on today's available demographic and psychographic data, which may itself be a year or more old. Even followers of trends must be aware that their data may be months old; this is a long time in marketing. The data, today's data, must be extrapolated to describe the consumer of the future; extrapolation does carry an element of guesswork that developers should be aware of.

Market research continues throughout the development process. It must continue throughout development because product development takes time. During this time customers' and consumers' tastes can change. They age, get married, move, and have children. Many new trends can emerge during this short period. What were once considered good ideas for products based on today's consumers may prove to have been short-lived fads by tomorrow when they are introduced into the marketplace. Consumers can be very volatile in their likes and dislikes. Information about such startling consumer changes is best learned early in the development process through continuing market research feedback before too much time and money are wasted. Products based on short-lived fads must be identified as such early in the development process.

The marketplace is replete with change, not all of which can be extrapolated with certainty to define trends despite all the surveys, for example, the sudden emergence of nutraceuticals in the health food market and government and medical concerns over the safety of some of them. At best, the studies merely underline the volatility of customers, consumers, and the marketplaces in which they are found. There are very few facts to be found in the marketplace; observations, yes, but precious few facts. Claude Bernard, a renowned French physiologist, wrote, "Observation is a passive science, experimentation an active science."

D. The Nature of Market Information

Chapter 2 includes a discussion of entities such as the average customer or average consumer. Entities making up groups that are averaged must have homogeneity or commonality. A casual observation in any of the various food marketplaces reveals that the customers and consumers have little in common except the fact that they were there at the time the observation was made. One must look for common traits before averaging becomes meaningful.

Nevertheless, there are attempts made to categorize consumers, to determine what traits in purchasing and in usage consumers have, or do not have, in common. This is a far cry from averaging the consumer and then trying to apply descriptive terms to this average, for example, "the average consumer is a white female between 23 and 35 years of age, etc.," which conveys very little useful information for product developers.

MacNulty (1989) described one such categorization. By classifying people, particularly those in the Western world, relative to their personal values and motivations, MacNulty arrived at three main groups:

- Inner directeds: These are "highly confident, self-determined individuals." They are best described as individualistic and as leaders.
- Outer directeds: According to MacNulty, "they want others to see how well they are doing." They need to be seen as having status. They are less confident and inclined to be followers.
- Sustenance driven: They are more safety and security conscious. They like the comfort of the traditional and avoid risk taking. They are very conservative in their buying habits.

New products are, therefore, directed toward the further characterization of the needs of the inner directeds, which MacNulty describes at length. The inner directeds will be the pacesetters and "will provide us a picture of what the Outer Directeds will be doing in the next year or two." In a little more time the sustenance driven will catch up to the outer directeds, who will in turn be trying to catch up to the inner directeds.

The British brewer, Bass Taverns, a conglomerate with many holdings in the food service industry, used a combination of fuzzy logic (analytical procedures used to interpret data that are approximations, i.e., are neither "yes" or "no" in nature) and search algorithms to locate their taverns and restaurants (Davidson, 1998). Bass Taverns's research distinguished eight consumer classifications:

- First tasters are young, affluent, and adventurous (perhaps identified with MacNulty's inner directeds).
- Blue-collar hunters, usually unskilled laborers, prefer a local pub with arcade games, bright lights, and canned music. Simple tastes run to draught ciders and ordinary lagers.
- Premium wanderers are singles on the prowl and out for enjoyment. They usually belong to the bottled beer brigade.
- The pint-and-pension group are an older, limited-income crowd who stick close to home for a quiet social evening.
- The student crowd, a young pint-and-pension crowd, are on a limited income and stay close to campus.
- Quality diners have good incomes and a love for good food and are willing to travel for it.
- The cards and dominoes group want only the quiet sociality of the local pub.
- Nightclubbers have money and want entertainment.

Yankelovitch Partners (2000) described the characterization of U.S. consumers:

- Up and comers: Young, upbeat, and childless "yuppies" (young, urban professionals), who lead the market and are not led by it. They tell marketers what to do.
- Young materialists: Single people who cynically equate money with happiness.
- Stressed by life: As expected these are people, especially parents, with heavy burdens.
- New traditionalists: Forward-looking, family-oriented upscale people who "set the agenda for boomers."
- Family limiteds: Families are central to these individuals.
- Detached introverts: Successful and moneyed but lonely "geeks."
- Renaissance elders: Wealthy, mature seniors with a zest for life.
- Retired from life: Uninvolved seniors who "hear you knocking but you can't come in."

These breakdowns of consumers reveal a diversity of market niches. Knowledge of the characteristics of these niches provides the direction for developing ideas and also for screening out ideas that would be unsuitable for the categories of consumers that research has found.

> Culture is smitten with counting and measuring; it feels out of place and uncomfortable with the innumerable.
>
> *Jean Dubuffet*

E. QUALITATIVE AND QUANTITATIVE MARKET RESEARCH INFORMATION

Market research information can be qualitative, quantitative, or both. Qualitative market information is based on the interpretation of focus groups and interviews. These tools require subjective interpretation. Quantitative market information is based on the statistical analysis of data obtained from surveys and questionnaires and direct measurements of what customers buy, when they buy, where they buy, and what combinations of purchases are made. Quantitative research techniques aided by computer-assisted data analyses are the norm. Techniques employed in researching the market (and there are many) require that the researcher clearly understand what information is wanted and that users of this knowledge understand its limitations.

1. Focus Groups

Focus groups are a commonly used research tool for developing product statements or concepts. A focus group is an assemblage of consumers sequestered in a specially designed and equipped room. Typically between 8 and 12 people, selected as representative of the target consumer that the new product development company wishes to reach, participate. Market research companies keep lists of consumers whose backgrounds are well documented. With little effort they can enlist consumers with any desired profile that the client wishes to have participate. The samplings of respondents are not randomly chosen but represent those the market research company has worked with and has on file.

The group is led by a moderator — a professionally trained discussion leader. The moderator leads the group through a discussion aided by a prop to stimulate a more focused discussion. The prop may be a simple description of the client's proposed product or descriptive artwork depicting the proposed product, or there may be a prototype product there for the group to see. The moderator elicits comments from the group about this proposed product, always probing their reactions, always gently pushing to get more focused attitudes to the prop. Sessions usually last 2 to 3 hours. Clients can investigate the impact a product concept might have on several different consumer profiles through these groups.

The rooms are equipped with cameras and voice recorders and are often paneled with one-way glass, allowing outsiders to note the reactions of the group without themselves being seen. Proceedings may be audio taped and filmed to enable consumer research companies to carefully analyze the oral and body language for hidden clues in the responses. Several focus groups (usually no more than three or four) with different individuals may be required before a clear, concise concept statement for a new product emerges that embodies what the product is and how it will meet the needs and expectations of consumers. Often representatives of the client company are present as observers.

There are as many variants of the focus group as there are market research companies. The technique masquerades under various, sometimes quite obscure, esoteric names such as real-time knowledge elicitation groups or simulated test markets (a form of focus group). They all have in common a trained discussion leader who focuses the discussion of a group of targeted consumers onto a food concept embodied in a prop. The leader later collates the reactions of the group to the prop and reports the findings to the client. Some consumer research companies use the same consumers over a 2-day session of discussions to get the desired concept statement.

The participants are paid, which introduces a potential danger of bias in the results. The participants may tend to please or agree with their moderator.

> I was a guest lecturer at Concordia University in Montreal for a course in communication. After a lecture on focus groups a marketing student related that she participated in focus groups regularly because they provided extra pocket money for her. As for her comments in these sessions, she said whatever it was obvious the moderator wanted to be said. It kept her on the recall list. This did not surprise me greatly after my personal experiences with focus groups conducted by a market research company in the San Francisco area.

Because of these experiences, I strongly urge companies to use any information derived from focus groups cautiously.

If done well, focus groups are intended to provide qualitative information about consumers' needs and interactions with the product concept. Done poorly, focus groups can be a waste of both time and money because they are expensive. The results can be distorted by group-think, with one dominant person swaying the other participants' opinions. They can be manipulated and interpreted to jolly along a client by the market research firm. Marlow (1987) and Cohen (1990a) discuss the value of focus groups, how best to use them, and the mechanics of using them in product development. A caution is in order: a food company interested in using the focus group technique for market research should seek out a professional market research company with an established reputation. As Marlow remarks, focus groups have value for suggesting direction for ideas but they cannot be the basis for business decisions. The main function of the focus group is to determine consumer reaction to product concepts and from this reaction to redesign the concept for products to be on target. Interpretation of the data is highly subjective.

2. Alternatives to Focus Groups

My opinions of the frailty of the focus group in providing reliable customer or consumer data are probably obvious. There are other techniques for eliciting the reactions of consumers. One of these is ZMET; another is neuromarketing. Both of these are described here from literature available from their developers.

a. Zaltman Metaphor Elicitation Technique (ZMET)

Olson Zaltman Associates (undated) distinguish two types of market research:

In Type I research the data is derived from descriptive research, structured surveys, and controlled experiments. That is, it is derived from focus groups, mini-market tests (done with a local social group), or taste test comparisons. This research attempts to answer questions such as: What variations of purchasing are there among consumer segments? Has a company's market share changed? Is a company's product preferred over its competitor's product regarding its qualitative attributes? Type I research "require[s] consumers to consciously focus on what researchers think to ask them." Therein lies a potential shortcoming of this type of research: the researcher introduces a bias.

Type II research attempts to understand the customers' and the consumers' needs, desires, and goals. The questions and the responses are more complex. Some typical questions asked are: What do my customers and consumers want and how and where do my products answer these desires? What does our brand or product mean to our customers and consumers? Or this more unusual question: Why do consumers use our product differently in different situations?

ZMET (a patented process) explores consumers' relationships with a brand or product. These complex relationships involve a multitude of constructs that form a mental model that Zaltman (2000a) refers to as a *consensus map*. To understand how consumers value a brand (or a product), this consensus map must be uncovered by an understanding of these various constructs.

In Type I market research, for example, focus groups, the researcher brings the stimuli to the subject. In ZMET testing the subjects brings their own stimuli to the researcher. Consequently, the subjects' responses are not couched in the nuances of meaning of the researcher's language nor in the need to develop such a language (Bone, 1987).

ZMET methodology is based on several established assumptions (Zaltman, 2000a). In practice, 20 to 30 individuals are first selected. They have expressed an interest in, intimacy with, knowledge of, or recognition of the product or brand under test. The participants are told what the subject matter is and are required to bring a number of pictures or to take photos (at least a dozen; camera supplied) that characterize their good and bad feelings and thinking about the topic.

The participants are given individual appointments for one-on-one interviews with skilled interviewers backed up by computer imaging specialists. The interview consists of several steps in which subjects relate their ideas and uncover hidden thoughts (metaphors) that describe very closely how they relate themselves to the product or brand. The constructs thus developed are measures of this relationship. The example in the reference, using data from an actual test, provides greater understanding of the process (Zaltman, 2000a).

b. Neuromarketing

The BrightHouse Institute for Thought Sciences in Atlanta, Georgia, has developed a technique they call *neuromarketing* for analyzing people's reactions to consumer products. They have used functional magnetic resonance imaging (fMRI) to study activity regions within the brain to study people's reactions to products, services, and advertising (Lovel, 2002). Researchers claim the fMR images reveal brain activity indicative of how the subject is actually evaluating a product, a service, or any advertising or promotion when viewing such objects. Thought Sciences researchers further claim this knowledge is a more accurate measure of consumer preferences than focus groups or surveys.

Prior to a test scanning, participants are surveyed to evaluate their responses (likes and dislikes) to a variety of food products, promotional materials, and other consumer goods. The subjects are placed under an fMRI scanner and then shown objects on a screen, whereupon a brain picture is taken. Comparison of the picture displaying brain activity and the survey results enables researchers to find the preference center of that participant's brain. With such data, researchers believe they can help clients develop better products, services, and marketing campaigns (Brighthouse, 2002).

c. Summary

Both ZMET and fMRI scans are new tools for measuring consumer valuations of products, brands, or advertising. With no experience of either process, I can describe neither advantages nor disadvantages for either ZMET or neuromarketing. Both may have shortcomings.

F. MARKETING'S WAR ROOM

Earlier in this chapter management's role in competitive intelligence gathering and its war room were described. It is in this war room that management gathers business intelligence about the general business climate, and about its competitor's activities and policies in particular, and formulates its own policies and counter activities.

TABLE 4.3
Getting to Know Customers and Consumers in the Computer Age

Tools	Description and Use
Cookies	A note (small text file) on user's computer dropped by a Web server. It allows third party advertisers to track the user via any site the third party advertises on. Hence they know the user's interests.
Web bugs	Invisible surveillance devices used by governments and big business to monitor Internet activity. They are 1-pixel GIFs (graphic image also referred to as 1-by-1 GIFs and invisible GIFs) that act like a banner advertisement that "talks" to cookies and reports back. The privacy concern is that they can track the user, but the user does not know they are present because the user does not see a banner ad.
Spyware	Software that is installed on a surfer's computer without the surfer's consent or knowledge. It is often part of free software downloaded from the Internet. The program reports back on the surfer's use of the Internet. Food companies do not use spyware but will possibly buy market information from companies that do.

From Hunt, J.T., *Natl. Post Bus. Mag.*, 48, Oct., 2000.

Marketing, in addition to its market research, also conducts its own war room activities, perhaps in a small corner of management's facilities. It gathers market intelligence about customers, retailers, and selling locations.

Customer shopping habits have changed due in part to the Internet and computer age. Selling, buying, advertising, and customer service (product maintenance) have all been influenced by these tools (Table 4.3).

A cartoon, *PC and Pixel*, by Thach Bui and Geoff Johnson depicted a customer reading a sign at the entrance to a supermarket. The sign read, "Here at MegaFood-Corp, you're more than just a customer … You're a completely predictable compilation of spending habits and product data." Privacy is rapidly becoming a thing of the past.

A dramatic and personal example of customer profiling or purchase profiling happened to me:

I received a telephone call from a person with my bank's credit card fraud department (I verified name and source). He asked first if I had my credit card (I had it), whether the night before I had visited a particular sushi bar where I must have been quite a bon vivant and spent over $700 on a meal, and if I had that morning spent over $300 on sunglasses. I had not done either. I was told to destroy my card at once and a new one would be in the mail immediately. The fraud department was alerted by the fact that *neither purchase fit my pattern of card use for purchases.* In addition, after the second purchase my card had been used, unsuccessfully, for a pay telephone call. The latter action was a sure giveaway that the credit card thief was trying to see if the card was still active. It was not: my bank had canceled it after two non-pattern-fitting purchases.

Courtesy cards, privilege cards, and client cards (for example, AIR MILES® cards) provide the card companies with tremendous customer information: what purchases were made, where they were made, when the purchases were made, as well as the addresses of purchasers and their credit ratings. All this information is requested when customers register for the card. One value of these cards is the discouragement of comparison shopping; customers are often held by their desire to accumulate points for redemption. With data-mining software a very complete picture of customers and their purchases can be obtained with very little effort. Service stations, food chains, and box stores requiring membership all can gain valuable customer purchasing data.

Web sites are doing this profiling for a profit, to sell a product or to target an audience. They maintain that this is actually to serve the interests of their customers, the Web surfers. They can tailor advertisements to suit the interests of their customers. Thus, they claim they serve the interests of Web surfers at the same time they better target potential customers for advertisers on the Web. This on-line profiling is justified as the ability to deliver the right message to the right people. It is no longer broadcasting a message but *narrowcasting* one.

Companies also use hotlines (1-800 numbers) and help lines to gain valuable customer and consumer information. As the last number is dialed, and even before it rings at the company's reception area, reverse directories have identified the caller's location and name while search engines pull up as much information on the caller as possible (e.g., previous calls, concerns, etc.). The receptionist is then armed with information about the caller to provide any product service, suggest new products, or reroute the call to others more able to help. Even if the caller routinely calls from an office telephone, the software can eventually associate that business number with the client.

G. MARKETING AND SALES DEPARTMENTS

The most important responsibility for marketing and sales departments is monitoring the marketplace for changes that might affect the course of development. Their vigilance in the marketplace alerts the team to new considerations in screening criteria, to new market opportunities, or to necessary changes in products under development, and most importantly to the activities of the competition during development.

Marketing and sales staff develop marketing and advertising strategies for products under development. With cooperation of technologists and the legal department (or an outside resource), they prepare label statements; print material for promotions; advertising copy for newspapers, radio, and television; and any product claims as well as recipe and usage suggestions. They oversee art work for labels and copy for advertisements and media campaigns that will be used in promotions.

Marketing personnel consider what impact new products may have on a company's established branded products. Will there be fragmentation of the market? If so, is this fragmentation good or bad? Within the larger market, there may be several smaller markets — niches — that might be profitably exploited, making fragmentation a good thing. The instant coffee market was profitably fragmented, for example, with the development and introduction of flavored coffees.

Marketing and, in particular, sales members of the team must evaluate what impact products may have on the retailer. The following example illustrates this concern:

A microbrewery (craft brewer) introduced a prestige beer with an old fashioned, wired-on cork stopper. Mini-tests were conducted in nearby campus pubs. Complaints from bartenders and waitresses poured in: Opening bottles was inconvenient; service was delayed; hand injuries resulted. Wires were a hazard underfoot, and the corks made good missiles. The beer was fine. The container was redesigned.

Much of this chapter has been devoted to getting to know customers, consumers, and retailers in order to generate ideas for new products that may satisfy the needs and expectations of all parties involved. Data generates information when interpreted. Information, in turn, generates ideas. Quantity of ideas, not quality, is important at the initial stage of the process. Critical screening will eliminate bad ideas later.

The generation of product ideas based on needs and desires, that is, the "I want," must come first from within either the customers' or the consumers' psyches, so to speak. This requires an intimate knowledge of consumers and their gatekeepers, the customers. The first cast of the net for ideas is cast wide. As both demographic and psychographic data about consumers are gathered and converted into information, the net is pulled tighter. A picture of the needs and expectations of a very specific group of consumers will emerge.

H. MARKETABILITY AND MARKETING SKILLS

Does the marketing department have the ability to market the product to the targeted consumer? Will the product require unique marketing skills to reach the intended consumer? For example, not all companies possess marketing skills to introduce new products into the food service or into food ingredient markets. They are simply not familiar with these markets or do not have the contacts for them. Inexperienced marketing resources within companies may make them unable to cope with new ventures. Companies should be aware of these dangers in unknown markets and should recognize their own shortcomings in these markets.

Can a market dominated by a single competitor be challenged? Is the marketing department able to penetrate a market with few major buyers or even a single buyer? How big is the market? Is it local? All these considerations should, in competent marketing departments, be uncovered in market research. Unfortunately, sometimes companies are not cognizant of their weaknesses. Yet these are very basic marketing considerations. All have been implicated, with the benefit of hindsight, as causes for new product failures (Anon., 1971). They are screening criteria that should have been applied. It may prove more profitable to sell a by-product of manufacture to an ingredient supplier to manufacture into an ingredient.

Products that lack marketability fall rather loosely into two categories: Either (1) they are so far ahead of their time they cannot be marketed (perhaps in time they will be) without an excessive amount of consumer education or marketplace reform or (2) they possess a difference from existing products that consumers can neither perceive nor appreciate or they are inferior to products already on the market or cost appreciably more than superior products already available. These products should be screened out early in development.

I. SUMMARY

Any of the criteria used to screen (i.e., marketability, technical feasibility, manufacturing capability, and financial criteria) could be reason enough to abort development at any stage. Two of these criteria, marketability and technical feasibility, must be applied carefully and dispassionately. Probabilities of success in both these areas are based on subjective assessments. Those responsible for new products can, and do, get emotionally attached to pet projects and are reluctant to accept the need to drop the project.

Being objective can be difficult for the product development team. Leadership must be enlightened and compassionate. They are, after all, in the "people business." At the same time, the leadership must be dispassionate in applying criteria in screening. It is here that leadership must be demonstrated.

5 The Tacticians: How They Influence Product Development

I. SCIENCE AND TECHNOLOGY IN ACTION

There is something fascinating about science. One gets such wholesale returns of conjecture out of a trifling amount of fact.

Mark Twain

Research and development, the engineering department, and the manufacturing department are the tacticians who carry out the product development strategy of the company, which as development progresses, becomes a winnowing of product ideas until only a select few that promise the greatest chance of success remain.

In many small companies, the only technically trained individual usually heads the quality control group. This person also heads up new product research and development if there is any such work in progress. The senior mechanic represents the engineering department, and the plant manager, the manufacturing department. Small companies usually do not have an active, ongoing new product development program. Large companies, on the other hand, may have amazingly complex infrastructures for organizing the tactical groupings of product development.

A. RESEARCH AND DEVELOPMENT: THE CHALLENGES

The development process began on paper when all the ideas were explored for their appropriateness for the company, their capability to bring in needed revenues, their marketability, and their technical and manufacturing feasibility. The technical research and development process really begins when the product concept is known. Food technologists require only rough concepts of products-to-be before they can provide basic recipes for them. Their sources are cookbooks, ingredient suppliers, or analyses of similar products in the marketplace. These basic recipes serve several purposes:

- Food technologists begin to anticipate the problems associated with maintaining quality and safety of the product that might be encountered during development and subsequent processing.

- These prototype recipes become props used by marketing personnel to refine concept statements for focus groups and serve as test samples for in-plant tastings.
- Some rough financial cost figures can be estimated through the purchasing department's efforts to find suppliers and cost materials. Engineering personnel get preliminary data of material properties and possible problems in handling these.

Nowadays, many large food companies have chef–food technologists on their development staffs. These combine the skills of chefs with food technologists well-versed in the sciences as these are applied to food.

Consumer opinions uncovered in market and consumer research, preliminary cost estimates, and foreseeable engineering difficulties of these prototype products eliminate some ideas, send others back for recipe tinkering, and cause rethinking of others. Eventually clear, concise statements of several possible product ideas emerge. Usually several products reach this stage, as companies often keep several products in development stages.

1. Recipe Development and Recipe Scale-Up

The first recipes often come from home recipes, cookbooks, analyses of similar products on the market, or more frequently from a *de novo* creation of a chef based on consumers' preferred tastes. These are used for stovetop samples only. They are inadequate for use in pilot plant preparation and certainly not suitable for large-scale plant production. Many owners of small food companies have difficulty understanding that their wife's spaghetti sauce recipe or Mama's chicken soup recipe are not readily adaptable to commercial production techniques without extensive modification of the ingredients and the processing steps.

I have experienced two instances of this lack of understanding of problems associated with scale-up of home recipes:

> In the first instance, a president of a medium-sized West Coast food processor insisted a particular canapé spread made by his wife for home entertaining could be processed as is — just increase the quantities. Against the advice of all including the plant manager, a trial was made. Much production time was lost cleaning and chipping out enamel-like gunk from the walls of the steam-jacketed kettle and getting the pump functional after the first disastrous trial run.

To commercialize a cherished family recipe and have it meet the needs and expectations of targeted customers and consumers (who are not family) at a price the customer is willing to pay requires much research and development for scale-up by food technologists.

In the second instance before the days of chef–food technologists:

> A large East Coast food company had hired a prominent TV chef to formulate several frozen main course items suitable for factory-scale production. The items were impossible to prepare in quantity on commercial equipment with unskilled manpower, with

TABLE 5.1
Comparison of Difficulties in Using Home Recipes For Commercial Production Runs

Commercial Recipes	Home-Cooked from Family Recipes
Safety and Stability	
Products must be safe with respect to all hazards of public health significance. Products must be stable with respect to their high-quality attributes from factory to table.	Home-cooked product is cooked and eaten within minutes of preparation with no, or few, hazards associated with packaging, storage, warehousing, distribution, or retailing.
Costs	
Commercially prepared product must be kept within well-defined cost limitations (labor, plant overheads, labels, packaging, advertising, and promotions).	Home-prepared recipes have flexible budget allowances (with the exception of low income families). Labor is free.
Ingredients	
Ingredients must be adaptable to mass production technologies for uniformity of processing characteristics (density, viscosity, particle size, thermal properties, etc.) and final product characteristics.	Ingredients and raw materials do not always have uniform characteristics. Home-cooked products often have a high degree of variability of size (e.g., of cookies), thickness and viscosity of sauces and gravies, and textures.
Volume of Production	
Large volume requirements of commerce require high-speed food processing equipment.	Equipment and processing technology used in the home could never produce the volume of product demanded in commerce.
Expectations	
Commercial products must consistently meet the needs and expectations of a wide cross section of customers and consumers.	Home-prepared products need only meet the expectations of family members and guests who have accepted that Mama's cooking is the best and that the family's way of preparing anything is the proper, traditional, and correct way.

readily available commercial ingredients, and within the cost parameters laid down for the products by management.

This failure was not the chef's fault. On the factory floor one does not have a multitude of trained sous chefs to assist in the preparation of dishes. Nor can a company use the quality of ingredients chefs would use in their restaurants. Factory operations must produce a consistently high-quality product rapidly and repeatedly at a price customers are willing to pay.

The need to modify family or cookbook recipes arises for several reasons (Table 5.1). Two points in this table deserve the technologist's attention. First, some manufacturers are beginning to realize that consumers believe products with uniform quality are synonymous with mass production and therefore are associated with being highly processed foods:

> I worked with a manufacturer of homestyle Italian food products who credited the wide popularity of their products to their variable, but always high, quality. They did not want uniformity in their pour-on sauces with its attendant glossy, gloopy smooth appearance. They cultivated nonuniformity by allowing a degree of syneresis and lack of gloss in their product.

These imperfections that gave their products the appearance of being homemade made perfection in their opinion. Lightbody (1990) makes the point that "there is evidence to indicate that uniformity of appearance, texture and taste within some manufactured food products can be judged by some consumers as unattractive and, in some cases, as a sign of 'excessive' processing."

Second, Worsfold and Griffith (1997) and Daniels (1998) have seriously questioned the safe and experienced hands of home preparers of food. In their studies, they have noted abysmal home food preparation techniques. It demonstrates the need for developers to take this consideration of a home hazard potential into their safety design for new products. Home food preparers might not be as skillful as thought.

If similar products exist in the marketplace, a basic recipe can be obtained by:

- Simple techniques such as sieving, filtering to determine quantities, and microscopy to identify plant material and perhaps even identify the herbs, spices, and starches used
- Ingredient lists in conjunction with the above techniques plus cost analyses of the quantities of ingredient found in the competitor's product to reveal much about a competitor's formulation, costs, and margins
- More sophisticated analytical procedures (mass spectroscopy, chromatographic techniques of separation, particle size enumeration, etc.) — and there are any number of commercial laboratories who can do these — to identify more complex constituents

Expert taste panelists as well as commercial flavor houses will supply information about flavors and are able to duplicate most flavors presented to them.

B. Spoilage and Public Health Concerns

With recipes based on the product concept in hand, food technologists begin to experiment with ingredients to get the desired quality attributes required for the final product. The food technologists also try to anticipate the expected or traditional avenues of product failure for that type of product. This is needed to determine the most suitable stabilizing system to be used to obtain a safe product of acceptable quality throughout the expected shelf life. This is accomplished by searches of the literature and food databases for data on similar products. Safety and quality must be designed into the product.

Products must be safe for moral and humanitarian reasons. Products must be stable, that is, remain unspoiled with their high-quality attribute intact throughout their shelf life, for economic and esthetic reasons. While spoiled products in the marketplace are a serious problem, safety with respect to hazards of public health significance for any food product is a far more important concern than spoilage.

1. Food Spoilage Concerns

Margaret Hungerford's well-known comment "beauty is in the eye of the beholder" could very easily be rewritten by food technologists: "Spoilage is in the eye of the consumer." The consumer decides whether a product is acceptable or spoiled. For consumers, spoilage is the failure of products to meet their expectations as promised by promotional material that convinced the customer to purchase that product. Spoilage in the consumer's mind is simply "you promised me this, but the product I got didn't fulfill this promise. It...." The manufacturer can finish the sentence with "bad smelling," "bad textured," "unsightly," "moldy," "didn't function as promised," "tasted nothing like ...," "wouldn't set," "cooked unevenly," and so forth to complete the consumer's reaction.

Foods are classified into three categories on the basis of their stability (McGinn, 1982):

- Highly perishable foods with a short period of acceptable quality shelf life that is measured in days. Typical examples are refrigerated fluid and flavored dairy and soy milks and yogurts; many fresh sausages and delicatessen meats; fresh meats (meat, poultry, and seafood), especially ground meats; fresh peeled and cut-up fruits and vegetables; mixed salad greens and delicatessen salads (slaws); fresh bread; and specialty cream- or custard-filled baked goods.
- Semiperishable foods with an acceptable quality shelf life calculated in weeks. Examples of this category are conserved meat products such as bacon, hams, some fermented and semidry sausages; some bakery goods (fruit cakes); dairy products (natural cheeses); potato and tortilla chips; and other snack foods.
- Highly stable foods with an acceptable shelf life determined in months or years. Dehydrated foods and food mixes, canned foods, many cereal products (flour, pastas, breakfast cereals), confectionery products (toffees, hard candies, and some chocolate products), and jams and jellies typify this group.

The shelf life of any food product and its safety are compromised in adverse storage conditions, if the integrity of the product's container is broken, or if abusive handling has occurred.

Defective product that is found in the marketplace can be devastating. Bad news spreads quickly, especially via the Internet, and may increase the awareness of spoiled or defective product through gripe sites (see Chapter 2).

TABLE 5.2
Categories of Changes in Foods with Examples

The Reaction Mechanism		
Physical	**Chemical**	**Biological**
Phase changes	Oxidation	Respiration
Evaporation	Reduction	Oxidation
Concentration	Hydrolysis	Autolysis
Crystallization	Condensation	Fermentation
Mass migration	Decarboxylation	Putrefaction
Irradiation	Deamination	
	Browning reaction	
	(e.g., Maillard reaction)	

The Sensible Changes in Foods		
Exudation	Off-flavors	Wilting
Separation	Off-odors	Softening
Precipitation	Discoloration	Discoloration
Clumping	Browning	Slime formation
Clotting	Fading	Clotting
Textural changes	Exudation	Exudation
Grittiness	Textural changes	Off-flavors
Staling	Container interactions	Off-odors
Toughening	Rusting	Toxin formation
Discoloration	Delamination	Senescence
Fading		Excessive cfu/gm
Opacity		

Note: cfu/gm = colony-forming units per gram.

Spoilage in food products, causing nutritive, textural, visual, and organoleptic alterations (Table 5.2), is a result of any, or all, of the three following causes:

- Microbiological changes resulting from the growth of microorganisms native to the particular foods, to the ingredients added to these foods, or present in the environment in which foods were prepared
- Biological changes caused by enzyme systems naturally occurring within foods that cause unacceptable quality changes
- Nonbiological causes resulting from chemical and physical reactions between the various chemical species within a food matrix that are promoted by processing, storage conditions, and packaging materials

Corrosion of aluminum containers by high-salt foods and detinning of metal containers by high-acid foods are well known to food technologists and easily avoided. Fluctuations in cold storage or freezer storage temperatures frequently occur in warehouses and in distribution channels. These changes cause a loss of esthetic appeal in frozen produce, with large ice crystal formation due to water migration within the package. Flavors must be carefully screened for their compatibility with plastic or plastic-lined containers and metal containers where the development of off-flavors and corrosion can occur.

Not all the changes noted in Table 5.2 are undesirable. For example, desirable changes occur in the aging of whiskey; fermentation and lipolysis in cheese making; the maturation of wine; or the fermentation of vegetables (krauts). These reactions are carefully controlled to allow the product to reach a desired state.

Measures used to control or prevent the changes (Table 5.2) may, in their turn, cause changes that are considered undesirable in the finished product by marketing departments or management; for example, homogenization of a pour-over plant-based sauce may cause undesired color and viscosity changes. Or a control measure may be considered too costly or inappropriate. The use of some control measures, such as irradiation, might raise sensitive consumer issues. The use of permitted chemical additives is also an unacceptable alternative if marketing personnel want to project a natural image for a product with a "green" label.

Where rigorous stabilizing processes such as application of heat are not applicable, alternative minimal processes may be used. Enzymatic activity is reduced or controlled by chilling or freezing, substrate removal, or use of substrate antagonists or specific chemical inhibitors of the enzyme in question. Separation of sauces is prevented with stabilizing agents if these are acceptable to management or by homogenization if this process does not cause unacceptable color and viscosity changes.

Chemical and physical changes that make foods unacceptable are somewhat more complex to treat. Some processing techniques used to delay or prevent chemical and physical changes are:

- Lower storage temperatures slow chemical and biological reactions.
- Homogenization prevents separation of oils.
- Agglomeration and crystallization prevents caking of powders and improves their solubility.
- A judicious choice of ingredients — for example, noncrystallizing sugars instead of crystallizing sugars — and the use of additives (doctoring agents) controls staling (e.g., moist cookies) and grittiness in products.
- Encapsulation separates and protects labile components from reacting during processing stresses and minimizes flavor losses.

Lowering the storage temperature slows the rate of a chemical reaction but decreases solubilities of dissolved solids, which may crystallize, causing cloudiness or sedimentation in liquids. As temperature is lowered further, phase changes can occur. Water becomes ice and oils solidify. Emulsions can be destabilized with extreme temperature changes.

In general, as the shelf life of products is lengthened the causes of instability alter from biological (largely microbiological) to physical or chemical. Short shelf life products spoil primarily from microbiological causes; for example, added-value products such as spiced, prepared hamburger patties or pasteurized flavored milks spoil for microbiological reasons, not usually for chemical or physical reasons. However, ultrahigh temperature processed milk has a long shelf life and is more likely to fail for flavor and textural (grittiness due to lactose crystallization) reasons. Beef jerky, a product with a long shelf life, fails due to fat oxidation causing off-flavors.

As the shelf life is increased further, chemical and physical interactions between ingredients in the food matrix and the package generally overshadow biological factors as determinants of change. These changes are accelerated by temperature fluctuations, light, oxygen, and abusive handling throughout storage and retailing that alter quality unacceptably. Biological factors in spoilage are minimal but always present. McWeeny (1980) observed that during long-term storage of flour, lipases and lipoxidases caused losses of "oven spring."

The expected duration of quality shelf life that is described in the product concept statement will determine which ingredients, raw materials, and preservative techniques are used to sustain the quality and safety of the product from the factory floor to the consumers' tables. Nevertheless, the quality attributes of any food product will deteriorate with time. Each attribute may deteriorate at a different rate according to changes in storage environment that occur over time. That is, the spoilage mechanisms have different rates of reaction (different Q_{10} values) (Labuza and Riboh, 1982; Labuza and Schmidl, 1985). At one set of environmental conditions, the color of a beverage may deteriorate faster than its flavor. At another set of conditions off-flavors may develop more quickly. Stresses likely to influence rates of deterioration are temperature changes, light (due to the wavelength and heat of incident light), vibration, and relative humidity (a major factor in plastic packaging).

Colored glass bottles or opaque plastic/paperboard/foil-laminated containers protect beverages from light damage but not from the radiant heat such incident light might cause. Laminated containers provide excellent protection but are regarded as environmentally unfriendly and even banned in some jurisdictions. Glass bottles are often part of recycling or reuse programs; however, retailers are not happy since such programs require a deposit and return-to-retailer system.

Tortilla chips spoil either by losing their crispness or by going rancid. Moisture-proof packaging will stop the former and elimination of oxygen, the latter.

> I worked with a manufacturer of an upscale tortilla chip snack. They had ascertained that rancidity was the limiting factor in the shelf life of their product, a determination based on an in-plant, expert tasting panel. Accordingly, the product was packed using a gas flush of nitrogen to combat oxidation. Rancidity was stopped, but loss of crispness became the limiting quality factor. Surprisingly, consumer complaints regarding flavor poured in. Consumers liked the hint of rancidity in the product. With that gone, they stopped purchases, leaving product sitting longer on the shelves and losing crispness.

2. Microbial Spoilage

Mossel and Ingram (1955) organized the determinants of microbial spoilage of foods as those due to:

- Intrinsic factors that are a characteristic of the food, its ingredients, its additives, and its processing, for example, pH, a_w, antimicrobial constituents, colloidal state or biological structure of the matrix, etc.
- Extrinsic factors that include all environmental factors such as temperature, relative humidity, or gases in contact with the food or its package
- Implicit factors that characterize the microorganisms involved, their population dynamics, with all the synergistic and antagonistic pressures they are affected by, and their nutrition and metabolism

It is the implicit factors that the developer must always be on guard against; does one know one's microbial enemies? There is always the danger of the introduction of a "harbinger of change in food safety" (Wachsmuth, 1997). For example, the emergence of a microorganism, *E. coli* 0157:H7, has challenged traditional barriers to food safety. Buchanan and Doyle (1997) discuss virulent strains of *E. coli* in detail.

Notwithstanding the above caveat, there are three broad approaches to predicting which stabilization techniques are suitable. The stabilizing techniques will be identified from:

- Knowledge of the chemical and biological properties of the food, its ingredients, the processing system used, and the susceptibility of these to microbial challenge
- The handling hazards (temperature variations, relative humidity, sensitivity to light, abusive treatment of the packaging material, etc.) that may be expected throughout casing, palletizing, warehousing, distribution, and retailing
- Knowledge gained from experience and from the technical literature on microbial hazards typical of the food and the growth characteristics of microorganisms implicated with these hazards. The growth characteristics used to control microorganisms are:
  - Optimal growth temperature
  - Optimal pH requirement for growth
  - Optimal reducing environment (Eh)
  - Optimum water activity relationship
  - Specific nutrient requirements
  - Sensitivity to competitive pressures of other microorganisms (Jay, 1997; Helander et al., 1997)
  - Sensitivity to antimicrobial agents (Helander et al., 1997; Roller, 1995), particularly interesting for the new antimicrobials derived from bacteria

That is, an understanding of the properties of the product under development and the microorganisms most likely to spoil the food allows one to prescribe the preservative techniques to be applied. Roller (1995), Jay (1997), and Helander (1997) discuss the role of microorganisms in stabilizing foods.

A simple example illustrates the application of this knowledge. Food technologists have determined that the expected cause of deterioration of a hypothetical new food product is an equally hypothetical microorganism that has the following characteristics:

- It grows poorly in pH conditions below 4.8.
- It does not tolerate water activity conditions below 0.9.
- It is a mesophile.
- It grows best in aerobic conditions.
- It is sensitive to sorbate as a preservative.

Food technologists can delay spoilage and increase the shelf life by:

- Acidification of the product to below pH 4.8. This requires reformulation.
- Reformulation with ingredients to lower the water activity of the product to below 0.9.
- Controlling extrinsic factors by presenting the product as a chilled product with refrigerated distribution, by vacuum packaging, by increasing the reducing potential of the food matrix, by considering modified or con- trolled atmosphere packaging, or by employing either the first two or the last two options.
- Addition of sorbic acid or its salts as a preservative.
- Heat treatment, irradiation, or ultrahigh pressure plus mild heat techniques to preserve the product.

If suitable combinations of techniques are used, several multiparameter methods of preservation are possible.

Not all the methods will be suitable for the product or acceptable to marketing personnel. For example, acidification may produce an undesirable tartness or may promote reactions with other ingredients; heating or high-pressure processing may cause unacceptable changes in the appearance of the product. Certain ingredients or additives may not be permitted by regulation in the class of food or their use may contravene a standard of identity for the food. The use of certain processing tech- niques, additives, or ingredients may contravene the company's ethical, environmen- tal, or internal policies.

3. Naturalness: Minimal Processing

Customers and the consumers they buy for are showing a desire for safe, healthy fresh looking, naturally full-flavored but still convenient food products. This chal- lenge has been met with minimally processed foods. These foods keep as much of their natural, fresh characteristics (flavor, nutrition, texture, color, disease-fighting phytochemicals) as possible. At the same time, they have a safe shelf life exceeding what consumers would normally expect.

The processes (Table 5.3) associated with minimally processed foods produce as little damage as possible to the quality attributes of a food, and the product's shelf life is extended. Often the techniques for minimal processing are used in

TABLE 5.3
Techniques Associated with the Production and Manufacture of Minimally Processed Foods

Controlled atmosphere storage
Postharvest treatments
Clean-room technology[a]
Protective microbiological treatment (Jay, 1997; Helander et al., 1997)
Hurdle technology (see Leistner, 1985, 1986, 1992; Leistner and Rödel, 1976a,b; Leistner et al., 1981)
Nonthermal processing
Mild thermal processing
Packaging
   Controlled atmosphere and modified atmosphere packaging
   Active packaging
   Edible coatings

[a] Clean room technology is not discussed in this book, as it is a somewhat limited, esoteric use and is mentioned only for reference sake.

From Ohlsson, T. *Trends Food Sci. Technol.*, 5, 341, 1994.

combinations. Ohlsson (1994) has reviewed minimal processing from storage and postharvest treatments of foods. The challenge for food technologists is stated deftly by Ohlsson: "Very short shelf-life products require preservation methods that will prolong their shelf life, while long shelf-life products require methods that reduce shelf life but improve quality."

Marechal et al. (1999) reviewed the application of thermal and water potential stresses mainly with references to the effects of pH lowering, pressure increases, and temperature decrease on the viability of microorganisms in order to develop optimal kinetics for minimal processing. Their goal is to optimize minimal processes to obtain maximum preservation of quality characteristics of foods while maintaining adequate safety of minimally processed foods.

C. Maintaining Safety and Product Integrity

Both safety and quality are designed into new food products. For simplicity, the expression "stabilizing system" will be used for processing technologies embodying mechanisms that ensure safety of foods with respect to hazards of public health significance and preserve their quality characteristics throughout their expected shelf life.

1. General Methods and Constraints to Their Use

Technologists use numerous techniques to stabilize foods (Table 5.4). The techniques can be used singly or in combinations planned for a particular product's quality characteristics and based on the food technologists' knowledge of the spoilage mechanics typical of the food and on the growth characteristics of the microorganisms of public health significance present.

TABLE 5.4
Traditional Techniques to Stabilize Foods

Stabilizing Stress	Possible Mechanisms
Thermal processing	
Temperature >100°C	Spore inactivation; vegetative cell destruction
Temperature <100°C	Vegetative cell destruction
	Enzyme inactivation
Chilling	
Refrigeration	Slowing of microbial metabolic pathways
	Slowing of chemical and enzyme-mediated reactions
Freezing	Immobilization of water (some microbial destruction)
Fermentation	Alteration or removal of a substrate
	Acidification
	Production of antimicrobial agents
	Overgrowth by beneficial or benign microorganisms
Control of water	(Partial) removal of water
	Humectants (control water activity)
	Freezing
Acidification	Hostile pH for microorganisms
	Suboptimal pH for enzymic reactions
	Preservative (specific property of acid)
	Fermentation
Chemicals	Specific preservative action
	Modification of a substrate
	Enzyme antagonist
	Specific chemical action (e.g., antioxidant)
	Control metabolic pathway
	Control redox potential
Control redox potential	Prevent senescence
	Prevent growth of some harmful microorganisms
Irradiation	Inactivate microorganisms
	Inactivate stages in metabolism
Pressure	Protein denaturation
	Disruption of cell organization

Limits to which techniques can be used were noted in previous sections. There are nontechnological constraints as well that limit which stabilizing systems are available to technologists. Freezing, for example, may well be unacceptable as a procedure for preservation if a company's manufacturing, marketing, and distribution capabilities cannot handle a frozen product. The alternative is to have the product copacked and distributed by reputable refrigerated handlers, but this adds a cost. There are packaging materials that are restricted for use by legislation because they are not recyclable; other materials may require a collection system for recycling. Some packaging materials may simply be unpopular with environmental activists (Akre, 1991). Customer or consumer resistance to irradiation, chemical preservatives, or genetically modified ingredients may deter the use of these technologies if marketing personnel or senior management are reluctant to use controversial or sensitive procedures. This reluctance effectively screens what ideas and products can be followed up.

Many new developments for stabilizing food products have emerged and technical advances in older, traditional technologies have improved their usefulness in making high-quality products. Some of these advances will be reviewed here, not because they are the newest or most novel but because they show promise in providing consumers with exciting new products in new market niches and even to supplant older, established products.

2. Thermal Processing of Foods

Technologists have long known that transferring heat rapidly into food and removing it rapidly results in less heat damage to the quality attributes of the food during pasteurization or commercial sterilization. The use of higher temperatures to inactivate heat-resistant microorganisms or spores led to shorter processing times — if the thermal conductivity of the food and its thermal diffusivity permitted the rapid movement of heat into and out of the product.

a. Continuous Flow and Swept Surface Heat Exchangers

The use of continuous flow plate or swept surface heat exchangers allows high temperature short time (HTST) or ultrahigh temperature (UHT) processes to rapidly heat foods to high temperatures. Rapid cooling in the heat exchangers followed by aseptic filling into sterile containers completes the process. This can be successfully used on many foods, except those with a considerable proportion of large particles in them. The required residence time in the heat exchanger for the more fluid convection heating part of the food is much less than for the thicker, conduction heating particles of the food. Large particles do not have a sufficiently long residence time in the hot portion of the exchanger and may not be adequately heat processed.

b. Agitating Retorts and Thin-Profile Containers

Thermal processes for foods are drastically shortened if heat penetration is speeded up by agitating the food while it is in its container in the retort. Many commercial retorts can now agitate cans in an end-over-end fashion, axially or in a rocking motion. As the container is agitated or rotated in the retort, the headspace as well as food particles in the container move through the food, mixing the contents and thereby assisting heat penetration to cold spots. Thus, a shorter thermal process is obtained with improved product quality. With the faster rate of heat transfer, higher processing temperatures can be used with still shorter process times.

In general, agitating retorts employing end-over-end can rotation or a rocking motion are batch-type retorts. Retorts that provide axial rotation to agitate the container's contents are usually continuous-type retorts. Production speeds and manpower requirements depend, therefore, on the type of retort used.

Altering the geometry of the conventional cylindrical can to speed up heat penetration will also minimize heat damage to the container contents. This altered geometry is accomplished with thin-profile (or low-profile) containers such as the (flexible) retort pouch, the semirigid container, or the larger institutional half-steam table tray. These containers present two broad surfaces separated by a shallow width for rapid heat penetration and cooling. Consequently, the contents of thin profile contents are subjected to less heat damage (Chapman and McKernan, 1963; Tung et al., 1975; Rizvi and Acton, 1982; Brody, 2003). Quality of the product with respect to color, flavor, nutrition, and integrity of particulates is greatly improved.

The retort pouch first became a very popular container in Japan for thermally processed sauce-based products such as curries, spaghetti sauces, and stews (Saito, 1983) and is well known in military rations (Mermelstein, 1978; Tuomy and Young, 1982; Lingle, 1989; Brody, 2003). It has not achieved great success in the consumer market in North America despite its many advantages (Mermelstein, 1978; but see Brody, 2003).

Developments in thin-profile containers and retorts able to agitate the can contents have given thermal processing a new appeal for the production of added-value, high-quality gourmet products that are gaining acceptance as main course items, particularly in the food service industry (Adams et al., 1983; Eisner, 1988).

3. Ohmic Heating

Ohmic heating occurs within an electrically conductive food by passing a low frequency alternating current through it. This heating depends on the electrical conductivity of the food. Since neither convection nor conduction heating play major roles in ohmic heating, there are no large temperature gradients between the more liquid portion of the food and any large particles in the food; these heat at roughly the same time. The size or other physical properties (e.g., thermal diffusivity) of particulate food have less effect on heat penetration (Biss et al., 1989; Halden et al., 1990; Selman, 1991). Packaging is, of course, done aseptically. Biss and coworkers describe the development of ohmic heating, and Sastry and Palaniappan (1992) discuss applications for liquid and particulate mixtures.

Effective ohmic heating requires that the product be an electrical conductor; this is not a problem since most foods have water contents in the 30 to 40% range, with dissolved ionic constituents present. Nonionized food components, for example, fats and oils, sugar syrups, alcohols, and nonconducting solids (bone and cellulosic material), are heated only indirectly by ohmic heating.

Ohmic heating promises to rival plate heat exchangers as a stabilizing system for rapidly heating foods containing nonuniform particulate material and liquid. Ohmic heating overcomes many of the problems (e.g., burn-on) encountered with continuous flow through plate heat exchangers. There is minimal damage to temperature-sensitive quality characteristics.

Several observations suggest that developers should proceed cautiously in establishing safe processes for products using ohmic heating (Halden et al., 1990; Parrott, 1992; Sastry and Palaniappan, 1992):

- The electrical conductivity of meats varied only slightly with temperature, but preheating increased the electrical conductivity of some foods.
- Processes must be designed to sufficiently heat treat the slowest heating, that is, nonconducting, food component.

- There were very specific plant tissue responses to ohmic heating as Halden et al. (1990) observed with aubergines and strawberries. These responses may be considered objectionable.
- Starch gelatinization caused a change in electrical conductivity.

Halden et al. (1990) caution that "electrical conductivity data from sources other than ohmic heating must thus be treated with care when designing an ohmic process."

4. Microwave Heating

Microwave heating, like ohmic heating, is an internal heating process, that is, food is heated from within. They differ in that microwave heating is by both conductive and dielectric heating, while ohmic heating requires that the food be (electrically) conductive.

There have been suggestions that microwaves have a sterilizing effect quite apart from their heating power. Mertens and Knorr (1992) and others believe the inactivation of microorganisms is primarily the result of the thermal effects of microwaves. Nevertheless they state, "If we assume [deleterious cellular effects] are real, it is difficult to imagine how these sub-lethal and long-term effects can be upgraded to a useful food preservation method."

Four very interesting presentations on the use of microwaves for pasteurization and sterilization were given at IFT's Food Engineering Division symposium at their annual meeting in 1992. Schiffman (1992) described the early history of microwave processing and posed reasons for its less than enthusiastic reception by the food industry. Harlfinger (1992) and Schlegel (1992), both representing commercial equipment manufacturers, described the basics of their respective companies' equipment, but it must be remembered that this describes equipment designed a decade or more ago. Datta and Hu (1992) review quality characteristics of foods processed with microwaves. Clark (2002) describes microwave equipment displayed at IFT's Food Expo.

5. Stabilizing with High Pressure

The history of, and developments in, the use of high pressures to stabilize food systems has been reviewed by Farr (1990), Hoover et al. (1989), and Hayashi (1989). The technique, nearly 100 years old, was first reported by Hite in 1899 for the preservation of milk (Hoover et al., 1989). The pressures used are in the order of 3,500 to nearly 10,000 atm (1 atm = 14.696 lb/in.^{2} = 1.033 kg/cm^2).

Changes occur in isolated proteins from 1000 atm and up, but Dörnenburg and Knorr (1998) discuss pressure-induced responses in plant tissue with pressures as low as 50 and 100 MPa (million Pascals; this range is roughly 500 to 1000 atm). My experiences using high-pressure processing to stabilize a fresh-pack salsa produced an undesirable glassy appearance in onion tissue and a faint but distinct flavor change that my client found undesirable.

Okamoto and coworkers (1990) studied the effect of different pressures using combinations of temperature and duration of pressure application on egg and soy protein solutions to produce gels. Softness and adhesiveness of the gels varied with the amount of pressure applied. As the pressure was increased, the hardness of the gels increased and their cohesiveness increased, but adhesiveness decreased. Based on these observations of coagulation of different proteins by heat and pressure, Okamoto et al. suggested that the mechanisms of gelation caused by the two techniques were different: Pressure-produced gels were softer than heat-produced gels. Taste and color were unchanged in the pressurized foods.

Messens et al. (1997) reviewed the effect of high pressure on milk, meat, egg, and soy proteins with process parameters such as pressure, time, temperature, protein concentration, pH, and the presence of salts to produce new textures and tastes.

Since changes are observed in protein foods, high pressures would be anticipated to produce alterations in the proteins of microorganisms. Metrick and colleagues (1989) studied the sensitivity of *Salmonella senftenberg* and *Salmonella typhimurium* to high pressures using two different media. Cell injury and death occurred in this pressure range and increased with increasing pressure. Inactivation was greater in the phosphate buffer medium than in the strained chicken baby food medium. The more heat resistant strain of *S. senftenberg* was more pressure sensitive than *S. typhimurium*.

Several caveats emerge for developers using high pressure to develop safe food products:

- High pressures will affect different proteins differently.
- The properties of the pressure-coagulated protein are different from those of heated proteins.
- The food matrix influences the survival of microorganisms. High pressures appear to be more efficient in inactivating cells in acid pH than in neutral pH (Hoover et al., 1989), but other factors in the food matrix may prevail (Metrick et al., 1989).
- Different strains of the same bacterium (Metrick et al., 1989) may have different sensitivities to the same lethal doses of pressure. That is, pressure lethalities (cf. z values in temperature studies) for different microorganisms vary.

Hayashi (1989) cites that the advantages of high pressure to stabilize foods are the avoidance of heat damage and the preservation of natural flavor, taste, and nutrients. Hydrostatic pressure is transmitted instantaneously and uniformly into food, unlike heat transmission where thermal conducting properties influence the rate of transfer. It can be used to kill insects as well. High pressure can also be used in conjunction with other stabilizing methods such as acidification, antimicrobial agents, or mild heat to stabilize foods.

In Japan, jams, with their high acidity and solids content, are further stabilized, and surimi from different fish sources is gelled with high hydrostatic pressure (Farr, 1990).

Knorr and colleagues (1998) improved the quality characteristics (flavor, lessened amount of thaw loss, color) by using pressure to freeze and thaw products. They describe the need for much more accurate instrumentation to verify the kinetic data on crystal formation as well as the effects of pressure freezing and thawing on the food systems to which it is applied.

In other work, Dörnenburg and Knorr (1998) reviewed the effect of high-pressure processing on the production of a plant stress responses, e.g., hydrogen peroxide, the triggering of anthocyanin synthesis, enzymatic browning, the presence of polyphenol oxidases, cell membrane integrity, and texture as a function of polyga-lacturonase and pectin methylesterase activation. Their substrates for these model systems were plant cell cultures of grape, potato, and tomato. This article is part of a series begun by Knorr (1994) to demonstrate the feasibility of using plant cultures to study the effects of processing stresses.

Information on an equipment supplier is available in Clark (2002).

6. Control of Water: Water Relationships in Stabilization

Water is necessary for enzymic reactions. It brings chemically reactive components together, brings nutrients to microorganisms, and permits the movement of motile bacterial species. Consequently, the control of water in a food is a tool to lengthen its duration of safe (with respect to hazards of public health significance), high-quality shelf life. My first experience with water relationships in foods was with research into the drip loss in poultry with the Department of Poultry Science at the Ontario Agricultural College (later to become the University of Guelph). At the Leatherhead Food Research Association (U.K.), I again worked on water relationships, studying water binding with various phosphate salts in meat curing and sausage manufacture.

Water is controlled by:

- Removing it from the food with dehydration by heat, filtering, pressing, or reverse osmosis techniques
- Immobilizing it within foods by chilling, freezing, or adding solutes that bind water

Jezek and Smyrl (1980) describe osmotic dehydration of apple slices using immersion of the slices in a 65% w/w sucrose solution to remove water, followed by a further drying by vacuum. Advantages were an increase in apple volatiles and improved appearance of the slices.

Tregunno and Goff (1996) used apple slices to study changes in the microstructure of apple tissue in a process they called osmodehydrofreezing. The slices were dehydrated with different sugar solutions and then rinsed and frozen. The different sugars used in osmotic dehydration did cause different microstructural changes to tissues — an effect developers should note.

Silveira et al. (1996) studied the kinetics of osmotic dehydration of pineapple wedges followed by air or vacuum drying. They claim data derived from their studies would assist commercial equipment designers.

As awareness grew of how the control of water in foods could stabilize those foods, the concept of water activity developed, making possible a whole new range of foods: semimoist foods or intermediate moisture foods (IMF). Early developments in IMF can be found in Davies et al. (1976), an excellent basic book on the subject.

Roser (1991) reviewed the use of the sugar trehalose in the drying of foods. Trehalose is found in high concentrations in cryptobiotic organisms. Cryptobionts have a remarkable ability to survive harsh conditions. Roser cites the cryptobiotic resurrection plant, which, when fully dry, can withstand heating to 100°C and megarad doses of irradiation. The secret appears to be the ability of trehalose (and other compounds with this ability) to dry as glasses rather than as crystals. Trehalose shows promise as an aid in producing superior dehydrated products.

MacDonald and Lanier (1991) reviewed the use of low and high molecular weight materials as cryoprotectants for meats and surimi and the mechanisms by which protection was obtained. Katz (1997) described innovations and application technologies of water binding and discussed the roles of minerals and carbohydrates, as well as plant-derived products such as raisin paste and oat bran.

Water activity and its measurement as a tool is comparatively new to many in food manufacturing, despite the technology's availability for two decades or more. Nevertheless, my requests to outside laboratories for water activity measurements still get answered negatively. For developers, water activity control is, as Best (1992) aptly described it a decade ago, a minimal processing concept that is undergoing a major upheaval; it is still awaiting full acceptance as a tool.

Slade and Levine (1991) revised general thinking on water relationships in foods when they introduced phase transitions in foods as factors in food quality and stability. Labuza and Hyman (1998) discuss at length the importance of moisture content and moisture migration to quality and safety in multidomain (multicomponent) foods. They define multidomain foods by example: at the macromolecular level examples of these are dry cereals with raisins or a frozen pizza crust with sauce, and at the micromolecular level these are water in a starch granule or water in baked goods. Moisture migration (loss of crispness in chips, staling of bread with loss of crustiness, or crystallization) is a function of the thermodynamics (water activity equilibrium) and dynamics of mass transfer (rate of diffusion of water) of the food system. The former is dependent on the water activity of the components in the multidomain food; the latter is dependent on, among other factors, pore dimensions within the components of the food.

Suggestions for developers in controlling moisture to improve quality of foods may be derived from the discussions of Labuza and Hyman (1998):

- The components (domains) should be chosen with water activities as close as possible to avoid moisture transfer between them.
- Processing techniques are recommended that assure that pore size is as small as possible and pore size distribution as narrow as possible.
- Viscosity within the components should be increased to inhibit diffusion and mobility of moisture.
- An edible barrier between components will deter diffusion.
- "Use ingredients with a high monolayer moisture content and high excess surface binding energy."

Taylor (1996) describes practical examples of the application of these control measures to the manufacture and packaging of sandwiches with extended shelf life for the vending machine and retail trade. Cauvain (1998) provides a practical discussion of these measures for frozen bakery products.

7. Controlled Atmosphere/Modified Atmosphere Packaging

Gases, used either singly or in precisely defined mixtures, provide a longer fresh shelf life to many foods both in bulk storage and in unit packages. The gases, usually mixtures of nitrogen, oxygen, and carbon dioxide, control plant metabolic pathways that lead to deleterious flavor and texture changes within the tissues and slow or stop the growth of some microorganisms on plant material.

This property has given rise to controlled atmosphere and modified atmosphere packaging (CA/MAP) to prolong the shelf life of many fruits, vegetables, and meats. An overview of CA/MAP for fresh produce in western Europe is provided by Day (1990).

Church (1994) reviews the different gases used in CA/MAP and various techniques of CA/MAP with their applications. He describes new development in what has been termed *intelligent* packaging. The new developments and their applications that are described are oxygen, water, ethylene, and taint removal; edible films; oxygen barrier; gas indicator; carbon dioxide release; and time-temperature indicators.

Despite its many advantages, CA/MAP has not been as popular in North America as it has been in Europe. Reasons for this were suggested by Day (1990). The reasons are important, as they highlight a shortcoming of CA/MAP foods and chilled or frozen foods in general: they are only as successful as the care and control that are in place in the national and local distribution systems.

Anthony as long ago as 1989 expressed concern about the adequacy of the available gas packaging technology, the distribution system to maintain proper cool temperature control, and customer and consumer acceptance of these products. Anthony also cautioned that product liability due to any abuse in the chilled food CA/MAP chain could result in serious product losses and the potential for risks of public health significance. Product developers should be cognizant of these concerns. Day (1990) echoed these concerns. Day (1990) and Anthony (1989) both stressed the need to control:

- Implicit factors such as the numbers of microorganisms present
- Extrinsic factors including the gaseous atmosphere and humidity in the package, storage and distribution temperature, as well as the film composition
- Intrinsic factors characteristic of the food itself

Geeson and coworkers (1987, 1991) successfully extended the shelf life of some varieties of apples but were unsuccessful in extending the shelf life of Conference pears using the same technology. This is another demonstration that developers need to understand that a preservative system successful for one product cannot, with certainty, be applied holus-bolus to another. Each product may have its own unique CA/MAP preservation system.

Fresh, sweet cherries, like many other fruits, would benefit greatly from extended shelf life both for marketing and for commercial processing. Meheriuk et al. (1995) studied the storage of Lapins sweet cherries that had been gas flushed with a mixture of oxygen, carbon dioxide, and nitrogen. They obtained acceptable quality characteristics for 4 to 6 weeks of storage at 0°C.

CA/MAP stabilizing systems are generally used in combination with other preservative techniques, for example, refrigeration. The greatest strength for CA/MAP might be for premium, added-value products.

8. Irradiation

> We cannot control atomic energy to an extent which would be of any value commercially, and I believe we are not likely ever to be able to do so.

That excerpt is from a speech by Rutherford, considered by many as the father of the study of the atom, delivered to the British Association for the Advancement of Science in 1933. Rutherford's remarks notwithstanding, Lieber (1905) had earlier taken out a patent for the preservation of

> canned foods, meat, beef extracts, and other manufactured or prepared foods, milk, cheese cream, and the compounds thereof, fruits, jams, juices, jellies, and preserves generally.

He preserved the foods by impregnating them with emanations from "thorium oxid," thus rendering the substances radioactive! Irradiation has come a long way since 1905 and 1933.

The mechanism for effectiveness of irradiation (at the levels used in food processing) in inactivating enzymes and reducing the counts of microorganisms is discussed by Robinson (1985). The effect irradiation (for some levels used see Table 5.5) has on plant or animal tissue is dose dependent (Giddings, 1984; Gaunt, 1985; AIC/CIFST, 1989).

TABLE 5.5
Dose Dependency of Irradiation on Plant and Animal Tissue

Food Product	Dose
Sprout inhibition	0.15 kGy to 0.2 kGy
Flour disinfestation	Up to 1 kGy
Spice cleaning	Up to 5 kGy for 2 to 3 log cycles of count reduction
Parasite elimination	Up to 6 kGy
Reduction of bacteria	Up to 5 kGy depending on level of reduction desired
Nonsporing pathogens	Up to 10 kGy

Collated from Giddings, G.G., *Act. Rep. Res. Dev. Assoc.*, 36, 20, 1984; Gaunt, I.F., *Inst. Food Sci. Technol. Proc.*, 19, 171, 1985; AIC/CIFST Joint Statement on Food Irradiation, Ottawa, 1989.

A joint expert committee of the United Nations determined that irradiation of food at doses up to 10 kGy presents no hazards, toxicologically, nutritionally, or microbiologically, of public health significance (AIC/CIFST, 1989). A steadily growing number of countries have accepted irradiation as a food process (Loaharanu, 1989). Developers of irradiated new food products for the export market should be aware of regulations respecting irradiation in the importing country.

Irradiation, for several years, has replaced fumigation to control infestations in fruit and vegetables, to prevent sprouting in potatoes, to sterilize sewage sludge for conversion into fertilizer, and to sterilize surgical equipment (Anon., 1981). Tsuji (1983) describes the use of low-dose cobalt 60 irradiation to reduce microbial counts in fish protein concentrate used as flavoring in the formulation of vitamin/mineral supplements in veterinary products.

The advantages of irradiation for product development have been glowingly summarized by Josephson (1984) and AIC/CIFST (1989) as follows:

- Irradiation can replace chemicals used for preservation and disinfestation of fruit, vegetables, and grains (see Giddings, 1989). A greater variety of improved quality tropical products will become available in the market-places of more northerly climates. It effectively eliminates parasites such as trichina in meat products, especially pork products, and cyclospora on imported raspberries (Strauss, 1998; Lund, 2002; IFST[UK], 2003); reduces or eliminates salmonella and other pathogenic microorganisms in poultry; and disinfests agricultural products (plant cuttings, fruit, etc.).
- Irradiation can be used to sanitize sensitive, labile products such as pharmaceuticals that cannot be sanitized by other techniques.
- Food can be prepackaged prior to irradiation. Neither the shape of the product nor the form (liquid, solid, frozen, or powder) is a major factor in irradiation, as these shapes and forms are in thermal processing. Freezing immobilizes ionic species, and chemical changes resulting from irradiation are minimized.
- Irradiation is competitive, in respect to cost and energy consumption, with other conventional techniques for food preservation.
- Irradiated foods provide convenience and versatility to meal components as well as snacks and reduce preparation and labor in the kitchen.

a. Controversy about Irradiation

Consumer reaction to and acceptance of irradiation in North America has been ambivalent. For example, irradiated Puerto Rican mangoes were enthusiastically received by customers in Florida according to press reports (Puzo, 1986; Bruhn and Schutz, 1989; Loaharanu, 1989). A comparison test of irradiated Hawaiian papayas and traditionally processed papayas showed that customers had no aversion to irradiation (Bruhn and Schutz, 1989; Loaharanu, 1989). In January of 1992, irradiated strawberries were sold in Florida (Marcotte, 1992). Despite protesters attempting to disrupt the market test, customers bought and were in general favorably impressed by irradiated strawberries.

Deep polarization between proponents and opponents of irradiation is well documented (see Giddings, 1989 for the pro side and Colby and Savagian, 1989 for the con side). There is slow acceptance of irradiation; customer resistance is declining (Pszczola, 1993; Demetrakakes, 1998a). Irradiation trials for test packs have kept the irradiation facilities busy at Vindicator Inc. (Lingle, 1992).

Bruhn and Schutz suggested as long ago as 1989 that the acceptance of irradiation was conditional on favorable answers to the following questions:

- Do alternative technologies offer greater or less safety for the food supply than irradiation?
- How does industry weight the costs of failure vs. the potential rewards of success in the introduction of irradiated foods?
- Will responsible media coverage accompany the introduction of irradiated foods?
- Have consumers been adequately educated to the value of irradiation for them?

In 2003 these questions still require favorable answers that food manufacturers, not just irradiation advocates, have not gone out on a limb to provide. Best (1989a) claimed customers and consumers do not see irradiation as an advantage to them, a gratification of their needs satisfied by irradiation. Frenzen et al. (2000) see irradiation as limited in scope unless consumers see increased value and safety of irradiated raw meat, poultry, and other products. At present, the opponents of irradiation argue that irradiation is an advantage to processors, who can ignore sanitary control features and zap product for sterility (Coghlan, 1998). Manufacturers fear that few customers will be willing to buy irradiated foods, despite favorable receptions, and, for fear of reprisals, are reluctant to undertake a public educational program. Processors and retailers must learn to handle protesters and the bad press they can bring with sound crisis management techniques. Unfortunately, some journalists still think themselves quite funny by feeding the public's reservations with comments that by eating irradiated food they will "glow in the dark," as I heard a journalist report.

b.   Irradiation Facilities and Costs: Inseparable Problem?

The very high capital costs can be better understood by a description of an irradiation facility. The U.S.'s first commercial facility for irradiating foods, Vindicator Inc., located in Mulberry, Florida, uses cobalt 60 as its source of gamma rays rather than x-rays from a machine-made source (Lingle, 1992). It is a wet-cell irradiator, that is, the cobalt 60 is housed in and shielded by an 18,000-gal pool of deionized water in a 28-ft deep well. Palletized food enters the chamber; the cobalt 60 is raised out of the water to activate the irradiation process. When submerged, the source of gamma rays is thoroughly shielded and workers can enter the chamber safely. Walls of 6.5-ft-thick steel reinforced concrete provide further shielding. This brief description provides some idea of the complex and expensive structures that are necessary to house these facilities. Both Demetrakakes (1998a) and Baird (1999) describe other commercial gamma irradiation plants.

Costs for irradiation vary according to the radiation source used. Comparisons can be made respecting the costs of irradiation since Tsuji (1983) estimated the cost of irradiation and associated handling at about 5 cents/lb and estimates made by Loaharanu (1994) and Frenzen et al. (2000). The latter two articles report costs of irradiation of meat or poultry of 0.5 to 1.5 cents/lb in a plant with a throughput of 100 million lb using an electron beam system, an enormous drop in cost. Irradiated food appears to be very price sensitive — a factor of major concern to developers.

Irradiation installations are very capital-intense facilities, and hence, irradiators will be few and widely scattered geographically. Only added-value, heat-labile products capable of benefiting from irradiation can bear the added costs. Some pilot plant–scale irradiation units mounted on trucks are available; they use electron beams or x-rays.

c. Technological Challenges

Since irradiation doses are restricted to pasteurizing doses (10 kGy or less) irradiation is used in combination with other stabilizing systems such as refrigeration and CA/MAP systems.

There are some cautions for developers using irradiation to design safety and stability into their product. Paster and coworkers (1985) used low-dose irradiation to pasteurize pomegranate kernels in combination with nitrogen gas flush packaging and refrigeration to extend their shelf life for industrial use. While spoilage microorganisms were greatly reduced, fungi were observed to be more resistant than bacteria to irradiation. The red flag for developers was the observation that the course of spoilage was altered. The dominant fungal contaminant common in nonirradiated kernels was no longer *Penicillium frequens*. In irradiated kernels *Sporothrix cyanescens* became the dominant fungal contaminant. Thus, one spoilage microorganism was removed only to be substituted by a different one.

Dempster and colleagues (1985) irradiated raw beef burgers vacuum packed in Cryovac® bags and stored at 3°C. The shelf life was extended by up to 7 days. Counts of microorganisms were significantly reduced, and shelf life was extended at refrigerator temperatures, but the irradiated burgers lost their redness, showed an increase in peroxide values in their fat, and developed a distinct, unpleasant odor during storage. Not only can the microbial path of spoilage be changed, but irradiation can change the chemical path of deterioration.

Grodner and Hinton (1986) noted similar alterations of conventional spoilage after irradiation. They irradiated sterile and nonsterile crabmeat to study the interrelationship of storage duration vs. storage temperature vs. irradiation (up to 1 kGy) on the viability of *Vibrio cholerae*. Both sterile and nonsterile samples of crabmeat were inoculated with *V. cholerae*. Three interesting findings are discussed in their work that should demonstrate the complexity of multicomponent stabilizing systems using irradiation:

- No *V. cholerae* were found in crabmeat pretreated by sterilization and then inoculated, or in nonsterile, inoculated crabmeat at dosages of either 0.5 or 1.0 kGy after irradiation. Irradiation at 0.25 kGy reduced the counts by several log cycles.

- Pretreatment of the crabmeat by sterilization removed competing microflora to *V. cholerae*'s advantage compared to nonirradiated crabmeat stored at each of the storage temperatures.
- Survival of *V. cholerae* was greater in crabmeat that had not been sterilized prior to inoculation in both irradiated and nonirradiated crabmeat stored at −8°C. Irradiation was, however, effective in reducing the counts. Grodner and Hinton suggested that pretreatment of the crabmeat caused the loss of some protective factor (perhaps a protein) against freezing, which contributed to the survival of *V. cholerae* in pretreated crabmeat at the coldest storage temperature.

In each of the above situations, the stabilizing systems either produced an unexpected result in the path of spoilage or altered something in the food's properties. The interrelationships between the application of the newer stabilizing systems, the product, and its microflora must be understood to ensure safety. Careless application of the newer technologies of stabilization may result in an alteration of the traditional courses of spoilage or of the microorganisms responsible for spoilage as these were recognized by food technologists. Unless this is understood, developers may find they have replaced the devil they know with the devil they don't know.

Irradiation offers great promise for developers of new, minimally processed foods with desired, added-value features for customers and consumers.

9. Hurdle Technology as a Tool for Product Development

By judiciously combining several stabilization techniques selected on the basis of a knowledge of the dominant path(s) of spoilage or of pathogenicity for a product, technologists can stabilize that product. Each of the individual systems selected inhibits or stops a spoilage vector in the food to which they are applied. They are by themselves not sufficient to wholly stabilize the food. But each, in combination, complements the stabilizing activities of other systems employed. In this way the high-quality shelf life and safety of food products can be maintained.

Leistner and Rödel (1976a) referred to this combining of stabilizing techniques as *Hürdeneffekt* (hurdle effect). Leister and Rödel (1976b) provide a further general discussion of the hurdle technique.

Hurdle technology (multiple stabilizing systems) has been practiced for many hundreds of years; it is recognized as the preservative technology behind many traditional foods such as:

- Sauerkraut: acidified through lactic fermentation; spoilable substrate removed through controlled fermentation; salt addition to control the course of fermentation and to encourage presence of highly competitive but benign lactic fermenters; lactic fermenter may also provide natural antibiotics (bacteriocins).
- Pemmican: sundried meat (lowered water activity); acidified by addition of high-acid berries.

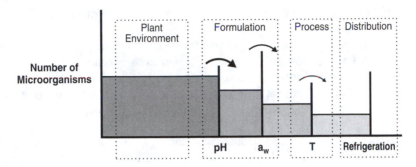

FIGURE 5.1 Pictorial representation of hurdle technology for a hypothetical food product formulated to have a low pH and low water activity.

- Smoked, fermented, hard sausage: dehydrated by heat of smoking treatment (lowered water activity); stabilization by smoke constituents (chemical preservative); acidified by fermentation (lowered pH and removal of a labile substrate); preservative action of the lactic fermenters (biological preservative); meat additives such as salts, spices, and herbs, some of which have preservative action (Beuchat and Golden, 1989).
- Cheeses (very similar to sausages in their stabilization systems); low water activity; fermentation to remove labile substrate; the presence of bacteriocins from lactose and other fermenters; competitive microflora; some traditional cheeses are also smoked.

Application of hurdle technology has been likened to an obstacle race. The course is set with hurdles of varying heights representing different intensities of stabilizing factors. The hurdles are designed to slow or stop chemical, physical, and biological spoilage reactions. Some microorganisms (the racers) fall at the first hurdle. Those microorganisms that pass that first hurdle are weakened for the next hurdle, where more runners are felled. Those that survive the first two hurdles are so further weakened when they face the third hurdle that they are in no condition to pass it or whatever hurdles may remain.

Hurdles can be synergistic. That is, a hurdle may complement the preservative action of another hurdle such that the combined preservative effect of the two systems is greater than the sum of the two separately. The principles of hurdle technology are pictured in Figure 5.1, where a hypothetical food is shown preserved by four systems:

- Rigidly adhered to ingredient standards; purchasing from reputable suppliers; effective plant sanitation, maintenance; hazard analysis critical control point (HACCP) programs; tight process control standards; and other plant support systems keep microorganisms in the plant environment and in the product low. These provide the first hurdle.
- Product formulation introduced two built-in hurdles into the product: a lowered pH and water activity of <0.90 that restricts the growth of spores

of Clostridia and provides a further check on the growth of many other microorganisms.

- Pasteurization eliminates vegetative microorganisms that can survive the lowered pH and lowered water activity and stops enzymic activities. The level of microorganisms is lowered even further.
- Finally, the finished, packaged product is stored, distributed, and retailed at refrigerator temperatures to slow chemical and physical changes.

The hypothetical product contaminated with the hypothetical microorganism discussed in "Microbial Spoilage" above, was stabilized with several hurdles: acidification, lowered water activity, chilled storage or vacuum packaging or both, use of a chemical preservative, and heat treatment. If all hurdles were used this would be overkill respecting stabilization.

The application of the concept of water activity, the enormous interest that developed for the hurdle concept, and its relation to food stability sparked the development of intermediate moisture foods. Intermediate moisture foods have moisture contents between those of dried foods and fresh foods, and their water activities are above those of dehydrated foods but below those of natural foods. They can be consumed without the addition of water.

Davey and Daughtry (1995) developed a predictive model (a linear Arrhenius predictive model; see reference for other predictive microbiological models) for modeling the growth of bacteria using three hurdles: temperature, salt, and pH. The model was based on a number of published growth curves of *Salmonella* spp. Cerf et al. (1996) developed a newer predictive model based on published data on *E. coli* spp. also using three hurdles: temperature, pH, and water activity. They suggest this model has many advantages over previous ones including the property of additivity; that is, "the effect of environmental factors on bacterial inactivation can be summed."

Vega-Mercado and colleagues (1996) demonstrated the use of combinations of pulsed electric fields, pH, and ionic strength on the inactivation of *E. coli* in a simulated milk ultrafiltrate.

Sous vide products, popular in the food service industry, use a multiple preservative system (a series of hurdles) for their shelf life. They undergo a partial cooking, which both pasteurizes and blanches the product. Vacuum packaging inhibits aerobic microorganisms and protects the product from further contamination during distribution and retailing. These products are refrigerated or frozen, which checks chemical and physical reactions and retards microbial growth. Product developers should, however, refer to the comments of Livingston (1990) and Mason and colleagues (1990) regarding the safety of this food service process.

a.   Some Cautions: Altering Traditional Courses of Spoilage

There are cautions to be aware of: Neaves and colleagues (1982), at the British Food Manufacturing Industries Research Association (now Leatherhead Food Research Association), used potassium sorbate, sodium benzoate, or ethyl p-hydroxy benzoate and combinations thereof; salt or propylene glycol and combinations thereof; pH; storage temperature; and pasteurization to inhibit the growth of *Clostridium botuli-*

num in a pork medium. Some combinations inhibited microbial growth and toxin production, but a lowered pH or reduced temperature were necessary to complement, for example, the salt plus sorbate stress. They found botulinal toxin could be present in a food when there was no obvious spoilage, and they feared inadequate preservation could result in the unwitting ingestion of toxic food by consumers. They believed the possibility of inadequate preservation was far more dangerous than no preservation in a food system.

This change in the course of spoilage due to the choice of hurdles was demonstrated by Webster and coworkers (1985), using a model system to test 72 hurdle interactions against a microbial cocktail consisting of *Staphylococcus aureus, Bacillus subtilis, Pseudomonas aeruginosa, Streptococcus faecalis, Lactobacillus casei* var. *rhamnosus*, and *Clostridium perfringens*. The variables were four levels of water activity, four pH values, and the presence of sodium citrate and sodium benzoate. Webster et al. observed that:

- When a lowered a_w was the sole hurdle, the predominant microorganism was *S. aureus*.
- By adding another hurdle such as citrate and benzoate to that of reduced water activity, the dominant microorganism became *S. faecalis*.

Developers must understand that the hurdles that are chosen can alter the expected course of microbial growth. Therefore, when applying different stabilizing stresses to foods in the course of improvements or reformulation with cheaper or newer ingredients, developers must understand the stabilizing system in its entirety since indiscriminate substitutions may simply alter the course of spoilage and not necessarily provide the safe, high-quality product the developer wishes.

Carlin et al. (1991) studied the effects of controlled atmospheres (CO_2, 0.03 to 40%; oxygen, 21 to 1%) and storage at 10°C on the spoilage of commercially prepared, fresh, ready-to-use grated carrots available on the French market. This temperature is higher than that recommended by the French government for such products but was the temperature at which such ready-to-use products were frequently sold. The growth of both lactic acid bacteria and yeasts was more rapid as the CO_2 concentration increased from 10 to 20% regardless of the oxygen concentration originally present. The changing composition of the atmosphere in CA/MAP products influenced the population dynamics of spoilage.

There are other concerns. Although a number of humectants (proteins, protein hydrolysates, amino acids, sugars, and sugar alcohols) are available to depress a food's water activity (see Table 2, in Chirife and Favetto, 1992), their use in many food products is precluded for flavor considerations because of the concentrations of humectant necessary to lower a food's water activity.

In addition to the ability of various solutes and humectants to lower water activity, Chirife and Favetto found specific solute and humectant effects, for example, (a) ethanol has an antimicrobial effect that is not due solely to its water activity–lowering ability and (b) sodium chloride and sucrose inhibit the growth of *S. aureus* around a value of 0.86, but when other solutes such as diols and polyols are used, growth is inhibited at a much higher water activity. Therefore, developers must be aware not only of the preservative potential of the hurdle that is imposed by a lowered water activity but also recognize the influence of the humectant or solute system that produced that level on the path of spoilage.

Leistner (1985, 1986, 1992) and Leistner and colleagues (1981) emphasize that the technology of the hurdle effect can be influenced by high microbial loads. Successful use of hurdle technology requires that high levels of quality control, plant sanitation, and personnel hygiene be exercised in processing. Good manufacturing programs and HACCP programs must be in place and exercised scrupulously. The hurdles established for any product are not designed to protect against high microbial loads: contamination introduced in the plant environment prior to packaging could swamp the hurdles.

The application of hurdle technology must, in effect, be preceded by a hazard analysis of the product prior to formulation with a description of which critical control points require management and must be built into the product. Developers must fully understand the properties of the food product under development and understand the possible destabilization mechanisms that could lead to intoxication and spoilage. Any stabilizing system is a balance of stresses that is unique for that product composed of those particular ingredients and processed precisely in the manner it was.

The final caution is that the developer must understand the purpose and function of each ingredient or additive in the product and record these in the product description. Each serves a purpose; the safety and quality of the product was determined based on these materials originally used in its manufacture and their characteristics and the process controls determined to be necessary. Substitution of any original raw material, ingredient, or additive with another could seriously alter the safety of the product. Defining these purposes alerts future developers faced with reformulation of the product to the importance of each ingredient or additive to the safety of the product.

10. Low-Temperature Stabilization

Chilled, prepared foods (refrigerated processed foods of extended durability [REPFEDs]) were hinted at in the preceding discussion of hurdle technology and are discussed further by D. A. A. Mossel (cited in Scott, 1987; Brackett, 1992). Their growing acceptance by consumers justifies a discussion of their potential for use in new products. As stated earlier, they are usually preserved using hurdle technology (Scott, 1987). The major concern with these products is the growth of pathogenic psychotrophs coupled with weaknesses in the distribution and handling systems, that is, in transportation to, and conditions at, the retail receiving and display level or food service outlet. Since many REPFEDs (e.g., *sous vide* products) may be used in institutions such as nursing or convalescent homes, controls to maintain their safety is of paramount importance.

In general, these products have four characteristics:

- A sound HACCP program must be clearly identified to prevent contamination in their preparation and packaging. A high level of quality control is demanded in their manufacture respecting sanitation and the maintenance of as low a level of microorganisms as possible.

- Packaging is usually done under vacuum (*sous vide*) or under CA/MAP conditions.
- A pasteurizing level of heat is applied either during preparation of the product (partial cooking) or to the product in the packaged state. If the product is heated during preparation, it is rapidly cooled and packaged as soon as possible when chilled to prevent any microbial growth.
- The products are stored, distributed, and displayed at refrigerator temperatures.

The design of stabilizing processes and formulation for chilled food products must be such that potentially harmful microorganisms are adequately controlled. The realization that several pathogenic and spoilage microorganisms can survive and grow at good refrigerator temperatures reveals an inherent hazard in chilled foods.

The shelf life of REPFEDs is, as one would suspect, highly variable. It depends on the intrinsic properties of the food, the stabilizing systems used in the product's design and processing, and extrinsic factors. REPFEDs have anywhere from 12 to 60 days of shelf life, with most in the 20 to 30 day range (Bristol, 1990). Fresh pasta at 40 days and their accompanying sauces at 60 days are among the most stable REPFEDs. The Institute of Food Science & Technology (U.K.) (IFST[UK], 1990) classifies chilled foods into three categories:

- Highly susceptible foods to be cooked immediately before consumption: fresh meat, fish, poultry, comminuted meats, sausage, etc. Recommended storage temperature is −1 to 2°C.
- Products often eaten without further heating that may have no preservative properties (chilling is the primary preservation): pasteurized products, cured meats, soft cheeses, prepackaged sandwiches, fresh doughs, milk and soy milk products, etc. Recommended storage temperature is 0 to 5°C.
- Foods not relying entirely on chilling for stability, which have some additional stability system: hard cheeses, yogurt, fermented sausages, etc. Recommended storage is 0 to 8°C.

Lechowich (1988) reported that low-acid foods (pH >4.6) with mild heat treatment and vacuum packaging or CA/MAP had approximately 14 days of shelf life. Acidic foods were less of a problem. *Sous vide* products had, in general, 2 to 3 weeks of shelf life if held at 2 to 4°C.

The short shelf life of chilled foods goes against their wider acceptance. Neither vacuum packaging nor CA/MAP alone plus refrigerated storage and distribution were sufficient to stabilize chilled foods adequately. This inadequacy has led Lechowich and Day (1989) to recommend that these stabilizing systems be supplemented with acidification, use of competitive microflora, addition of preservatives, or heat processing.

Waite-Wright (1990) recommended a five-point program to ensure safe, high-quality, chilled products:

- Raw materials of the highest quality. This requires procurement from reliable and dependable suppliers plus audits of all supplies and suppliers. This is expensive but necessary.
- A safe controlled process employing all the plant's internal support systems. This involves purchasing standards for all raw materials, active quality control and HACCP programs in place, and close attention to good manufacturing programs.
- Strict vigilance against cross contamination caused by workers or by poor separation of cooked and raw materials and their associated equipment. (This and the following recommendations are echoes of the preceding one.)
- Close adherence to sound plant hygiene and sanitation programs.
- Employee training in sanitation and personal hygiene.

All are necessary for safe chilled foods and all can contribute significantly to cost overheads. The development of REPFEDs as a new product venture requires a cadre of skilled, well-trained, and disciplined workers. Peck (1997) reviews concerns about the safety of REPFEDs with respect to *C. botulinum*.

a. Distribution and Handling: A Weakness

Weaknesses in the distribution and handling system for REPFEDs are well documented. As early as 1980, Slight in a study of U.K. storage and transportation facilities for chilled foods found that the inspection and maintenance of refrigeration systems for chilled foods were poor and none of the transport refrigeration systems studied were operating properly. In 1981 Bramsnaes discussed how the high-quality shelf life of frozen foods was affected by poor storage temperature and poor control of temperature of retail cabinets. Similarly, while studying the shelf stability of a newly introduced chilled Mexican burrito in southern California in the mid-1980s, my colleagues and I found that chilled food display counters in supermarkets would at times reach temperatures of 9°C (48°F) and that they held that poor temperature for extensive periods of time. Light and coworkers (1987) in the U.K. found temperatures to fluctuate from −1 to 10°C in chilled food vending machines. They also noted that some machines could not maintain the desired temperature range of 0 to 5°C for even 50% of their working cycle. Clarke (1990), for example, found that multideck display units in retail outlets had day-to-day temperature variations as high as 15.8°C and as low as −1.2°C. The situation has apparently not improved with time. Audits International, reported in Brody (1997), found temperatures of delicatessen products in retail display cabinets across the U.S. to range from 14 to 71°F with a mean of 47.1°F.

The amorphous glassy state and the importance of the transition temperature at which a food passes into this state during freezing has been stressed by O'Donnell (1993) for its effect on the quality of foods (see also "Control of Water: Water Relationships in Stabilization" and references therein).

11. Other Stabilizing Systems

New nonthermal stabilizing systems include pulsed electric fields, oscillating magnetic field pulses, and intense light pulses. These and several other nonthermal processes are reviewed by Mertens and Knorr (1992), Ohlsson and Bengtsson (2002), and Leadley (2003).

a. Pulsed Electric Fields

High electric field pulses kill microorganisms by causing dielectric cell membrane rupture. Mertens and Knorr cite two industrial applications: one to improve fat recovery from animal slurries from slaughterhouses and the other to stabilize pumpable foods. Gauri Mittal, at the University of Guelph (*The Gazette, Montreal*, Feb. 28, 1996) demonstrated the pasteurizing of milk products, fruit juices, and the brine used in smokehouses to cool cooked meats, which allowed the brine's reuse to prevent it becoming an environmental hazard. Vega-Mercado et al. (1997) discuss developments in pulsed electric field. Products successfully stabilized are orange juice, milk, yogurt, and pea soup. All are pumpable, fluid foods. Details for the design and construction of a pulsed electric unit are provided.

b. Oscillating Magnetic Fields

Mertens and Knorr (1992) provide no commercial details about the use of oscillating magnetic field pulses. They describe the principle of oscillating magnetic field thus: "When a large number of magnetic dipoles are present in one molecule, enough energy can be transferred to the molecule to break a covalent bond." This rupture of the bond in any essential enzyme, DNA, or protein within microorganisms would disrupt its metabolism and reproductive capability. Pothakamury et al. (1993) review in detail the effects of this technique on microorganisms, cell membranes, and malignant cells. They conclude that (a) there is minimal heat damage to nutritional and organoleptic properties, (b) energy requirements are low, and (c) product can be treated already packaged in flexible film packaging.

They caution that very little is known about the death kinetics of oscillating magnetic fields. Leadley (2003), however, concludes that evidence of oscillating magnetic fields for microbial inactivation is inconclusive.

c. Intense Light Pulses and Other Systems

Mertens and Knorr (1992) suggest that the inactivation effect of intense light pulses may be a combination of photochemical and photothermal effects. They also discuss carbon dioxide treatments; the use of chitosan (deacylated chitin) as an antifungal agent; antimicrobial enzymes; and biological control systems that could be used in conjunction with the hurdle concept of food preservation. Roller (1995) also discusses the use of enzymes, bacteriocins, and microorganisms as preserving systems.

The main advantage of nonthermal techniques is the minimization of undesirable changes to flavor, color, texture, and nutritive value that thermal processing often brings about in foods. For a developer seeking information on minimal processing techniques for a delicate product, reference should be made to Ohlsson and Bengtsson (2002).

12. A Summary and a Caution

The foregoing review of procedures for stabilizing food products has demonstrated many opportunities for producing new products, in particular, for minimally processed food products. In the main, the references have been chosen to make developers, first, aware of these newer technologies and, second, aware of some cautions to be observed in choosing which of these techniques to use singly or in combinations. The danger in their indiscriminate use to produce novel foods is that they may alter generally recognized spoilage pathways of familiar products or expected (through experience with similar product formulations) spoilage pathways of new products. Something has changed in the food. Some hitherto unsuspected microorganism may emerge as a dominant cause of spoilage (Paster et al., 1985; Webster et al., 1985; Metrick et al., 1989) or a technique or techniques applied to a product may cause physical changes that are esthetically unacceptable to developers and consumers (Dempster et al., 1985; Grodner and Hinton, 1986). Great care must be used in selecting stabilizing systems for new products.

Many countries have introduced food laws regulating the introduction of novel foods. Their concern is with consumer safety. The observation that some combinations of novel techniques may leave overt and recognizable signs of spoilage (visible mold growth, loss of color, or an off-odor) unseen by consumers is worrisome. No obvious signs of spoilage would suggest to consumers that no spoilage has occurred. Nothing forewarns them of a problem. Rates of spoilage of some quality or safety attributes or both may proceed at different rates as different sets of stresses applied to a food are changed. However, unseen and unwelcome changes may have occurred and be hazardous.

II. THE ROLE OF ENGINEERING IN THE DEVELOPMENT PROCESS

The time required for completion of any new product development process would become impossibly long if it were a serial operation, with one step following after the other. Once some prototype formulations that meet the objectives of the company and the perceived needs of consumers have been developed, the members of the team and the skills they represent are interacting. Research and development and engineering (as well as other groups) advise one another on foreseeable problems respecting fragility of raw materials, capabilities and limitations of in-house equipment, and possible sources of new equipment should these be needed. As soon as prototype formulations have been obtained, all team members begin to interact in earnest.

A. THE ENGINEERING DEPARTMENT

1. General Activities

Engineering personnel serve many functions in new product development. The nature and complexity of their tasks vary with the product under development. For me-too products, their input is comparatively simple. For more exciting, novel products their input may be on a vast scale if they are required to develop and design equipment for new processes for which standard off-the-shelf food equipment is not available. Engineers identify whether existing plant equipment can be modified to handle the new product, design novel equipment for any innovative processes involved, or investigate equipment suppliers for suitable machinery if new or innovative processes are key to the product's manufacture. They, with the production department, identify for the team process requirements on the basis of product throughput, water usage, and energy requirements.

Besides investigating equipment suppliers, the engineers conduct pilot trials where possible in the suppliers' test facilities. They write equipment specifications for new equipment and bargain for the best delivery dates and prices for needed equipment. New production equipment costs or delays in delivery dates of equipment must be communicated to the team early. Marketing plans for launch dates are determined by all the delays that are encountered.

2. Process Design

Engineers, with production personnel and food technologists, have two pathways when faced with the need to design a safe process *de novo* for plant-scale production of products needed for consumer taste and use studies:

- One is to use theoretical models, that is, mathematical modeling based on the best available heat and mass transfer data.
- The second option requires a vast amount of experimenting, analyzing data statistically, and developing a process from this information.

The former is prone to errors because most food processes are complex and cannot be easily modeled. Food is not uniform; pieces are not always regular in shape, density, or composition; raw material varies by variety, by season, and by weather conditions. The second procedure is obviously time consuming, laborious, and hence expensive. De Vries et al. (1995) describe the design of a computer model for baking ovens; they were preparing biscuits of the Marie type.

The first step for the engineering department is to produce a flow diagram of the expected process, identifying as closely as possible the unit operations and processes along with the materials, services, and energy required at each step. From this, all members of the team help identify critical points and the processes to control them. These critical operations can then be studied in isolation for developing design data. It is apparent that this is a combination of the two pathways above. The behavior of food material within these critical processes (changes in viscosity, heat transfer properties, or phases) will be studied in order to describe the process as closely as possible. This permits mathematical simulation of changes that occur within the process.

3. Scale-Up

Products prepared on a large scale in these pilot plant trials will not resemble the kitchen-prepared products on which the original concept trials were conducted. They do however provide the preliminary data necessary to begin design and optimization of the process. Each level of product scale-up brings its own need for screening to get the product as close as possible to that which satisfies the consumer's need.

Taylor (1969) suggested the following valid problems generally associated with scale-up and their causes:

1. Much of food technology lacks a sound scientific basis or, as Taylor puts it, technology is running ahead of the science. There certainly does seem to be a gap between food technologists' understanding of the basic principles underlying many novel processes and their effect on microorganisms and food components and the equipment to take advantage of these with surety. When this occurs, "scale-up is not very soundly based." The caveats mentioned in the previous chapter illustrate this; for example, some of the newer minimal processes such as high pressure processing and oscillating magnetic fields lack a sound scientific basis. Although this situation is slowly improving, for many of the newer processes there is no standard, off-the-shelf equipment, and in some processes there are no sound scientific principles to assess inactivation of microorganisms.

2. Raw materials used in food processing are highly variable from one variety to another of raw produce, from supplier to supplier, from season to season, and during the season from one geographical area to another. One thing engineers dislike in designing a scaled-up plant for new products is variability. Overcoming variability in processes requires operator skills, which means relying on the judgment of operators.

3. Pressure mounts to cut corners in new product development to meet launch date targets. "There is regularly pressure by marketing to eliminate any intermediate scale-up on the grounds of time saving. Such elimination seldom saves time in the long run" (Taylor, 1969).

4. Engineers and technologists want pilot plant studies to get design data for scale-up. Product is secondary. Marketing personnel, however, see the need for product for test market or consumer research studies as paramount and the technologists' demand for ever more data as dithering around. These are the seeds for conflict. If potentially unsafe or unstable product is placed in the hands of consumers, repercussions could be severe.

5. Engineers tend to specify process designs based on the available data. Such designs require unique purpose-built equipment, which is expensive. Over the life cycle of the product, costs may not be recovered fully or may be a total loss if the development project is a failure. Consequently, standard, off-the-shelf equipment is used and scale-up efficiencies are thereby compromised.

6. Echoing Items 1, 5, and to an extent 2, the state of process control in the food industry is still not highly developed. Off-line controls and batch operations are the norm in many processing plants. Many have on-line controls; in-line controls delivering data in real time are only beginning to be developed to their full potential.

7. "The present food industry is still fairly labor intensive and in some parts of the country the labor force is unskilled with a high turnover rate"

(Taylor, 1969). The conservatism of the food industry and its lack of skilled personnel able to cope with new technologies is discussed by Demetrakakes (1998b).

Taylor's remarks made nearly four decades ago and intended for the U.K. food industry apply equally well today in North America and in most major food processing areas. However, equipment technology and process control systems are rapidly improving in sophistication and versatility.

4. In-Process Specifications

Engineers need to know the sensitivity of food products to any treatments to which they are subjected, for example, heating, freezing, or size reduction. Where thermal processes required for safe product are required, engineers work with the research and development group and professional thermal process authorities to develop safe thermal processes. The process designed by engineers with assistance of the team must adequately meet all safe process requirements and maintain the final product's quality characteristics as defined in the concept statement. All treatments that have an effect on quality must have tolerance limits specified (i.e., upper and lower temperature limits, mixer speeds, pump and flow rates), and equipment must be designed to maintain these conditions. When this has been done, engineers will finalize their activities by signing off on a product flow document identifying every unit operation and process, the equipment to be used, and the conditions under which these will be used. This includes:

- Temperatures of the product at each stage in the product flow; times at these temperatures, flow rates, and tube diameters in heat exchangers need to be calculated to maintain this residency time; product viscosity changes require monitoring to prevent settling and turbulence in pipes, and amount and size of particulates present, etc.
- Cooling. Rapid cooling to reduce temperatures below critical values is often a requirement as food is transferred within the plant. Heat balance studies are required. Recovery of heat as an environmental necessity for workers and for reuse in preheating the product is important.
- Shearing action on the product as it is transported through the plant. Data are required for flow rate, pumping action for minimum product damage and mixer speeds, pipe diameters, energy consumption of motors, etc.
- Pressure changes on the process and the product. This plus rheological properties are factors in extrusion cooking and supercritical carbon dioxide extraction.

Any operation that might have an effect on a product's character must be described together with its safe (with respect to hazards of public health significance and maintenance of quality attributes) operating limits.

III. THE MANUFACTURING PLANT

A. THE MANUFACTURING PLANT: A STUMBLING BLOCK OR AN ASSET?

The manufacturing plant often appears as a Jeremiah crying in the wilderness; so often must it take a wet blanket approach to dampen the enthusiasm of the rest of the development team. Plant personnel see the need for new products, but they must also keep a plant running to make the money for product development. They must tell the others of their ability or lack of ability to manufacture the product or inform them of how disruptive the new operation will be.

1. Concerns: Space, Facilities, Labor, and Disruptiveness

Late recognition of the increased burden self-manufacture of new products will put on existing processing lines, availability of labor, and the physical plant can be disruptive of normal plant operations and of the timeliness of the new product's introduction. This can be disastrous for the launch of seasonal (warming soups in the wintertime or leisure foods in the summertime) new products.

As early as possible the production unit must communicate whether they have the necessary skilled labor and physical plant available to produce the new product within the cost constraints and quality parameters required by marketing. Together with engineers they evaluate available processes and determine what modifications may be necessary for the new process. They identify production costs and estimate the disruption that a new product introduction will have on normal plant operations; all these considerations impact the course of development and ultimately the product's launch scheduling. The production department, after all, makes the products that make the money for the company to be able to undertake research and development. New products cannot destroy the goose that is laying the golden eggs and paying the bills.

New lines mean new equipment and the finding of space to put the new lines or the juggling and movement of regular lines to accommodate the new production facilities. It may also mean a new plant.

Production staff work with marketing staff to review their sales volume projections and to coordinate personnel requirements for the processing lines. Their knowledge of local labor markets and the availability of skilled labor become factors in determining the new product's costs and ultimately serve in screening deliberations. If a high level of technical skills is required for the manufacture of the product, then outside skills may have to be bought.

2. Copack Partnerships

Inability to manufacture new products is not sufficient reason to reject product ideas. It is the least consequential of the criteria, but it may have consequences for the anticipated financial success of the project. Manufacturing and packing capabilities can always be had through copackers. There is, of course, a price to pay. Copackers levy a fee per case packed. If new products are price sensitive, companies will not have as much flexibility in pricing structures because of the added cost per case. Lower profits result.

A frequently overlooked cost associated with employing copackers is the extra vigilance required for quality control. Companies will require quality control staff to be in the copackers' plants when their products are run to ensure that they are manufactured according to specification. This is necessary even when the most amicable and trusting relationships exist between the two parties. Copacked products require the same vigilance as self-manufactured products.

Copacking costs can be a small price to pay to get products launched. If products should prove unsuccessful, no capital expenses or extra staff have been acquired. If products meet or exceed sales forecasts, then plans can be made for either plant expansion to undertake self-manufacture or acquisition studies to purchase the necessary manufacturing capability.

For many reasons, use of a copacker can be a very attractive route to new product development. Development costs and time might be telescoped if a copacker skilled in manufacturing the particular type of product can be found. Copackers with experience with similar products have a more accurate assessment of their development and manufacturing costs and initially may be able to manufacture a product more efficiently, and copackers have an experienced work force.

B. THE ROLE OF THE PURCHASING DEPARTMENT

Two aspects that impinge heavily on the success of new product development will be considered under the umbrella of the manufacturing plant: the purchasing department and the warehousing and distribution department.

1. The Purchasing Department's Activities

The purchasing department plays a role in product costs by having the task of finding inexpensive yet reliable (with respect to availability, delivery schedules, and consistency of quality) sources of the raw materials, ingredients, and packaging materials specified by technologists. Obviously, if the purchasing staff can obtain materials and ingredients meeting specifications cheaply, product costs will be lower. If purchasing can negotiate delivery cycles from its suppliers and still maintain favorable terms, then the impact on warehousing will be lessened and warehousing costs reduced.

The food technologists should have identified and described the necessary characteristics of all raw materials, ingredients, and additives in their written product standards. Where technologists have insisted on restrictive or unusual specifications, purchasing departments will require suppliers to submit samples of their products that most closely fit the requirements of the technologists. They must research several sources to ensure that continuity of supplies will not become a problem in the future if the product is a success.

The more exotic and stringent the specifications are, often the higher are the costs. I have always been bemused, and frustrated, to note how many technologists establish standards for ingredients. Suppliers send them samples to use in their formulations. There is nothing wrong with this, but if that ingredient works satis-factorily, the technologists often use that supplier's specification sheet as the ingre-dient standard. This limits the source to one supplier and one cost. There is rarely an attempt to characterize what unique properties of an ingredient are essential to the product's quality; these and only these essential characteristics are required in the purchasing standard. Better price, availability, delivery, servicing, and quality may be found with other suppliers.

Here is an example of what I mean:

> I worked with a tortilla manufacturer that produced its own masa dough. The lime used was builder's grade and not food grade. The company had discovered that builder's grade lime met all the specifications of food grade lime (indeed surpassed them) and was considerably cheaper. For protection, every batch of lime purchased was sampled and analyzed for a wide range of chemical and microbiological components by the quality control department and by an outside U.S. government approved consulting laboratory. It was still cheaper, even after factoring in laboratory expenses, to buy builder's grade lime.

Purchasing personnel with the traffic department balance transportation costs with geographic availability and the reliability of the source with the item's cost. This interplay between geography (transportation costs from supplier), availability and cost (they fluctuate inversely), reliability of supply and supplier, and quality (adherence to specification of ingredients) for material, ingredients, and additives is one in which great cooperation and interaction are required between suppliers and purchasers.

Reliable, quality suppliers often assist in the development process for any appli-cation the developer wants. This is especially true with flavor houses, which own their flavor formulations. Developing and maintaining a working relationship with suppliers is discussed by Williams (2002). Practical advice and cautions are described (see Chapter 8).

Ingredient, raw material, and packaging costs contribute significantly to any product's pricing structure; as such they serve as a deterrent in markets where price sensitivity is important. For example, for the military market, special packaging that meets military specifications may be required; for vending machines, a size and packaging material that is appropriate for machine dispensing is needed; for big box stores and warehouse clubs, still different forms of packaging are needed. Customers will reject a high unit cost if quality, novelty of the product, and its added value do not justify that elevated price in the eyes of the customer and the consumer. Pur-chasing personnel's success or failure in obtaining supplies at an economic price is important to the success or failure of the project.

C. THE IMPACT OF WAREHOUSING AND DISTRIBUTION

Not many in the food industry realize how expensive warehousing is; every square meter of space can cost several hundreds of dollars per year to maintain. Therefore warehousing needs for both raw materials and the finished goods can be a factor in the costing elements for screening. A warehouse can be unheated and subject to the vagaries of summer and winter temperatures and humidities. It can be a chilled or frozen food warehouse and maintained at specific temperatures suited to the product — and expensive — or it can be a controlled atmosphere warehouse for storage of fresh fruits and vegetables, controlled to inhibit ripening. For example, popping corn and chipping potatoes require special temperature and humidity storage. Many fruits and vegetables require controlled atmosphere storage to keep adequate supplies of fresh produce on hand for processing.

Warehouses can also be leased or held by a contracted third party warehouse-cum-distributor. Warehousing considerations are usually overlooked, or at best the structure is considered to be only a buffer between the end of the production line and the distribution system. If this latter attitude is taken, it will soon be realized by all that it is a very expensive buffer. Warehousing and its associated distribution function are major factors in product development. Warehousing and distribution staff are important in the development process because of the cost factors they introduce:

- How many days of production are planned to be held in storage?
- Will production have a seasonal peak? Will it be necessary to store excess production off site?
- Can production and distribution be phased to complement one another? Is it necessary to ship orders from stock in the warehouse or can it be shipped from production?
- What are storage requirements for raw material and packaging material? Are there environmental considerations concerning packaging and container materials?

Special warehousing or distribution channels (chilled or frozen foods) or special environmentally controlled storage may be required for raw materials, ingredients, and finished products. If these are not available in-house or locally, then additional storage and delivery costs must be factored into financial planning.

Products with short shelf lives require a returns pick-up system for products kept beyond their expiry date and a means of destroying or recycling out-of-date product. Recyclable or reusable returned containers require a pick-up service and a cleaning operation before reuse.

D. SIMPLIFYING THE WORK: USING COMPUTERS

The tacticians, research and development, engineering and manufacturing (especially purchasing and warehousing) groups rely heavily on computers to accomplish their tasks. Computerization of operations usually starts in the company's accounts department, for which there is an abundance of available software applications, then progresses to warehousing and distribution for use in stock keeping and product tracing, into the manufacturing plant (Mermelstein, 2000) and eventually finds its way into technical departments and, at last, into new product development arenas (Gaisford, 1989). The ready availability of software programs capable of performing a myriad of tasks gives small companies the same computational ability and power as larger companies. O'Donnell (1991) describes the use of computers in research and development programs in the food industry in the following:

- Costing of ingredients in formulations (see also Mermelstein, 2000)
- Nutritional computations for formulations
- Experimental design to reduce the amount of experimentation

Developers, equipped with their computers and appropriate software, are now in a position to formulate a product to have a given nutritional value, meet a nutritional standard, and to reach these goals with the least number of experimental tests.

Computers (and the software to use with them) serve three basic functions in new product development:

- Management (storage and retrieval) of information. Search engines coupled with external databases provide access to technical literature as well as market information to assist product design.
- A "number crunching" capability that allows the use of sophisticated statistical analytical programs to reduce experimental trials, analyze sensory and consumer data, and provide least cost formulation techniques for optimization of ingredient usage.
- Graphics programs that permit manipulation of three-dimensional solids such as graphs to complement statistical studies, to assist package and packing designs, or to create and design food labels.

These functions or variants of them provide developers with tools to reduce the work load, to manipulate data for more efficient information retrieval, to retrieve and classify data to see trends more clearly, to develop expert systems, and to communicate data and information more rapidly and efficiently.

1. Number Crunching

Statistical software packages are available that have been a boon to developers in the manipulation of numerical data in diverse fields from consumer preference studies to obtaining response surfaces from formulation trials. Software programs permit rapid analysis of multiple variables to efficiently extract all the information buried in the data.

The analysis of sensory data has been effectively accomplished with the aid of appropriate software. McLellan and coworkers (1987) describe an early use of computers in the collection and analysis of sensory data. They combined a computer system with software for sensory analysis, an optical card reader, and cards specially designed for collecting data from a multitude of sensory test procedures: triangle difference tests, rank analysis, category scaling, hedonic scaling, paired preference/difference tests, and magnitude estimation. In about 10 minutes they could enter a day's set of data from a sensory laboratory running 4 panels a day, each panel consisting of 30 judges doing category scaling assessments. A review of this now very dated paper accentuates the progress that has been made in the analysis of sensory data in real time.

Thomson (1989) describes software for generalized Procrustes analysis (*Procrustes PC©*) for sensory methods relying on consumers rather than on trained panelists. A program called *REST©* (Repertory Elicitation with Statistical Treatment) is also described, with an explanatory example, as a structured method of qualitative market research based partly on the repertory grid method and generalized Procrustes analysis. A more detailed description of the software used in these studies is given by Scriven and coworkers (1989), who applied the technique to study the context (time, manner, place, or circumstances) under which consumers drank a variety of alcoholic beverages. Gains and Thomson (1990) also used generalized Procrustes analysis with the repertory grid method to study under what contexts a group of consumers used a range of canned lagers. Such data is invaluable in defining market niches for products and opening up new market opportunities. GAP analysis is a crude and unsophisticated variant of the above.

By far the greatest use of computers is associated with statistically based experimental design software. These software programs allow developers to take calculated shortcuts in the number of trials dictated by classical statistical experimental design. For example, using a factorial design, the number of experiments mushroom rapidly as the number of variables and the levels at which each variable is to be tested increases. Thus, the number of trials required in a factorial design where v is the number of variables and L is the number of levels at which each variable is to be tested, is L^v. Astronomical, and costly, numbers of trials are required to test four ingredients of a product formulation at three concentrations.

Mullen and Ennis (1979a,b) describe the design for applying a linear equation process to a computer program to produce a six-ingredient hypothetical product that got 10% of its calories from protein, 35% from fat (high by today's standard), and 55% from carbohydrate. Mullen and Ennis (1985) later refined their program to handle 15 variables but reduced the amount of experimentation by using fractional replication. The procedure is described in a detail that clarifies the assumptions used in the shortcuts that underlie many of the statistical software programs available for experimental design.

Optimization designs are particularly useful, as these permit the developer to investigate the optimum levels of ingredients needed to maximize a particular quality feature or to alter a process to get a maximum effect (or a minimal undesired effect). Two techniques are used: response surface methodology designs and mixture designs. Henika (1972) used the example of improving the wettability and flavor of an instant breakfast cereal product to compare and explain classical testing vs. the response surface methodology approach in getting answers more quickly and cost effectively. Hsieh and coworkers (1980) developed a synthetic meat flavor using response surface methodology. Mixture designs treat the unique problem of formulations whose proportions of all ingredients must equal 100%.

The history of experimental design, as well as an explanation of the principles of experimental design as an aid to product development, are described by Dziezak (1990). She describes screening and optimization designs using response surface methodology and techniques based on mixture designs and lists available software to accomplish these techniques for product development. Statistical software has proliferated, and sources can be readily found in journals, technical and business magazines, and computer stores.

Further examples of response surface methodology in the optimization of processes can be found in work of Bastos and coworkers (1991), who upgraded offal processing by extrusion cooking to produce a finished product with good solubility and emulsifying capacity. King and Zall (1992) used a model system to study low temperature vacuum drying using a design with three variables and three levels.

Skinner and Debling (1969) describe the application of linear programming techniques to management decisions in food manufacturing. They describe with examples three problems in food manufacturing:

- The allocation problem faced when several products use the same commodities, which are available only in limited supply in their formulation. Which product to make? Their example concerned fruit salad vs. fruit cocktail.
- The blending problem that arises when ingredients of a particular product can be blended either to meet a quality standard or to achieve a cost standard (applicable to least cost formulations). Which proportions to use? A sausage formulation problem was used which is akin to a reformulation situation.
- A simultaneous blending and allocation problem exemplified by a pork vs. beef sausage example.

The examples are worked through to demonstrate the principles, but as the authors state, a statistical software package designed for linear programming could have been used.

2. Graphics

Computer graphics capability has grown immensely as anyone interested in the subject of virtual reality can attest. Architects now plan a building in their computers with suitable software. They can figuratively stroll through the building noting the views from different aspects within the simulated structure. This application has been a boon to package designers both to create new packages and to minimize waste in packaging.

Dziezak (1990) describes some of the three-dimensional and contour plots that can be accomplished to view response surfaces to rapidly assess the most rewarding avenues of investigation. Rotation of these surfaces reveals information about areas of optimum and minimum values.

Bishop and coworkers (1981) make the case for three-dimensional graphics in food science applications. They discuss programs they developed and used in their applications. Floros and Chinnan (1988), using optimization techniques based on response surface methodology, demonstrated the application of graphical optimization to the alkali peeling of tomatoes. Product quality was improved by studying response surface plots to select the optimum concentration of alkali and temperature.

Roberts (1990) demonstrated the use of computer generated three-dimensional graphics in predictive microbial growth modeling. While predictions may not be precise, he claimed that knowing the trend of growth would be highly important in designing stabilization systems for new products.

An obvious application of computer graphic techniques is in the design of packages and labels. Lingle (1991) describes the use by several food companies of computer-based systems to control package and label design. The advantages cited are:

- Ability to manipulate designs in any desired fashion for application to line extensions or redesign packages
- Ability to store designs more easily on disk than as paper art
- Ease with which images can be communicated to others for decision making

Graphics software allows a company to bring package design in-house, provided the company has talented graphic artists on the payroll.

IV. TECHNICAL FEASIBILITY

There comes a time in any development process when the tactical unit of the development team must ask itself, or will be asked by the strategists, what the chances are for the technical and marketing success of the project.

A. THE LOOP: THE INTERCONNECTIVITY OF QUESTIONS WITH INDEFINITE ANSWERS

Technical feasibility is measured by the project's chance of successful development and introduction, the time needed to reach successful development, and the costs of that success. The more qualitative attributes demanded in a product, the more research and development efforts are obviously needed to attain these desired features. Greater effort in research and development, in turn, lengthens the time between the start of the development process and the final technical success of the project. Costs rise as a consequence. The foregoing is straightforward and uncomplicated.

Marketing personnel need to have firm time commitments so that promotional material, labels, and associated artwork for advertising are ready for the appropriate launch date, but before any launch date can be decided, the distribution channels need to be filled by distribution personnel. Prior to the distribution channels being filled, manufacturing has to know whether it can produce the product. Even before this, any special equipment has had to be designed or modified by engineering or specifications had to be written for new equipment to be purchased. Financial interests should not be overlooked here. Accountants need to have financial estimates for the project in order to fund it.

It is much like the children's nonsense song:

There's a hole in my bucket, dear Liza, dear Liza
There's a hole in my bucket, dear Liza, there's a hole.
With what shall I mend it? dear Liza, dear Liza,
With what shall I mend it? dear Liza, with what?

Liza gives many suggestions to her companion, each one requiring a subsequent step until finally the last step requires a pail of water, whereupon the entire song commences again.

Technical development of new products depends on the probability that the company's technologists or those contracted by the company can succeed in matching claims implied in product concept statements with safe, stable products. Development teams now enter into the realm of trying to estimate probabilities and produce guesstimates. Products with little or no creativity, products that are imitations of existing products, or products that are simple line extensions can usually be brought on stream with a high probability of success in a comparatively short time. However, no development project is simple and without hitches.

Each step in product development needs to be analyzed for its chance of success, its cost contribution to the company, and its time to success. Any one step can present an insurmountable hurdle if chances of success are impossibly slim, if the time to success is too long, or if the project is too costly; one step can thus stop a project's chance of moving forward.

B. THE ART OF GUESSTIMATING

Determining the probability of certain events happening is familiar to every student of statistics, to every gambler, and to planners of outdoor events; we live with probability every day. For example, the probability of getting at least four heads when tossing seven pennies or the odds of picking a red ace from a deck of cards are common problems described in textbooks. Also familiar to students of statistics are problems associated with calculating the probability of a particular event occurring as the result of a sequence of events when the probability of each step of the sequence is known (readers unfamiliar with probability statistics should review Bender et al., 1982 and Parsons, 1978 for a concise readable account of probability statistics).

If going from A to B in some sequence of events has a 9 out of 10 chance of success (0.9), there is a high probability that B will be reached. If there is a third stage, C, and the probability of going from B to C is also 0.9, then the probability of going from A to C is, as any student of statistics knows,

$$0.9 \times 0.9 = 0.8$$

Therefore, the likelihood that C will be reached from A is still high but somewhat diminished. If more steps are added, even though each step has a high likelihood of success, the chance for success becomes less and less from the original starting point A. Instead of likelihood of success of an event or reaction, one could easily have substituted processing yields (Malpas, 1977). Thus, if in this simple processing sequence, a 90% yield was anticipated at each step, the yield of C would be 80%.

$$A \xrightarrow{a} B \xrightarrow{b} C \xrightarrow{c} D \xrightarrow{d} P$$

	a	b	c	d
Probability	0.5	0.8	0.3	0.9
Cost	$W	$X	$Y	$Z
Time	T_1	T_2	T_3	T_4

FIGURE 5.2 Probability, costs, and time as factors for consideration during development stages for a hypothetical food product.

The *a priori* probabilities associated with tossing coins or picking cards from a deck of cards are either readily calculated or are determinable by a long series of trials. They can be established.

Problems arise when developers attempt to assign probabilities to phases of the development process. Objective probabilities determined from coin tosses and picking cards no longer apply. There is no history of observations from which one can state, on the average, that such and such an event will happen with a specific probability. The development team must work with situations in which the probability cannot be calculated. Rather, the developer is forced to assign probabilities that "are arrived at by considering such objective evidence as is available and, in addition, incorporating the subjective feelings of the individual" (Parsons, 1978).

Subjective probabilities assigned by developers to the various phases of development must be realistically based on the best available information. They must not be unrealistic probabilities based on an enthusiastic over-assessment of the technological skills of the development team. There can be no gut feel.

The development process for a hypothetical product has been broken down into a simple sequence (Figure 5.2). To proceed from a starting raw material, A, to the final desired product, P, requires three intermediate stages, B, C, and D, and four intermediate steps, a, b, c, and d. The steps could be key processing steps to provide a desired characteristic in a product; they could be the likelihood of getting a change in legislation for a permitted additive; they could be steps to undertake the necessary change in some product's standard of identity; or they could be the possibility of penetrating a particular market. They can be represented as logical steps on the way to products or decisions or events for which probabilities have been assigned.

Each step can be given a cost figure for its accomplishment. The sum of the costs, $(w + x + y + z)$, for each recognized phase in Figure 5.2 represents the total developmental costs to go from A to the final phase P. It should be noted that these costs refer only to the costs of the processes involved. The impact of development on other areas in the company such as sales or production cannot readily be factored in.

The time to accomplish this sequence is estimated to be (T1 + T2 + T3 + T4), the sum of the subjectively assessed time requirements for each step. The probability, the expected cost, and the time expected to go from A to P can then be assessed. It must be remembered that they are all subjective estimates in the screening process.

What is the probability of success? The phases (Figure 5.2) range from a more than moderately difficult one, D, with a probability of 0.3, to the very easy last phase, D, estimated at 0.9. The overall success of the entire sequence, the probability of reaching P, is a disappointing 0.1 arrived at in the following fashion:

$$0.5 \times 0.8 \times 0.3 \times 0.9 = 0.1$$

This poor probability of success in association with the cost and estimated time to success may suggest that abandonment of the project is the wisest move. Much depends on the company's objectives and its strategy to get to these goals.

If the product, P, is highly desired, the technology team may be tempted to tackle the difficult d process first. This may be the most economical approach to the problem for ingredient developers; it avoids the input of time and money in solving the initial phases if it should be determined that the project is not feasible within the timeframe of the company at the C to D process (Holmes, 1968). Probability analysis does serve a useful purpose.

Again, if these were percentage yields in the manufacture of some new food ingredient and not probabilities of success for processes, one would anticipate only a 10% yield for the entire process. Such a low yield may be acceptable if, for example, the product is a highly desirable ingredient for which customers will accept the high cost.

New and improved products are almost a dead certainty to be successfully developed. For example, a breakfast cereal can be improved in several different ways. The probability of better flavor is 0.5, better crispiness is 0.7, higher fiber content is 0.8, and longer shelf life is 0.6.

To improve this breakfast cereal with respect to one of the above quality characteristics — but without specifying which one — the chance of success is 1 (complete success) less the product of all the probabilities of failure or:

$$1 - (1 - 0.5)(1 - 0.7)(1 - 0.8)(1 - 0.6)$$
$$= 1 - (0.5)(0.3)(0.2)(0.4)$$
$$= 1 - (0.01) = 0.99$$

The product is almost certain to be a new and improved one.

V. SUMMARY

Only in very large companies can there be such a separation of duties as presented here. As stated previously, in small companies, individuals wear several hats. The quality control manager is usually also the research and development manager and consequently has responsibility for design of the HACCP program, formulation of the product, and preparation of test samples. This person also does the taste testing, analyses of the product, costing of ingredients, writing of ingredient specifications, and product specifications.

Trained mechanics (usually one mechanical and one electrical), often with drafting skills, substitute for engineers. They prepare specifications and where necessary submit their work to professional engineers for signing off on. The trained mechanics and the plant manager, often one and the same person, combine most of the functions of the engineer (except where a professional engineer's input is required).

In the small company then the duties of product development as related here are not distinct but merge. This gives greater flexibility to the development team but may also limit the horizons of the team because they have fewer resources for marketing, technical, and engineering skills.

Because of the scale of a small company, personalities, usually that of the company president, may dominate, and without the openness and frankness to voice opinions against the dominant personality, which is more usual in large companies, development decisions may not be wisely made.

6 The Legal Department: Protecting the Company, Its Name, Goodwill, and Image

I. INTRODUCTION

Two very different departments protect the safety of the company's customers and consumers, the company itself, and its real and intellectual properties. I have chosen to call these the support groups, as they play supportive roles for the other departments and the development process. They also play a very active role. These groups are the legal and the quality control departments (or whatever name the latter function goes under). Areas that require protection are:

- The consuming public must be protected from any injury or intoxication from preparing, using, and eating the company's food products.
- The development team must receive assistance in setting standards and controls to ensure the stability of the high-quality attributes of the new product and its wholesomeness throughout its shelf life.
- The developers must be advised and monitored to ensure adherence to customer protection legislation.
- Label nomenclature, advertising statements, and promotional claims must adhere to government-established norms.
- The company's rights in contracts, leases, and partnership transactions must be protected.
- Trademarks must be registered and protected; copyrights or patents must be protected from illegal use.
- The company and its products must be protected from fraudulent damage claims.

It very much depends on the size of the company as to how these responsibilities are divided between quality control and the legal departments. Some are more obviously one than the other, but there is a murky middle ground. In many smaller companies it is the quality control manager who advises on label statements and claims (particularly nutrient claims), and many smaller companies do not have legal departments in-house but resort to legal firms specializing in food legislation. Advertising is often handled by outside agencies who also have knowledge in the field of food advertising.

The topics, legal matters and quality control, are different enough, however, to be treated in separate chapters.

II. THE LAW AND PRODUCT DEVELOPMENT

Legislation, government regulation, and food have had a long history of association. The ancient Code of Hammurabi (Hammurabi: 2123–2081 B.C.) contained legislation regulating food standards and trade; parts of the code are believed to predate Hammurabi himself. The Emperor Shun in China is reported to have controlled the production and distribution of grain around 2000 B.C. (Spitz, 1979). Food and government regulation have been intimately entwined:

> The role of the state in building up and controlling grain reserves is nowhere better illustrated than in ancient Morocco, where the same word — *mahkzen* — was used for both granary and government (Spitz, 1979).

There are sound political reasons for government intervention in food matters. For example, governments could use their public treasuries to buy up surplus grain when grain was abundant and cheap and thereby maintain a stable fair price. Farmers did not suffer, and customers did not grumble. When grain was scarce and expensive, the government intervened to prevent starving by doling out grain from its stores, and the people did not suffer. Farmers were happy, and the urban population was kept fed and happy also. Most importantly for politicians, nobody became restive and rebellious.

Government, through its laws and regulations, influences the food microcosm in many ways (see, e.g., Table 1.4); many of these ways have a direct bearing on new foods, new food product development, and the expansion of markets into new geographic areas. In Table 6.1, attention is drawn more closely to legislation that can be expected to affect foodstuff, its production at the primary source, its subsequent manufacture into food, and the product development process itself.

To understand this intervention into and regulation of product and ingredient development more fully, a quick overview of how food legislation is developed and influenced is useful as a tool for developers to anticipate possible regulatory developments that can have an impact on the progress of development (cf., repercussions from the ban on saccharin). A simple overview of a generalized food legislative system is shown in Figure 6.1.

Groups that influence the policy-making process and hence indirectly influence the development process are in the upper half of the figure. They exert their influence, which can be formidable, by lobbying elected representatives, by presenting briefs at hearings called by policy makers, with overwhelming attendance of opponents at such hearings, with organized write-in campaigns to elected representatives, or by overt disruptive demonstrations. The ultimate result is that any decision of the

TABLE 6.1
An Overview of the Extensive Reach of Legislation and Regulations Pertaining to Foodstuff

Area of Impact	Elements of Foodstuff Regulated and Impact on Food Development
Agriculture and fisheries	Siting of farms and fish corrals: restricted land use for industrial, agricultural, or residential use.
	Restricting the use of pesticides, herbicides, and fertilizers, as well as regulating the use of antibiotics, pharmaceuticals, and feed nutrient supplements: inspection costs.
	Establishing quotas for raw materials: controlling availability and cost of materials.
	Requiring odor abatement at farms in or near municipal areas; demanding environmental controls at primary production and manufacturing sites.
	Restricting or preventing waste water or run-off into lakes, rivers, and streams; restricting agricultural practices for factory farms and waste treatment from these: indirect impact on raw material costs.
Processing	Establishing zoning requirements restricting site location for plants.
	Requiring waste water recycling at plant sites and noise and odor abatement programs at food plants in municipal areas: costs for environmental control affect plant overheads.
	Specifying design of, and requiring the use of certain construction materials for, plants: certain product types require specially designed and constructed facilities.
	Requiring adherence to processing codes of good manufacturing practice.
	Imposing import and export permit control that affect both sales and availability of goods: government imposed restrictions on imports limit material availability and marketing plans for sales abroad.
	Imposing traffic regulations on plants situated in or near residential areas. Local regulations may limit manufacturing hours for plants.
	Challenging the safety of many new foods and innovative processing techniques: onus on manufacturer to establish safety.
Product	Establishing commodity grades for raw produce. Grades influence pricing schedule of produce grade.
	Establishing standards of identity for produce and some products. May restrict use of certain ingredients and additives. Formulation changes for cost reductions are restricted.
	Establishing lists of approved additives or lists of restricted additives.
Package	Regulating package sizes. This imposes inflexibility of container sizes and could have a possible impact on marketing plans.
	Restricting composition of packaging materials. Some packaging materials are prohibited because of environmental concerns.
	Imposing weights and measures controls.
	Establishing proper nomenclature for products; requiring product. Information for safe use and nutrition information. Product nomenclature may have an impact on marketing plans.

TABLE 6.1 (*Continued*)
An Overview of the Extensive Reach of Legislation and Regulations Pertaining to Foodstuff

Area of Impact	Elements of Foodstuff Regulated and Impact on Food Development
Package (*continued*)	Requiring label declaration for ingredients. These lists could bring possible consumer backlash concerning ingredients, additives, serving sizes, and nutrient content (or lack thereof).
Marketing	Establishing guidelines for advertising claims respecting nutritional and health benefits.
	Establishing guidelines to prevent misleading promotional tactics respecting product and claims for it. Severe consumer backlash (bad publicity) and penalties for misleading claims.
	Restricting advertising targeted for children. This restriction applies in many countries.
	Restricting the display of some forms of advertising, particularly the siting of advertising.
International trade	Establishing trade alliances and treaties with foreign nations.
	Applying tariffs and nontariff trade barriers.
	Proclaiming antidumping regulations that are attempts aimed at supply management.

Adapted from Fuller, G.W., *Food, Consumers, and the Food Industry: Catastrophe or Opportunity?*, CRC Press, Boca Raton, FL, 2001. With permission.

legislative body bears the imprint of these groups. The departments that are influenced occupy the bottom half of the figure.

A. NONGOVERNMENTAL ORGANIZATIONS (NGOs)

Marketing boards (which shall be considered NGOs but are really quasi-governmental agencies) regulate the supply and price of food commodities; they limit the numbers of suppliers and the amount suppliers can produce (i.e., supply management). They set the commodity prices to further processors. Through supply licensing arrangements, they control who is allowed to further process the commodities. They control, for example, the supply of industrial milk to boutique cheese makers in Quebec; unfortunately, the latter cannot afford the licenses and therefore there is deep resentment of this board. Boards influence governments to impose quotas on imports of foreign commodities or on products containing the commodities that the boards control. Product development with such controlled raw materials results in more costly added-value products.

Consumer associations and trade associations representing food processors invariably oppose the higher prices that are forced upon them by the activities of marketing boards or by supply management policies. These associations have their own agendas respecting food legislation. Consumer associations have influenced regulations requiring the uniformity of the size of containers, the provision of

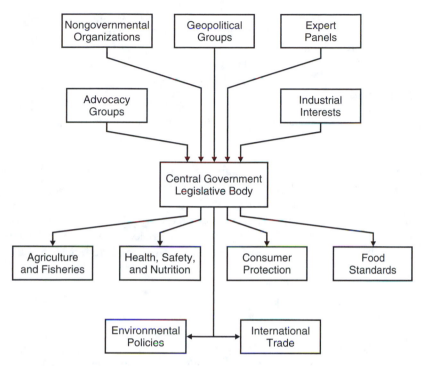

FIGURE 6.1 Generalized overview of a food legislative system with the bodies influencing it. (Adapted from Fuller, G.W., *Food, Consumers, and the Food Industry: Catastrophe or Opportunity?*, CRC Press, Boca Raton, FL, 2001. With permission.)

nutritional labeling, and truth in labeling legislation. Trade associations are often solicited for their opinion on many proposed food legislative issues.

B. ADVOCACY GROUPS

Advocacy groups are often more militant than NGOs and manage to polarize groups of people: animal rights activists are pitted against farmers using factory farming techniques; those for food irradiation are opposed to those against food irradiation. There are groups that demand food laws that restrict the movement and importation of food products manufactured by socially, ethically, and environmentally irresponsible companies or countries, or laws that restrict the importation or sale of products that do not meet certain religious laws. The strongest weapon of these groups is their ability to organize and mobilize vociferous segments of the population to sometimes violent demonstrative action that gets reported widely or to organize their supporters to undertake simple letter writing campaigns to their representatives.

C. GEOPOLITICAL GROUPS

These groups are vested interest parties, perhaps united because they have in common closely defined economies based on their geography or natural resources in their regions. For example, maritime regions have a fisheries industry as a key economy, the Prairie Provinces in Canada or the midwestern states in the U.S. have cereal crops and livestock production as major industries. These groups have specific regional interests that they defend fiercely. Such partisan activities influence food legislation.

D. EXPERT PANELS

Expert food panels are made up of prominent food scientists, nutritionists, biochemists, and agronomists, indeed all those associated with food, its production, its manufacture, its retailing, and its consumption. They provide "informed opinion" on matters of health, safety, nutrition, and agriculture. In this manner, governments expect to receive a rational basis for any legislation that is dependent on scientific issues. There is an inherent weakness in this: science cannot prove the absence of harm, and therefore absolute safety cannot be guaranteed. Food scientists are guided by risk analysis (an intensive investigation of the risks as they see and interpret them) and the so-called precautionary principle (a euphemistic term for Murphy's Law, i.e., if it can go wrong it will, and is also the underlying principle of the hazard analysis critical control point program).

Where there are conflicting scientific opinions or disagreement over the interpretation of scientific data, governments have two options:

- They delay enactment of legislation until a clear opinion can be gotten. This is an option that satisfies no one and certainly intensifies feelings between the parties concerned. The consuming public is simply left confused by the scientific authorities unleashed by both sides.
- They pass legislation based on the best available information even if it proves later to be wrong. This option, too, satisfies no one. Those whose expert opinion it went against are not happy, and laws, once passed, are difficult to overturn (cf., the Delaney Clause).

Interpreting what is the best available information evokes fierce debate in lobbyists and other vested interest groups.

No lesson seems to be so deeply inculcated by experience of life as that you never should trust experts.

Lord Salisbury

These, then, are the major influences on the legislation and regulation of the food product development process. The anti-this and anti-thats, the pro-this and pro-thats, the vested interest groups, all attempt to see their causes espoused and all are voters. Governments hear and enact on the basis of their own self-interest; they wish to stay in power. These groups should be recognized for their influence on food and their potential impact on novel foods and food processes in development. Development takes time, and developers should recognize turmoil within the legislative arena that might raise the ire of some group.

III. FOOD REGULATION AND THE DEVELOPMENT PROCESS

A. LEGISLATION, REGULATIONS, AND SAFETY: A DILEMMA

At best, legislation and its regulations supply little assurance of a safe and wholesome food supply. A government publishes regulations that:

- Establish limits on the presence of toxic or proscribed chemicals or agricultural residues in a food. This requires inspection and analytical facilities with the ability to test for *known* hazards; new, novel, or unknown hazards are impossible to predetermine. The latter is a major concern in the development of new products developed using novel technology. Both inspection and analysis are an economic burden for governments.
- Set limits on the presence of microbiological hazards in foods. Again marketplace inspection and subsequent analysis are required, and bureaucratic expansion and costs grow.
- Set limits for the presence of extraneous matter in foods. The result is more growth in bureaucracy for inspection and analysis.
- Define the composition of traditional, standardized foods.

These are end-product standards and have little to do with safety, quality, or wholesomeness. A consumer eats the contaminated food, becomes ill or hurt, and then the legal system swings into action, but that consumer was not protected. These are production standards so to speak; foods that contravene these standards should not have been produced in the first place. If defects are discovered by food inspectors in the marketplace, the manufacturer will be fined or its management jailed for negligence. These faults represent gross incompetence or negligence or mismanagement on the part of the manufacturer. Yet these errors do occur.

To overcome the shortcomings of these standards respecting safety, food law extends to the regulation of the manufacturing process by creating codes of practice for the safe handling, manufacture, storage, and sale of food. These laws establish that foods must be processed in such a manner that they are not contaminated during processing or ensuing distribution, storage, and retailing. Legislation extends to the premises where food is prepared or stored. Building codes, processing regulations, and codes of practice are attempts at ensuring that the environment in which food is prepared keeps food safe from contamination during production and that the preparation or processing was designed to assure a safe and wholesome product. These codes assure that the purveyors of foods (or their distributors) in the many different marketplaces receive a safe product. Customers and consumers should receive safe food products if the rules are followed.

Governments also protect the customer and the consumer from fraud through legislation:

- They provide standards of compositional identity for commonly defined foods.

- They establish grade standards for commodities and many other processed foods.
- They regulate weights, measures, and package sizes to eliminate deception and regulate profusion of container sizes that will make difficult the calculation of unit costs and eliminate false impressions of contents through overpackaging.
- They provide product information with proper and standard nomenclature of foods, itemized lists of ingredients in descending order of magnitude, and nutritional data.
- They publish advertising and promotional guidelines to prevent misrepresentation of products and their properties to customers.

By enforcement of these practices through inspection and analysis, food is considered to be safe. Or is it? Countries with strict food legislation or with a long history of such legislation or that employ strict food regulatory and enforcement agencies do not necessarily have food supplies that are free from problems of public health significance. Food legislation, no matter how much of it there is or how strongly it is written or how vigorously it is enforced, cannot guarantee either the maintenance of quality or the safety of all food manufactured and consumed within any country's borders.

For example, in the U.K. in 1997 more Britons suffered food poisoning than had been recorded since records were first kept (Coghlan, 1998). Britain has a long history of and an excellent library of food legislation. The U.S. Agriculture Department in April 1999 ordered a recall of meat products including hot dogs, luncheon meats, and sausages made by an Arkansas company; the products were declared unfit for human consumption. The U.S. probably has the largest government organization in the world dedicated to food safety and inspection as well as one of the most comprehensive bodies of food regulations and legislation.

Food legislation can only be developed to answer to known hazards. Knowledge of food hazards is based solely on processes whose history of maintaining food safety is known. Nevertheless, the process of making salami, a food with a long history, is being questioned concerning its safety against new variants of an old microbiological hazard. Existing regulations cannot safeguard the public:

- Against new hazards that may be associated with novel or innovative foods or from microorganisms that have mutated into more deadly forms
- Against food processes with an unknown history of safety or that lack an established theoretical basis for safety
- Against microorganisms that have developed more virulent and more resistant characteristics

The lack of history of long-term safety of genetically modified foods is one argument that its opponents have used.

Legislation alone cannot ensure safety, nor can adherence to codes of manufacturing practice make food safe; the need for developers to design safety into their products is paramount.

B. THE ROLE OF LAWYERS

The overt and covert legal implications in new product development, and there are many, cannot be overlooked by the product development team. Good advice from a lawyer or a specialized resource familiar with food legislative matters early in the process concerning the implications that food legislation or pending legislation might have on the development process will prevent costly surprises later. This resource can advise the team against wasting time and money pursuing developmental goals that are not legal, for example, development of a product that is similar to a standardized product but contravenes the standard. The implications for marketing are enormous; if the standard for permitted ingredients is not followed, the product may have to be labeled "imitation product" or have some quirky computer-generated name found for it. The imitation label may frighten customers off, while the latter option requires education of customers about the meaning of the new name.

The legal department serves in two essential areas in product development:

- Interpreting the law and its regulations: They review labels and advertising copy for misleading statements (Anscombe, 2003) and examine ingredient statements. They investigate the appropriateness of product nomenclature. They take care of patents and copyright matters and advise on the infringement of these (Garetto, 2003; Newiss, 1998).
- Overseeing contractual arrangements: Any contracts with copackers, contracts for consulting services, negotiating partnership arrangements, or licensing arrangements for new products; or sales of technology are best left to lawyers. They also prepare employment contracts for executives brought in to manage new divisions, etc.

Another important service provided by lawyers that is not directly related to the new product development process is the defense of the company against litigation.

C. LEGISLATING QUALITY AND SAFETY

The quality of food cannot be regulated, at least not directly. Regulations providing grade standards for foods such as meats, fruits, and vegetables and standards of identity for composite foods identified by a common name are merely statements that the product meets the stated standards for that product or may be called by the common name for the standard of identity. This is not quality; it is avoidance of fraud. Since most new composite products (multicomponent foods) do not have a grading system or a standard of identity, most foods are not included. Neither developers nor consumers are protected if novel, innovative food products that adhere to *existing* regulations are introduced (European Union, 1997; Huggett and Conzelmann, 1997). Most countries do not have regulations for new foods; it is up to developers through their lawyers to determine what constitutes a new food.

Quality is associated, in both the customer's and consumer's minds, with some characteristic, for example, quality with respect to color, flavor, chunkiness, creaminess, or nutritive content, and quite often with respect to a brand. It is the manufacturer's brand that identifies quality in the eyes of customers and consumers.

Labeling regulations permit only statements that the product within the container adheres to certain minimum characteristics specified for that product. In essence they are standards of commerce. Few manufacturers would exceed these minimal standards because their costs would increase over the costs of their competitors.

Ingredient and additive usage is often regulated; therefore, such regulations have an impact on formulations. Since food regulations vary from country to country a product developed for export markets may have as many formulations as there are foreign destinations. Meal replacements as a new product category, for example, must meet nutritional standards of the meal they are replacing in the country they are destined for. Product formulations must be screened against all legal restrictions in their destinations. Legal departments can provide such information.

There are legal implications involving patent or copyright infringement as new equipment is designed or old equipment and processes are adapted during development. Conversely, the team must also recognize when equipment or processes they develop are patentable and file for these patents through their lawyers. Thermal processes must be verified for safety before being filed with the proper governmental agencies.

There are a host of packaging regulations. These range from the use of only environmentally friendly materials to recyclable packaging to the composition of laminated films and the regulations restricting transfer of the packaging material to the food it protects. Governments, very sensitive to environmentalists, have legislated against the use of certain types of packaging. Some governments have decreed that all packaging be recyclable or require a pick-up program for empty containers.

Marketing departments work closely with their legal departments on label statements, packaging design, and trademarks; on the guidelines for promotions and advertising; and on the legality of the claims they wish to make. There are unwritten laws too. Not only can advertising not be misleading, but it must also not be sexist or racist — this invites protests and cancellation of expensive campaigns. Legal advice is necessary to walk the thin line and avoid bad publicity during product introductions. There are laws regulating package sizes.

Names of new food products cannot be misleading. Goldenfield (1977) presents an excellent example of how the impact of food regulations plus company-based restraints can influence product development. Goldenfield uses as an example a refrigerated whipped fruit-flavored puree that the marketing department insisted contain only fruit pulp and fruit juice so that advantage could be taken of declaring all natural ingredients. Naturalness was key to the concept. It was to be marketed under the product name FRUIT FLUFF. Problems began when the technical team attempted to keep within cost parameters established by the strategic development team. The FRUIT FLUFF Orange Dessert Whip underwent a metamorphosis from FRUIT FLUFF (Natural) Orange Flavored Dessert Whip to FRUIT FLUFF (Natural) Orange Flavored Dessert Whip With Other Natural Flavor to finally end up as FRUIT FLUFF Artificially Flavored Orange Dessert Whip as technical and budgetary considerations played havoc with the original market concept.

IV. ENVIRONMENTAL STANDARDS

Environmental legislation affects not only food packaging (e.g., Akre, 1991) but also the operation of food plants. Food plants are heavy polluters, largely because of the waste and water produced in cleaning and preparation of fresh produce, the considerable amounts of water and detergents required in plant sanitation, and the odor and noise produced. They must operate night and day to process seasonal crops, which can be an annoyance to local residents. Plant and animal waste must be sent off-site, the animal waste to rendering plants and plant waste to landfill sites. This is expensive. New products should not contribute unnecessarily to the already heavy burden plants must pay to remove or treat waste in an environmentally friendly manner or lead to the company being condemned as a bad corporate citizen.

Legal departments advise companies on available alternatives for environmental compliance programs. These alternatives are expensive and often shift the onus of environmental compliance elsewhere, for example:

- Sending the waste away, which involves transportation, handling, and tip charges. This does not solve the disposal problem, as suitable dump sites must be found.
- Recycling the waste to recover any valuable by-products or converting the waste to a marketable product (e.g., mulch or fertilizer). This too involves an expensive, parallel development process and leads the company away from its core business.
- Where possible and permitted, using the waste as fuel for heating and power generation.
- Where possible, preprocessing of produce at the farm level to minimize processing waste at the plant. This simply moves the disposal problem to someone else's backyard.

Boudouropoulos and Arvanitoyannis (1998) discuss at length the impact on food industries of environmental standards published by the International Standards Organization (ISO 14000), which have been adopted by many chemical and automobile industries.

V. SUMMARY

New products present an interesting legal dilemma for developers. New processes for which there has been no historical establishment of safety will always be challenged in the law courts by groups who have real or imagined grief with products made with the new process. New ingredients (e.g., for nutrified foods) will be challenged. Environmentalists will challenge, indeed are challenging, biotechnological applications used in traditional food production systems. Companies urgently require a legal presence within their development teams to be aware of the legislative climate as they plan to introduce innovative, never-before-seen products derived from unconventional sources.

Edible, adj., good to eat, and wholesome to digest, as a worm to a toad, a toad to a snake, a snake to a pig, a pig to a man, and a man to a worm.

Ambrose Bierce

Edibility is apparently something for the law to interpret. Somewhere in Bierce's chain someone is sure to object.

7 Quality Control: Protecting the Consumer, the Product, and the Company

Look beneath the surface; let not the several quality of a thing or its worth escape thee.

Marcus Aurelius Antoninus

I. INTRODUCTION

There has been much discussion about, and hence confusion over, the nomenclature regarding this function within food manufacturing. It was referred to first as quality control; this was then followed by quality assurance; then it became quality management; then total quality management, or TQM. Somewhere in between these terms there was product integrity, a term that I preferred. However, by quality control, I refer to that function in a company responsible for assuring that all processing, product, environmental, and worker safety standards are adhered to and that all reasonable and practicable precautions to protect the product from hazards of public health significance have been taken. I include sanitation, worker training, worker hygiene, pest control programs, observance of good manufacturing practices, and establishment and observance of hazard analysis critical control point (HACCP) programs.

II. THE EVER-PRESENT WATCHDOG

The quality control department acts much like an internal policing unit satisfying itself that all the safety precautions designed into the product to protect its wholesomeness, integrity, and quality are adhered to within the specified written limits. They must, as the above quote says, look beneath the surface. Their role should not be adversarial either in their day-to-day functions or in their contribution to product development. They must, on the basis of experience, present the development team with all their concerns regarding the sensitivity of the process and the product to hazards and provide guidance in putting in place practical and practicable solutions.

Engineers and research technologists prepare flow diagrams of the process and describe operational standards at critical points; they recognize limitations of plant equipment and identify modifications that will be required in the plant. With the cooperation of the production staff, developers must establish specifications for raw materials and ingredients. It is now up to the quality control department to:

- Select reliable and rapid analytical procedures, preferably usable in-line and in real time, to monitor the desirable quality attributes designed into the product and to determine whether safety practices for the product have been followed.
- Verify the safety of the sterilization processes and determine the shelf life of the product under anticipated storage conditions.
- Develop a practicable HACCP program based on data supplied by the engineering and research departments that provides the desired level of product integrity.
- With the production department, train staff in new analytical procedures and, if needed, teach new skills to line personnel if the new products require processing technologies and ingredient handling for which there is no prior history in the plant.
- Conduct sensory evaluation tests and microbiological analyses throughout development to monitor organoleptic values and effectiveness of stabilization processes. These tests may be used for off-line control procedures.

A. Sensory Analysis in Product Development

One of the major requirements of food products is that they taste nice (i.e., flavorful), typical of the product, or as advertised. All the sensory characteristics associated with a particular food must be met. To ascertain these characteristics in a new product, its sensory appeal must be measured. Herein begin problems in measuring something as subjective as sensory appeal.

1. Sensory Techniques

Peryam (1990) described the historical development of the understanding and application of sensory evaluation to product development. Sensory appeal is difficult to measure. An excellent review discussing sensory analysis is presented by Piggott et al. (1998). Sensory testing must be carried out rigorously with proper experimental design and correctly selected and trained sensory panelists to minimize the errors that can creep into any scientific trial; *this is why technologists knowledgeable in the techniques should carry these out.* Data from organoleptic tests require special statistical skills for their analysis and interpretation; *this is why sensory measurements should not be left to those who have had no training in the field of sensory analysis.* Some populations of tasters, such as children, present unique problems in assessing sensory appeal; *this is why special skills are required in sensory analysis.* It should be apparent that sensory evaluation measurements and their analyses should be carried out by technologists trained to conduct them properly. At the very least, a company needs a member of the research and development staff familiar enough with the technology to be aware of his or her own limitations in conducting the tests and the limitations of the tests themselves.

There are four superb references, any one or all of which sensory technologists or any person contemplating undertaking sensory evaluation measurements should have as a *vade mecum* for easy reference:

- *Laboratory Methods For Sensory Analysis of Food* (Poste et al., 1991) is a compact handbook written in an easy to assimilate style.
- *Sensory Evaluation Guide For Testing Food and Beverage Products* (Sensory Evaluation Division, IFT, 1981) is a concisely written guide describing the tests that can be used and providing references that provide further details.
- *Tasting Tests Carried Out at the Leatherhead Food Research Association* (Williamson, 1981) is a very readable document giving many of the descriptive terms used in sensory analysis and providing concerns for the safety aspects of tasting panels (expert panels) and the need to have a clearly defined statement of policy on the use of tasting panels.
- The Bureau de Normalisation du Quebec (Ministere de l'Industrie, du Commerce et du Tourisme, Gouvernement du Quebec) (Quebec Bureau of Standards, Ministry of Industry, Trade, and Tourism, Government of Quebec) has published an excellent series of documents covering:
    Vocabulary (BNQ 8000-500; 84-03-06)
    General methodology (BNQ 8000-510; 84-03-06)
    Sample preparation (BNQ 8000-512; 84-03-05)
    Scaling techniques (BNQ 8000-515; 1982-10-08)
    Triangular tests (BNQ 8000-517; 84-03-06)
    Paired comparisons (BNQ 8000-519; 84-03-06)
    Designing a sensory analysis location (BNQ 8000-525; 1982-08-27)
    Determination of taste acuity (BNQ 8000-560; 1982-10-08)
    Methods for determining flavor profile (BNQ 8000-570;84-03-05)

There are two types of sensory evaluation:

- Objective sensory evaluations: These are all generally difference methods, that is, can the tester pick out the odd sample or the difference between samples?
- Subjective sensory evaluations: These are descriptive tests usually used with trained panelists. Piggott et al. (1998) identified the flavor profile method, the texture profile method, quantitative descriptive analysis, and the spectrum method.

Each serves quite different purposes; they cannot be interchanged. Far too often, the purposes for which the sensory tests are meant are either misunderstood or are ignored. Companies will use the results of an objective test as indicating a sensory preference for one product over another. This is wrong and can lead to incorrect decisions in product development that can be disastrous.

2. Objective Sensory Testing

Objective sensory evaluation tests are used for just that, to get an objective evaluation of some sensory appeal. Other names used for objective tests are more descriptive of their purpose: analytical sensory tests, expert panel tests, and difference tests. The questions asked of panelists are:

- Does a difference exist between the samples?
- How would you rank the samples on the strength of some sensory characteristic?
- How would you describe and rank the sensory characteristics you can identify in a sample (profiling)?

Objective tests are used to determine if there is a difference in some sensory quality between products or between a reference sample and a test sample. For example, Jeremiah and coworkers (1992), using a procedure called consensus profiling, followed the changes in chilled pork loin packaged under carbon dioxide and with vacuum and stored at $-1.5°C$ for up to 24 weeks. From a pool of 12 highly trained flavor/texture panelists, at least 6 at any one testing profiled each sample individually. In subsequent discussions a consensus profile was developed. Details of their procedure of analysis are in the reference.

Objective testing requires trained panelists. Poste et al. (1991) provide advice on the selection and training of panelists. Panelists should be trained to recognize different sensations, to be discriminative, and to quantify these sensations against recognized standards. Like wine tasters, panelists must develop a common vocabulary with which to describe sensations. Rutledge (1992) describes what she refers to as an accelerated program for the training of descriptive flavor analysts. Her program requires up to 9 weeks of training during which novice flavor analysts are coached, guided, and tutored by established analysts, much like an apprenticeship program. In an earlier paper, Rutledge and Hudson (1990) describe a general method for the training of descriptive flavor analysis panels. Powers (1988) describes the use of statistical procedures such as multivariate analysis for the screening and training of sensory panelists.

Many food companies select their panelists from staff; this practice can have dangers. Sensory analysis is obviously not the only job of those selected. Indeed, sensory analysis may intrude upon the panelists' day-to-day company responsibilities. Sensory specialists must be careful with selecting panelists, keeping them in top tasting form, and not intruding on their workloads, so that they are not irritated and do not feel threatened by their supervisors if normal workloads are tardy. Panelists, on their part, must like what they are doing, be in good health, have no genetic or psychological sensory biases, and not be harried into taking part in panels. Williamson (1981) reviews at length the cautions to be observed in selecting panelists.

Sensory technologists should keep a large pool of panelists. This will compensate for the usual absenteeism and unavailability that occurs. Another reason for having a large pool is that not all panelists are suitable for testing all sensory characteristics; there are genetic factors influencing taste acuity. It profits sensory specialists to keep profiles of each panelist, listing availability, threshold levels of discrimination, and sensory record in the various testing sessions the panelist has attended (see Powers, 1988).

3.   Subjective or Preference Testing

In preference testing, also called subjective or affective testing, panelists are presented with a choice of samples and must state which sample is preferred. The word *preferred* is understood to mean *most acceptable*, *tastes best*, *looks best*, *would buy*, or any other expression indicating greater satisfaction.

There are three main variants in the way preference testing can be carried out:

- The focus group
- The central location test
- The in-home test

Focus groups were discussed in Chapter 4. A small number of panelists representative of the targeted consumer are asked to test the product and fill in a questionnaire. This process is repeated several times and always with selected consumers.

For a central location test, the new product development team may take the product under test to some central location. Here the team will be assured of a broad cross section of potential testers. Many companies select church groups, veterans' associations, shopping malls (if the surrounding district is appropriate to the target), ethnic clubs (with consideration of who the target consumer is), or social organizations for their testing. These groups can be selected from locations all over the country. A carefully prepared questionnaire is used by the new product team to evaluate consumer preferences.

Focus type tests and central location tests provide developers with control over the product. That is, samples are prepared uniformly and in a controlled fashion. Qualified personnel carry out the interviews or are at least available for interpretation of questions on self-administered questionnaires.

In the third variant, the in-home test, preselected consumers (there is some control of the consumer) are sent coded samples of product that are to be prepared at home. Developers have no control of the preparation of the samples. Nor is there control of the environment in which the test is carried out. How well were the instructions followed, with all the distractions that are possible in the home? Were the time, place, and other circumstances in the home conducive to a good test? These are unanswerable questions for the survey takers. Consumers then fill out and return the accompanying questionnaire or they are interviewed by telephone.

Preference testing techniques are compared in Table 7.1.

4.   The Panelists

Choosing who should be panelists is influenced by what kinds of data are wanted and what the data are to be used for. Trained panelists (including company personnel)

TABLE 7.1
Three Main Variants of Preference Testing

Test Variant	Characteristics
Focus groups	They require 8 to 12 carefully selected, usually untrained, participants. Tests are repeated two or three times.
	They can be easily repeated, but overrepetition of qualitative data can be unrewarding.
	They require a professional leader to conduct tests.
	Control over display and preparation of product is good.
	Results are obtained quickly.
	Only qualitative data is obtained, and it is not projectable.
Central location tests	These tests involve larger numbers of people (i.e., social clubs, church groups, etc.).
	There is poor selectivity of respondents, but the test can be easily repeated and comparatively large numbers of respondents can be reached.
	They are somewhat more expensive to conduct; they require a well-prepared questionnaire, but testers can explain the questionnaire.
	Control over product preparation is excellent.
	The results are quickly obtained. Quantitative and qualitative data are obtained, but data are projectable with caution.
In-home tests	These can involve several hundred respondents and there can be good selectivity of respondents.
	They are usually a one-time test (a mini-test market) and they are more expensive to conduct than either of the other two tests.
	They require intensive follow-up with a well-prepared questionnaire; there is poor control over how product is prepared and over the circumstances under which the product is used.
	Results are slowly obtained (hence intensive follow-up needed); both quantitative and qualitative data are obtained that are projectable with caution.

should not be used for preference (subjective) testing; they are simply not at all representative of consumers the company wishes to target.

An example of how an expert panel missed the mark on consumer tastes has already been described in Chapter 5, where the trained, expert panel of a tortilla chip manufacturer had determined that an incipient taste of oxidative rancidity was offensive in an established tortilla chip product. The expert panel was proven wrong: consumers preferred that flavor.

Unfortunately many companies often use their expert panels, panelists gathered from among their staff, to get data on the acceptability of a product. This is wrong. Company employees, whether from the plant floor or from the offices, have too intimate a knowledge of their products and what they expect their products to taste like and may be expected to be prejudiced. The preferences of knowledgeable but untrained panelists have no marketing validity with respect to the population. It has been my experience that while these people are not trained members of an expert panel, they do have an expert knowledge of what taste image their company's products have or should have; this may bias their opinions. That is, they are tasting a brand, an image, that they know and are familiar with. They are not typical consumers. They have a bias related to a pride in their work. They are, in all respects, experts on that brand image. Comments such as "This isn't our company's flavor" or "I wouldn't want our company to put out a product like this" will prevail and confound whatever results are obtained.

These testers know nothing of the company's objectives in this test. The company may be working on developing another distinctive taste image in a very different marketing niche. Where, however, companies have a strong brand image, these same panelists may have a better idea of what the brand can carry than does management. The point must be made strongly that company workers, particularly the workers in the plant handling the product, can bring a very definite brand (image) bias to preference sensory testing.

MacFie (1990) and Gutteridge (1990) discuss sensory techniques affecting consumers. Macfie describes characteristics affecting consumers' choices and explains "free choice profiling" and preference mapping. Gutteridge discusses the technique of repertory elicitation with statistical treatment (REST®; Mathematical Market Research, Ltd. Oxford, U.K.; see also Thomson, 1989) for finding the most appropriate market niche for products.

5. Other Considerations in Sensory Analysis

Before sensory testing begins, there must be a clear notion of what is required from the test. Does the development team want to determine whether they have produced the best formulation of a product with respect to some criterion? What is that criterion? Or do they want to determine whether this particular company product is as good as, or better than, the competition's product? That is, are the results of technical product development under investigation or are marketing personnel trying to understand the product in its full marketplace context, where branding may be a factor? Answers to these questions about product characteristics have a decided impact on how the tests will be conducted and who will do the testing.

Martin (1990) broke product characteristics into three components:

- Physical characteristics of a product, which are well recognized by the consumer as poor, acceptable, or high quality; the technologist is trying to formulate only the high-quality (ideal) attributes into the product.
- Image characteristics. Martin recognizes these as most dominant in perfumes, fragrances, and tobacco products. They are not unknown in food products, for example, liqueurs, liqueur-flavored instant coffees, exotic gourmet sauces, etc.
- A combination or interaction of physical and image characteristics.

This interaction is interesting since it introduces a problem. Martin (1990) describes it as follows: when testing similar products "blind" (for example, various brands of beer) in sensory tests, consumers would place the product ideal (the "best" product) some distance away from the test product and with other unbranded competitive products. However, when branded products are evaluated (that is, the test is no longer conducted blind), all products are ranked closer to the consumer's ideal product. Branding is an important factor in sensory testing.

6. To Test Blind or Not?

Should a product be tested branded or unbranded, based on the foregoing? After all, the product will have to face other similar products in the marketplace eventually, and these will be branded. Brands have uniqueness: they communicate images, a product persona. Therein lies a brand's value. A brand is comfort, known values, and security to consumers.

There is overwhelming evidence that branding does influence tasters in ranking similar products and in picking a preferred sample. Martin (1990) provides evidence of vastly different assessments of various characteristics of ciders, beers, and chocolate confectionery when the products were tested branded or blind. When one beer was tested blind Martin found that it was rated higher than when it had been tasted identified.

Moskowitz and coworkers (1981) describe an interesting analysis of consumer perception, magnitude estimation scaling, carried out on chocolate bars. They found that branding encouraged a product's acceptability; furthermore, for some products, it could be branding and not a product's quality that lifted a product's acceptability. (One must be aware of a difference here: branding influences acceptability; branding is the image, a persona, a product has. It does not influence the evaluation of objective characteristics posed by questions such as "Is product A smoother than product B? sweeter? sourer?")

Schutz (1988), Scriven et al. (1989), and Gains and Thomson (1990) describe sensory techniques used to evaluate consumer attitudes about foods and determine the circumstances under which consumers would use or serve particular foods (contextual analysis). Such techniques are excellent tools to guide marketing personnel in determining market niches for products.

The questions must be repeated: What is the purpose of the sensory test? Does the company want to know whether it has the best formulated product or does the company want to know whether it has a product that is preferred over the competition's product? This represents a technological capability question vs. a marketing capability question, technology vs. psychology. As Martin (1990) aptly put it:

> It may not be necessary to develop a clearly superior product in sensory terms, if reputation can sufficiently enhance one which is the equal of the competition.

Hardy (1991) is more adamant and maintains that superior tasting food and beverage products are no surety of a loyal consumer base.

When faced with unbranded popular products, most tasters cannot distinguish between the unbranded products or even successfully choose their favorite brand. For example, Hardy (1991) found that most tasters not only could not distinguish consistently between competitive products but did not improve with experience without formal feedback (i.e., training). Sensory testers must decide what the purpose of the test is and consider whether the biases that branding may introduce will adversely affect this purpose.

7. Are There Differences among Tasters Affecting Discrimination?

Through the application of their magnitude estimation scaling technique to the survey of chocolate candy preference, Moskowitz and colleagues (1981) note — at least for chocolate candies — that there could be age differences and sex differences in the response to branding. Age and sex of tasters influence how they perceive a product. Clearly this suggests that great care is required in the selection of panelists. Whom are the developers targeting?

There are other factors to be accounted for in tasters. Reaction to the genetic taste marker, 6-n-propylthiouracil, marks tasters. Some cannot taste it; some have a slight ability to taste it (medium tasters); some have violent reactions to the taste (strong tasters). Drewnowski et al. (2000) found a statistically significant relation between strong and medium tasters who disliked cruciferous and other raw green vegetables and nontasters who liked them. Duffy and Bartoshuk (2000) found that women's liking for sweet and high-fat foods declined as their ability to taste the genetic marker increased. Men did not have this taste preference.

Ideally, sensory panelists should be representative of those for whom the new product is targeted. Ideally too, all geographic areas where the target consumer can be found should be represented in consumer research tests; taste preferences vary geographically. However, the ideal panel tasters can seldom be gathered without assistance from specialists.

Consequently, food companies enlist the aid of product testing/consumer research companies. These groups keep extensive files of potential panelists categorized by age, sex, ethnic background, economic status, and other characteristics that may be important to product developers. McDermott (1990) discusses recruiting for sensory testing, problems encountered with setting specifications on recruitment, and how these involve cost considerations.

Money limitations may provide constraints on both the number of subjective tests that can be carried out and the size of the panels used in them. When to test, how to test, what to test, and how big a test to conduct are decisions companies undertaking product development must be prepared to make.

8. Using Children

Products designed for young children present unique problems. There is first the difficulty of working with young children, communicating with them, and determining how to measure their preferences. Skilled panel leaders are required, and it is virtually a necessity that the relationship be one leader to one child. Such testing is expensive; extra skills are required. Then there are decisions concerning the test itself. How big should the test be? How often can the test be conducted?

Kroll (1990) describes another difficulty with testing children: what scaling system to use. She tested children using one-on-one interviewing, a self-administered questionnaire, and three types of rating scales. She found that while all three scales discriminated at the 10% level, a simple in-house scale used by Peryam and Kroll (Peryam & Kroll, Chicago) was better than a traditional hedonic scale and (surprisingly) better than a face scale.

B. Using Electronics: The Perfect Nose?

The advent of gas chromatography to analyze volatile or volatizable materials must inevitably lead to the use of the "electronic nose" to analyze the components that make up flavors. Warburton (1996) described one commercial unit and its use in identifying nut varieties and in distinguishing good from bad walnuts. Several types of units are described in detail in a review by Schaller et al. (1998). They describe their use with meats, grains, coffee, beer and other alcoholic beverages, fish, fruit juices, and soft drinks.

Neugebauer (1998) discusses the electronic nose in detail, comparing it with the human olfactory system, neuronal networks, and sample recognition. Advantages and disadvantages of the noses in quality control and sample recognition are discussed with comparison to classical methods.

C. Shelf Life Testing

Developers often speak in terms of *high-quality shelf life*, *acceptable quality shelf life*, or *useful storage life*. These terms have meaning to other developers; they have no meaning to customers and consumers.

A product's shelf life is a verification of the stabilizing systems designed into the product. Food companies cannot release new products into a market, especially a test market, without knowing how stable that product will be. The company must have some idea of what the product's shelf life is. This seemingly simple requirement, to provide a shelf life estimate, is fraught with all sorts of difficulties Curiale (1991) discusses with particular reference to microbiological shelf life. His and other points to consider in shelf life determinations follow.

1. Selecting Criteria to Assess Shelf Life

First, some criterion that changes and can be measured perceptibly and that is appropriate for the product must be selected. This criterion must change gradually with time so the onset of the change can be measured. One that is sudden or abrupt without some measurable antecedent or precursor is not satisfactory. The criteria that may be chosen are:

- Microbiological changes (Curiale, 1991): Where appropriate, total plate counts, psychrophilic counts, or counts of specific microorganisms of public health or economic significance may be monitored to estimate shelf life.
- Nutritional change: The loss of a nutrient such as vitamin C might be chosen. This nutrient should be one for which the food product is a significant source.

- Undesirable change: The loss or change of color or the production of breakdown of color compounds can be followed. Other changes might be exudation or drip loss, moisture transfer, shrinkage, and malodor production.
- Change in a functional property: The loss of a functional property for which the product is noted can be used to follow shelf life, for example, its ability to whip, to color, to flavor, to foam, to leaven, or to set.
- Undesirable textural change: Hardening, softening, staling, loss of crispness, development of graininess, etc. might all be suitable criteria to follow the course of shelf life.

Selection of a criterion for stability presents different degrees of complexity for developers. It is comparatively simple to follow the destruction of a nutrient like vitamin C in a food. But if the product is not an important or even significant source of vitamin C, then vitamin C is not, in all likelihood, a useful criterion for monitoring shelf life. However, if the loss of vitamin C correlates closely to the loss of a major but difficult to measure quality characteristic of the new food product, then vitamin C is a good standard.

Rarely is only one quality characteristic of a complex system such as a food the sole determinant of its shelf life. It is more likely that several characteristics will break down concomitantly; for example, color, texture, and flavor will all degrade over time. And frustratingly, they will degrade at different rates. It is also possible that the breakdown of one characteristic may accelerate the breakdown of another characteristic; for example, the breakdown of some artificial colors can accelerate the development of off-flavors in some foods. Therefore, choosing the correct criterion or criteria to follow during the determination of shelf life stability becomes very important. Figure 7.1 depicts three quality attributes that deteriorate at different

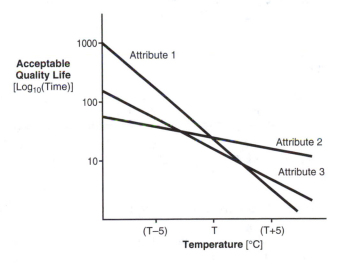

FIGURE 7.1 Comparison of different rates of deterioration of quality parameters for a hypothetical food product with increasing temperature.

rates with increasing temperature. The problem is to choose the most apt one (perhaps the one most characteristic of the product) to follow.

Second, there must be some decision made about how much loss of quality characteristics can be accepted (and by whom). How much loss of a quality characteristic can be accepted before spoilage is declared? If the criterion is color, then how much color loss is acceptable? The loss of a nutrient cannot be seen or tasted, but if a label declaration has been made for that nutrient, unacceptability acquires a new meaning, that is, a label violation. Losses of color, flavor, and texture are assessed by consumers. The result can be dissatisfied consumers.

What constitutes an acceptable degree of instability? Is it the loss of 60%, 50%, or whatever percentage of the vitamin C content or the redness or the crispness of the product? Or must the total plate counts of microorganisms reach a particular level? At 10^5 to 10^8 microorganisms per gram of product there will be a distinct malodor for most products and obvious slime production as well.

In jurisdictions where statements such as "best before ..." or "best quality before ..." are required by label legislation, a misstated shelf life has serious economic implications. A conservatively stated shelf life will cause retailers to return wholesome product that is past its stated expiry date, believing it to be either deteriorated or unhealthy. This constitutes a loss to the manufacturer. If an exaggeratedly long shelf life is projected, a large number of consumer complaints may arise from failed product.

A rough rule of thumb that is frequently used, although many may deny that they do, is the two thirds rule. The rule works as follows: shelf life tests show that a particular packaged product has a good quality shelf life of 90 days at refrigerator temperature. According to the two thirds rule, the shelf life is stated as 60 days. I have never found any practical or scientific justification for this, but I have found that it is used. Since the publication of the first edition of this book in 1994, several people have commented that they too have used this rule. No one has suggested any scientific basis for it.

2.   Selecting Conditions for the Test

A final decision: under what conditions will the shelf life test be carried out? If ideal temperature conditions of frozen storage for a frozen new food product or ideal refrigerator temperature conditions of storage for a refrigerated food product are used, the resultant shelf life will probably bear no likeness to what will happen in the real world of consumers. For frozen or chilled foods, this real world may include the following:

- Temperature changes encountered in factory warehouses that abuse the product.
- Temperature changes during transport and transfer to a wholesaler's or a retail chain's warehouse, storage in this warehouse, and subsequent transport and transfer to a retail outlet (Slight, 1980) may all contribute to abuse of products.

- Storage at inadequate temperatures in the retail store abuse the product. A short period of temporary storage in the retail outlet until shelf stocking starts follows transport. This storage is often in a refrigerator set at one temperature that may not be optimal for the great variety of products held in it. Finally, there may be shelf display in refrigerated cabinets stocked above their limit line and frequently improperly maintained at unsuitable temperatures.
- Storage in poorly maintained vending machines can damage frozen or refrigerated goods (Light et al., 1987).
- Further temperature abuse results from transport and in-house handling by customers and consumers.

Is the product to be tested under nonabusive conditions, that is, following recommended conditions of storage throughout, or under abusive conditions such as those encountered in the distribution chain? This, then, is the dilemma to be faced in determining the shelf life of a product.

Three factors help developers to determine which test to use for the quality shelf life of a food product:

- The preservative systems designed into the food product, which are protected by packaging material selected for this ability
- The physical abuse that food handlers — the people involved in the warehousing, distribution, retailing, purchasing, and in-home storage of the food — can give the package of food
- The environmental abuse that the product and its package may encounter from manufacturing and packaging until it is opened by the consumer and consumed

A fourth possible factor has been omitted — the microbiological load on the food after processing and as it is packaged. It is assumed that good manufacturing practices and a sound HACCP program were in place during manufacturing. However, for chilled foods and minimally processed foods, such an assumption is not valid. The initial microbiological load is an important factor in an acceptable shelf life. A sound HACCP program is critical to these foods.

There can, therefore, be several factors that affect a product's shelf life adversely from factory door to consumer's table:

- Interactions between the chemical components of the food that can alter the quality and the safety of the product. These were (it is hoped) anticipated during formulation development and developers designed stabilizing systems to prevent them.
- Interactions between the food and its package that can damage the integrity of the package and alter the quality. These, too, were anticipated.
- Adverse environmental effects (temperature changes, relative humidity changes, irradiation) to which the package is subjected during these

interactions. Abusive treatment cannot be anticipated and may confound the integrity of the safeguards.

For example, temperature influences the rate of chemical reactions, as every secondary school student knows. The familiar Q_{10} tells us that for every 10°C increase in temperature one can anticipate a two- to fourfold increase in the rate of a chemical reaction. This influence of temperature can be seen in the work of Labuza and Riboh (1982), who reviewed the influence of abusive temperature treatment on Arrhenius kinetics with respect to nutrient losses used as the predictor of shelf life of foods.

Temperature changes will cause phase changes in foods. Gels and emulsions will break down; ice will thaw or ice crystals can grow and damage the structural integrity of soft foods. Texture or tackiness of foods will alter (Slade and Levine, 1991; Goff, 1992).

Temperature changes affect biological reactions. Such changes can alter growth rates and growth patterns of microorganisms and the activity of enzymes (Thorne, 1978; Williams, 1978).

Temperature changes in warehouses can be caused by doors being opened and closed during receiving and shipping. Heaters in warehouses can warm product stacked near them. Changes occur during transportation in refrigerated vehicles with improperly set or maintained thermostats. Retail cabinets with their defrost cycles impose further temperature changes. Humidity changes closely associated with temperature changes can cause sweating of packages, resulting in rusting of metal containers or label damage outside the package. Incident light striking exposed containers on ship decks can cause large temperature changes in foods.

3. Types of Tests

Product development technologists have traditionally used three general approaches to determine shelf life:

- Static tests in which the product is stored under a given set of environmental conditions selected as most representative of the conditions to which the product will be subjected
- Accelerated tests in which the product is stored under a range of some environmental variable (e.g., temperature)
- Use/abuse tests in which the product is cycled through environmental variables

At intervals during these tests, samples are taken and subjected to sensory, chemical, and microbiological assessment.

To these there should now be added another. Good estimates can also be made for the shelf life of foods by examining the technical literature describing the shelf life dynamics of similar products. An excellent review of shelf life and techniques for predicting it can be found in Robertson (2000).

a. Static Tests

Static tests have obvious shortcomings. First, it takes too long for the tests to produce noticeable changes in many foods. Second, because of this fault, the tests are too costly to undertake given the paucity of information they provide. A static test can be likened to a one-point viscosity measurement or a one-point moisture sorption curve: it tells nothing of the behavior of the product under other stresses. It provides no kinetic data.

Okoli and Ezenweke (1990) used a static test to determine shelf life for a new product, pawpaw juice, for consumption in Nigeria. Although freshly sliced and peeled pawpaw was well known and accepted, pawpaw juice was unknown there. Their storage conditions were 30 ± 2°C for the test bottles of pawpaw juice and 10°C for the comparison controls. Samples were subjected to sensory, physical, and chemical analyses at intervals over 80 weeks. This is much too long a test period for practical research and development and would certainly frustrate the marketing program of most companies.

b. Accelerated Tests

Most researchers prefer accelerated tests since they provide much more information about products and the kinetics of their deterioration. A range of conditions of an environmental variable (such as temperature) are carefully chosen to cover the range that could be anticipated in distribution. The packaged product is stored under each of these temperature conditions and analyzed at intervals for the loss of a particular quality characteristic. If the variable chosen is temperature, a simple application of the Arrhenius equation allows researchers to demonstrate graphically the relationship between temperature and time in days until an undesirable degree of loss of quality occurs. Then researchers can calculate the number of days of good shelf life to be expected if one assumes similar storage conditions to prevail. Factors to be considered in accelerated tests are well documented by Labuza and Schmidl (1985).

Conditions of the accelerated test must be selected with care (Labuza and Schmidl, 1985). If one considers only temperature as the environmental variable, one must remember that ice can thaw and water can freeze; fats can melt or solidify; suspensions (emulsions) can degrade; and rheological properties can change drastically. These changes can seriously skew data obtained by this sort of kinetic modeling.

Obviously, there is a limit to increasing the storage temperature of frozen foods, but this limit might be lower than one might expect. Water in a food system can remain liquid well below the freezing point of pure water. One is no longer studying the behavior of a frozen product but the behavior of a liquid product.

Ideally the variable chosen for accelerating should not alter the normal or anticipated path of spoilage, that is, the path of spoilage to be expected in normal nonabusive conditions. The purpose of an accelerated test is primarily to determine the shelf life of a packaged new food product under normal marketing conditions, whatever these may be, and only secondarily to study the reaction kinetics of its deterioration. (The eternal question in product development: Is the purpose to satisfy the needs of the research technologist or the marketing needs?)

Labuza and Schmidl (1985) provide a table of recommended storage temperatures for frozen foods (−5 to −40°C for the control), for dry and intermediate moisture foods (0 to 45°C for the control), and for thermally processed foods (5 to 40°C for the control).

Some quality characteristics of foods vary widely with temperature. That is, at temperature T°C (Figure 6.2), attribute 2 of a hypothetical product is the limiting quality characteristic. At a higher temperature (T + 5)°C, quality attribute 1 is the limiting quality characteristic for a good quality shelf life. At a colder temperature (T − 5)°C attribute 3 limits acceptability. Again, reference to Labuza and Schmidl (1985) will provide guidance in determining both storage times and intervals between samplings.

Temperature is not the only accelerating factor that can be used, but it is the usual one. Chuzel and Zakhia (1991) used adsorption isotherms at three temperatures to derive an equation describing the shelf life of gari. (Gari is a semolina prepared from cassava that has been fermented, cooked, and dried and is popular in West Africa and in Brazil, called *farinha de mandioca*.) The equation considers the shelf life in terms of:

- Its equilibrium moisture content
- Its safe storage moisture content
- Its initial moisture content when packaged
- The permeability of the package and its surface area
- The fill weight of the package
- The slope of the product isotherm

Their suggestions for further work are to include the initial microorganism load and quality characteristics such as color, flavor, and crispness.

Hardas et al. (2000) describe accelerated stability studies on encapsulated milkfat. They studied peroxide value, hexanal production, fatty acids, and emulsion droplet size distribution.

c. Use/Abuse Tests

Use/abuse tests are different from either static or accelerated tests. They are included here because many technologists use them to assess the shelf life of the food and its package as a unit. They are as varied as an imaginative mind can make them.

Frozen food developers commonly cycle new frozen food through the temperature range of −10 to +20°F. This range is intended to duplicate the freeze-thaw cycles of frost-free frozen food cabinets. Cycles are set to correspond to those that could be anticipated under store conditions. Some developers go so far as to purchase freezer display cabinets, stack them in the manner seen in most supermarkets (with a good portion of the product above the recommended fill level) and thus simulate real supermarket conditions.

In one instance I am aware of, a company's frozen product was stored in freezer lockers set to cycle at the temperatures that their experience had taught them would be encountered from factory warehouse to frozen wholesale warehouse through to the retailer. In between these stages, product was removed to simulate the transfer from storage to ambient temperature to frozen transport to dock transfer at the next stage and so on.

In another example of a use/abuse test, a pallet of cased product was dispatched on a journey around the country by truck and by train. The rigors of transportation and the influence of temperature and humidity changes due to weather on both the condition of the product and the effectiveness of the package could be studied on the pallet's return to the plant. Application of this data to shelf life calculations can be limited, but information on what one can expect of the product and its package in handling and distribution can be useful in protecting the validity of one's shelf life statement. Perhaps this expectation of abuse was the origin of the two thirds rule?!?

Cardoso and Labuza (1983) studied the moisture gain and loss of egg noodles packaged in paperboard, polypropylene, and polyethylene, typical packaging materials for this product. The products were cycled under controlled but varying temperature (30 to 45°C) and relative humidity (11 to 85%). From their data they developed a kinetic model to predict moisture transfer, an important factor in product stability.

Porter (1981) discusses the unique problems the military has with shelf life prediction in uncontrolled environments, a predicament the military knows only too well when shipping food from controlled temperature facilities to Arctic bases or to arid desert conditions such as in the recent Iraqi conflict. The military also has the problem of packaging material to survive air drops. However, food manufacturers face many of these same problems in exporting their products to foreign countries with vastly different environmental conditions as well as warehousing and shipping conditions. Temperatures in container shipments can reach levels that can be very stressful to a good quality shelf life.

Static tests and accelerated tests challenge the first and partially simulate the last of the three factors that determine the shelf life of a food product, that is:

- Its preservative system and the package protecting it
- Abusive handling within the factory, from factory to consumer, and by the customer until use
- The environment the food, and its package, encounters throughout its shelf life

Use/abuse tests cover all three factors. It is difficult to predict any estimation of shelf life with absolute certainty from such data, and no tests can simulate all the circumstances that may befall a product from manufacturer to consumer.

Abusive or unusual treatment a food might encounter during storage or transportation cannot be anticipated. No use/abuse test can simulate all the stress that may be heaped on the food product and its protective packaging:

How can one duplicate the military mind that pinholed film-wrapped food packages so they would fit better into the ration cartons? Or how can one simulate the damage done to the package and ultimately to the product by retail stock clerks who slash food containers as they open cartons or practice basketball shots with packaged frozen chicken?

The above abuses, which I have personally encountered, can be eliminated only by educating the food handlers in the safe handling of food products. Food manufacturers have relied on retailers to do this, but unfortunately many retailers, with part-time, temporary help, find no profitability doing so and rarely undertake to train staff. Nevertheless, abusive treatment of food products can be minimized only by training and education.

Transport of product can result in abusive treatment. Transport of product in containers that are exposed to the hot sun during a trans-Atlantic crossing have been already mentioned. Temperatures inside such containers can exceed 120°F.

Vibration caused by transport results in product changes. The engines of a ship produce a gentle vibration that can destabilize some food suspensions. Surface transport, with its gentle rocking action, has been known to produce subtle, and in one case unexpected but desirable, changes in a chocolate product.

A chocolate *couverture* supplier shipped bulk chocolate by tank car from its factory in the east to a customer on the West Coast. Demand was so good it was considered advantageous to all parties to produce this product closer to the customer in a new facility. This step turned into a disaster when the customer complained about the loss in quality of the product, although the satellite plant rigorously followed the parent plant's procedures for making the *couverture*. Investigations revealed that the extra conching action caused by rocking of the tank cars developed the better flavor that the customer preferred. The problem was solved with longer conching times.

4. Guidelines to Determining Shelf Life

Shelf life is an important characteristic of a food product that may be required by legislation or by contract between a copacker and a buyer or insisted upon by a retailer. Consumers, at the end of the distribution chain, certainly expect food to have a good quality life until it is consumed, no matter how long and under what potentially abusive conditions they may have stored it. At the start of the chain, manufacturers of sensitive food products, chilled foods for example, are at the mercy of:

- Distribution and warehousing companies: Manufacturers depend on them to store and handle products under nonabusive conditions.
- Retailers: Retailers need to practice a stock rotation system in their warehouses and on their shelves. This is not always the situation.

It is also unfortunate for manufacturers of sensitive products that many retail food chains misinterpret the product statement that reads that a particular product has, for example, a 40-day refrigerated shelf life. They assume that a 40-day refrigerated shelf life starts when they decide to stock their shelves or to plan a promotion.

A client who was a packer of a fresh-pack salsa had shipped product to a retailer for a special promotion. The retailer delayed the promotion until the product was beyond its expiry date. Product was returned. My client was devastated.

Few manufacturers, especially small manufacturers, would dare to refuse to take back product beyond its expiry date for fear of losing future orders. These are the facts of life that developers determining shelf life must deal with.

Estimating shelf life is a guessing game. Many scientists who model kinetics of spoilage reactions may take umbrage at this description of estimating shelf life: it is, nevertheless, true. The kindest that can be said is that the stated shelf life stamped on any product has a strong element of guessing in it along with the element of hard data that food scientists can add. Neither consumers nor retailers know what previous treatment a product has had when they read the "best before" date. As data on the deterioration of foods are collected and knowledge of growth mechanisms of microorganisms in foods progresses, this guesstimate can be refined.

Let it be clearly understood: no one can say what the shelf life of a particular product will be. The best that can be done is predict what the shelf life should be if.... That "if" encompasses all the precautions that should be taken in manufacturing, distribution, warehousing, retailing, and home storage and preparation. This becomes a big if.

The following points should be considered. First, there are no tests that can absolutely be relied upon to predict the shelf life of a given food product. *All such tests can do is provide an approximation.* Experience with similar products in the same product category (e.g., chilled foods) can help provide some initial estimates. Data in the scientific and technical literature can help refine these initial estimates. Audits of competitor's similar products drawn from the retail showcases will provide more data to complement one's own findings. Here too, the complaint records of a company can provide information on a product's stability that may be applied to another product in development. This marks another reason for documenting the complaint files. The development team should not waste their time looking for the perfect test.

Second, shelf life tests should ideally be carried out only on finished product manufactured on the line (and equipment) to be used in regular production and packaged in the container that will be placed on the shelves for consumers. Product prepared in test kitchens or pilot plants does not simulate the product prepared in a plant at the height of the packing season with adjacent packing lines running products capable of being a source of cross contamination. Bailey (1988) discusses these problems of scale-up from the pilot plant to the production floor and describes attempts using predictive techniques to minimize the discrepancy between the pilot plant and the manufacturing plant.

The following examples illustrate the fallacy of relying heavily on test kitchen samples for reliable shelf life data.

A potato processor wanted to develop a line of chilled prepackaged, prepeeled potatoes. A shelf life of the product was determined to be an excellent chilled shelf life exceeding 30 days. I had determined this on pilot plant–prepared samples made from potatoes purchased on the retail market as part of my graduate research studies. In full plant production trials, a shelf life of 10 to 15 days was the norm. The microbiological load between the samples prepared in the well-kept, sanitary test pilot plant and those prepared in the plant factory under full production with field grown produce was significantly different.

In another case,

the unavailability of the desired canister with metal ends from a supplier and the pressure exerted by marketing "to get on with it" led researchers in one company to substitute a canister with plastic ends for a spiced and herbed bread crumb mix for shelf life trials. The supplier suggested that the plastic substitute would provide a more rigorous test "since plastic breathed." Tests were successfully conducted and shelf life determined. Consumer complaints began to pour in when the final product used the canisters with the metal ends. Off-flavors were noted and rusting was observed. Constituents (principally citral) in the spice and herb blend reacted with the metal ends to cause a breakdown of the flavor and initiate detinning.

Factory produced product is different from product prepared in the pilot plant. Scale-up from the test kitchen to the pilot plant to the factory floor has always produced changes, subtle and not so subtle, in a product. The manner in which heat is applied is different; stirring action is different, and therefore heat transfer can be altered; pumping action may be different, and the distances the product has to be pumped in the factory may alter its temperature, its rheological properties, and hence the shear stress the product receives. Two products, one factory-produced and one pilot plant- or kitchen-produced, should not be expected to have the same storage properties.

Third, once the shelf life of a product has been determined, any change in the recipe, in the suppliers of the ingredients, in the water treatment system in the plant or in the water used in batch preparation occasioned by plant relocation, or any other change can have a major impact on the acceptable quality shelf life of a food product. For example, the mineral content of plant waters can have a profound effect on the flavor of a product, not only immediately but over a period of time.

It was found that the flavor of marinated, fried peppers was adversely affected when the supplier of the frying oil, a fractionated peanut oil, changed from one antioxidant for their oil to another. Both the user and the oil supplier were surprised at the flavor change.

Fourth, extrapolation of data obtained from accelerated storage tests should be treated warily. Extrapolation of data for chilled foods in particular can be misleading if due care is not given to the types of microorganisms contributing to the spoilage: spoilage can be overt with obvious indications of slime, off-odors, loss of color, etc., or it can be covert with no obvious outward signs of spoilage but with undesirable increases of microorganisms of public health significance or even of toxins.

Fifth, lack of acceptance by the customer, the consumer, and even the retailer is the ultimate determinant of a product's shelf life. Nevertheless, in basing shelf life on sensory analysis by consumers — that is, on taste — their interpretation of spoilage and their ability to detect it varies considerably. Dethmers (1979) discusses the use of sensory panel evaluation for failure criteria for open-dating. Curiale (1991) and Beauchamp (1990) both point out some of the shortcomings of the use of sensory panels for shelf life determinations. Beauchamp, in particular, cites intra-individual and inter-individual variations that can confound the use of sensory panels as an evaluation tool. Therefore, the use of taste panelists in determining shelf life poses health hazards for the panelists and may not produce reliable estimates of shelf life. A clear understanding of the criteria used to assess the end of acceptable (or high quality) shelf life must be established.

Finally, determining shelf life involves measuring the differences between control samples and test samples that are subjected to stress over time. Wolfe (1979) discusses the advantages and disadvantages of different reference standards, especially in sensory studies, that are equally valid for shelf life studies. See also Labuza and Schmidl (1985) for recommended storage temperatures for control samples.

5. Advances in Shelf Life Considerations

There is a need for predictive techniques in microbiology as food manufacturers provide more added value to food products. Added value, as Williams and coworkers (1992) see it, alters traditional paths of food spoilage and intoxication. To alleviate this problem, predictive techniques to properly assess duration of quality and safety are needed.

Some exciting new thinking is emerging in the study of the kinetics of food deterioration and in the development of predictive techniques in microbiology. Instead of observing a food spoil over time under predetermined conditions, could one, from a knowledge of spoilage mechanics and a product's composition, accurately predict its expected shelf life? That is, kinetic information about the spoilage mechanics ought to make it possible to estimate shelf life duration.

The use of predictive models is not new. For example, the botulinum cook established for the safe thermal processing of low-acid, high-pH foods is an application of such a model; it is an inactivation model of the destruction of spores of *Clostridium botulinum* by heat (Gould, 1989). The value of predictive models of microbial spoilage is that they provide a data gathering tool to apply to many types of food stabilized by various techniques. Based on this data, rapid predictions of duration of quality and safety are possible with a greatly reduced need for testing and therefore with reduced costs (Roberts, 1989).

Roberts describes two types of predictive models: the probabilistic model and the kinetic model. Each serves a different purpose. Probabilistic models, as their name suggests, predict the probability of an event, for example, probability of toxin development in a food, however they provide no information about how quickly the toxin develops or the amount produced. Roberts (1989, 1990) describes a mathematical model able to predict botulinal toxin production as a function of salt, nitrite, thermal treatment, storage temperature of the food, and the presence or absence of preservatives.

Kinetic models predict the rate of growth of microorganisms and therefore are useful for anticipating time to microbial spoilage or time to growth of critical numbers of food-intoxicating microorganisms. Roberts (1989) discusses several examples of kinetic modeling. Figure 7.2 is a hypothetical depiction that shows how

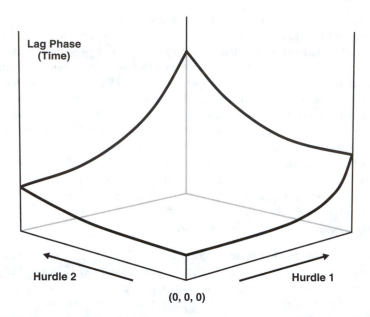

FIGURE 7.2 Three-dimensional representation of the influence of concentration of two hypothetical hurdles on the lag phase of a hypothetical microorganism.

the lag phase of microbial growth behaves under the stress of increasing concentrations of two different hurdles. Choosing the conditions that provide the longest lag phase of microbial growth gives the greater security. This visual presentation is valuable for technologists designing stabilizing systems for new products. It would be advantageous to select concentrations of Hurdle 1 and Hurdle 2 that extend the length of the lag period of microbial growth (vertical axis) if these are consistent with other quality factors such as taste. Likewise, one could determine which stabilizing system causes the slowest rate of microbial growth, and this would again contribute to the product's stability. With the aid of predictive models, technologists are able to design hurdles into the product or its process, and from this the length of the lag phase or the rate of growth could be predicted.

The Arrhenius equation has been used for assessing both nutrient losses with temperature (Labuza and Riboh, 1982) and microbiological growth with temperature (Gibbs and Williams, 1990). The Ratkowski square root equation has also been used:

$$r^{1/2} = b\,(T - T_0)$$

where
    r = growth rate constant
    b = regression coefficient
    T = storage temperature (K)
    T_0 = a temperature at which the growth rate is zero

Gibbs and Williams describe the use of this equation for plotting the growth of *Yersinia enterocolitica*.

New foods can spoil not only because of temperature variations. The demand for low-salt, low-acid, nitrite-reduced, sulfite-free, or sugar-free variants of food products have multiplied the number of deteriorative routes of conventional products (Williams et al., 1992). As multiparameter stabilizing systems are used to preserve these foods, new risks are introduced. There is an increased need to understand the mechanisms of these systems and to be able to predict quality changes as well as microbial activity in a complex matrix of variables (see, e.g., Chuzel and Zakhia, 1991).

In his review of predictive food microbiology, Buchanan (1993) attributes the growing interest in mathematical modeling to three major factors:

- Easy access to powerful, number crunching software programs and computers to process the data.
- A growing desire of consumers for minimally processed, just-like-fresh foods. This desire usually favors the chilled foods category.
- The need to organize quantitatively and systematically the wealth of microbiological data on the vast array of foods potentially at risk as either economic or public health hazards.

Buchanan discusses and describes various classifications of models. Examples of the use of Arrhenius models and Ratkowski and Gompertzian equations to model growth are available (Gibbs and Williams, 1990; Buchanan, 1993).

Walker and Jones (1992) explain predictive microbiology in general and describe the coordinated research program then current in the U.K. This program, under the direction of the Ministry of Agriculture, Fisheries, and Food had a host of participating scientists at various laboratories. In addition, private food companies were invited to contribute data. Campden Food and Drink Research Association would house this U.K. Predictive Food Microbiology Database. This program has had a rather checkered history. Commercialization was undertaken by Leatherhead Food International; then the intellectual property rights were sold to the U.K. Food Standards Agency. In a press release dated June 16, 2003, this agency describes an international collaboration among itself, the Institute of Food Research, and the U.S. Department of Agriculture to provide free access to the database: "The new common database, called ComBase, already contains around 20,000 growth and survival curves and 8,000 records containing growth rates."

In the U.S. the database is available at http://wyndmoor.arserrc.gov/combase/ and in the U.K. at http://www.ifr.ac.uk/combase. Such a database resource is useful to all in product development for predicting food safety and quality.

Kinetic studies and predictive modeling techniques have been applied to measuring the deteriorative rates of food quality or specific food components. Labuza (1980) used water activity and temperature to measure quality losses in foods and assess water activity's effect on reaction rates. Rockland and Nishi (1980) also used water activity as a variable affecting food product stability. Andrieu and coworkers

(1985) used an Oswin type relation ($X = f\{a_w, T\}$) to model pasta drying. Norback (1980) discusses modeling in general for optimization of food processes.

Lund (1983) cited three major reasons for generating models to describe food processes:

- Models allow developers to optimize processes with the minimum amount of costly trial and error.
- Models provide better understanding of processes, which leads to better processes and safer new products.
- Models permit better prediction of shelf life and quality changes in foods.

Descriptions of modeling techniques and applications to food quality deterioration can be found in Lenz and Lund (1980), Hill and Grieger-Block (1980), and Saguy and Karel (1980). Heldman and Newsome (2003) review papers on microbial inactivation kinetic models presented at the Institute of Food Technologists second Research Summit held in Orlando, Florida, in January 2003. These papers focus on microbial survival during processing. Erkmen (2000) used high-pressure carbon dioxide to develop a predictive model for inactivation of *Listeria monocytogenes*.

III. DESIGNING FOR PRODUCT INTEGRITY

The safety of products with respect to hazards of public health significance as well as hazards of economic significance is established only by designing safety into the product from the start of development. This concept has been well established in instrument manufacturing. Mayo at AT&T (1986) argues that quality by design applies equally to products and to services. There are four elements to the program:

- Design to the correct requirements of the customer. If the customer's needs have not been clearly identified, product design will be faulty.
- Design using the right technology. This influences the cost of the product and the customer's satisfaction with the product.
- Design for manufacturability. As Mayo states it, the product and its process should be designed "to be insensitive to 'noise' such as conditions of customer use, drift in components, or variations in the manufacturing environment."
- Design for reliability.

Coincident with development's progress, feedback mechanisms should continually monitor and analyze the quality and safety of the product. This is Mayo's design for reliability.

Huizenga and colleagues (1987) at the Perkin-Elmer Corporation put forward six steps for cooperation and communication in safe product design:

- The needs of both internal and external customers must be identified.
- Quality characteristics must be identified.

- Means to measure these quality characteristics must be obtained.
- Quality goals that satisfy customers and suppliers at reasonable cost need to be established.
- The process to attain the stated goals must be set in place.
- Processing capability must be verified.

The steps described by both Mayo and Huizenga and colleagues pertain to instrument design and not specifically to food products. Pearce (1987) appears to have rewritten Mayo's and Huizenga and colleagues' elements to make them food oriented. Pearce's rewrite describes six principles that are involved in a design assurance policy for food products:

- Design work for quality must conform to marketing concepts and regulatory needs.
- Design work for quality should conform to properly established procedures and standards.
- All design work should be properly documented for quality with changes both recorded and regulated.
- Challenges of the design should be carried out at each stage of scale-up including production trials.
- Third party review of design work for quality is required at critical stages before advancing to subsequent stages.
- A feedback system must be prepared to collate all activities and planning that support manufacture of new products.

Pearce (1987) further broke these principles into 12 subsystem requirements and developed a responsibility matrix for product design assurance with primary responsibility delineated.

Wilhelmi (1988) considers the following requirements to be important in product design:

- Product composition
- Safety considerations for the product
- Regulatory compliance
- Knowledge of product stability
- Packaging considerations
- Marketplace considerations

A comparison of Wilhelmi's points and Pearce's shows some obvious similarities.

Indeed, all the above authors have essentially stressed the same points concerning designing for quality. That is, they emphasize formulation for quality, stability, conformance, and safety, based on known spoilage mechanisms and anticipated abuse and protected by suitable packaging. Such design is a built-in safety net; it is error prevention in new food product development. Technologists must incorporate safety and quality design into their products through the judicious use of ingredients, processing, and packaging technology.

A. SAFETY CONCERNS

Equally important to maintaining desired quality characteristics are control systems to monitor safety with respect to hazards of public health significance and to maintain the integrity of these systems throughout processing, storage, distribution, and retailing. Monitoring systems should already exist in the plant's good manufacturing practices, quality control/quality assurance systems, and HACCP programs; these should be in place for the company's other products.

All these existing systems for maintaining the integrity of existing products must be reevaluated every time a new product is introduced into a food plant. Each product needs an HAACP program unique to itself. New products may introduce unsuspected new hazards through the introduction of new raw materials and ingredients. The result is that the safety of all the company's products and processes may be jeopardized.

B. FOOD SAFETY DESIGN CONCERNS

Two major questions need to be answered when prototype recipes or products have been obtained:

- What hazards of public health significance are, or could be, associated with these products during manufacture; during warehousing, storage, and distribution; and with their home use? In short, a complete HACCP program must be developed.
- What desired quality attributes are to be built into these products that require stabilizing or maximizing?

This begs other ancillary questions:

- What are the major (probable) spoilage routes of products of this particular nature and composition, and what health hazards are associated with them?
- What duration of acceptable quality shelf life is desired for these products?

These desired attributes could be organoleptic attributes, specific dietary or nutritional attributes, functional attributes, or convenience of preparation.

An HACCP program begins with product design. Elements of quality and safety must be designed into products at their inception.

Concern about the safety of any food product centers around all of the following:

1. Possible presence of food-intoxicating microorganisms exceeding recognized norms in the ingredients and raw materials forming part of the product or in the product itself that, when ingested by susceptible consumers, may cause illness or death.
2. Presence of preformed toxicants from a biological source (enterotoxins, mycotoxins, or compounds such as domoic acid of algal blooms), which may have entered with the ingredients or be formed during processing.

3. Presence of chemical hazards (pesticides, herbicides, growth stimulants, or even fertilizer uptake) resulting from carry-over in the food product or in any of the ingredients.
4. Development of chemical hazards from processing through intercompound reactions within the food.
5. Presence of miscellaneous extraneous matter such as stones, glass fragments, metal pieces, or wood in the food product, which can cause serious injury if ingested.
6. Presence of insects and insect parts, an esthetic hazard that often has a shock effect on consumers, resulting in illness.

Items 3, 5, and 6 should be prevented from entering the food or removed from the food by a total quality management program (Shapton and Shapton, 1991). Item 4 may be circumvented by a more judicious choice of ingredients and processing conditions or both.

Many processing steps (trimming, cleaning, and blanching) plus plant support systems (good manufacturing practices, HACCP programs, cleaning and sanitation) reduce the numbers of microorganisms that jeopardize the safety and stability of products (see, e.g., Shapton and Shapton, 1991). For minimally processed products, extreme attention to HACCP programs is essential.

C. New Concepts of Safety

The concept of safety is evolving; its interpretation is broadening and coming under closer scrutiny from regulatory agencies in all countries. Safety concerns have traditionally centered on susceptible consumers, that is, the very young and the elderly.

But the concept of the susceptible consumer must be expanded now to include immuno- or health-compromised individuals. Brackett (1992) describes these as

> people with underlying chronic health problems such as cancer, diabetes, or heart disease; individuals taking certain immunocompromising drugs, such as corticosteroids; and individuals with immune deficiency diseases such as acquired immunodeficiency syndrome (AIDS).

The latter group is a growing proportion of the consuming public who are more susceptible than uninfected individuals to food-intoxicating microorganisms. They may fall ill after ingesting much smaller numbers of infective microorganisms than would harm other consumers. Archer (1988) reports that AIDS-infected male patients are 300 times more susceptible to listeriosis than are AIDS-negative males.

This new susceptible consumer becomes an additional consideration that shapes technologists' thinking in designing safe stabilizing systems for new products.

Another consideration is the emerging recognition by food microbiologists of the ability of some exotic and some well-known microorganisms to grow and become health hazards in stored chilled foods. One such exotic microorganism of concern to food microbiologists is *Mycobacterium avium* subsp. *paratuberculosis*, thought to have a role in Crohn's disease (Williams, 2003). Indeed, as the limits to growth and the viability of food-associated microorganisms in multiparameter stabilizing systems (e.g., hurdle technology) are studied, some anomalies of accepted knowledge of the limits of growth and viability are appearing. Pathogenic psychrotrophs capable of growth around 5°C are *Listeria monocytogenes, Yersinia enterocolitica*, type E *Clostridium botulinum, Vibrio parahemolyticus, Vibrio cholerae, Bacillus cereus, Aeromonas hydrophila, Staphylococcus aureus*, enterotoxigenic *Escherichia coli*, and some Salmonella species (Farber, 1989). Any temperature abuse of susceptible products in the chain from processor to consumer could be potentially dangerous.

Developers must also consider the origin of contamination. Konuma and colleagues (1988) determined that the most likely sources of contamination by *B. cereus* in meat products in Japan were meat product additives and not the meat itself. Beckers (1988), reviewing the incidence of food-borne diseases in the Netherlands (1979 to 1982), found *B. cereus*, Salmonella sp., *Campylobacter jejuni, S. aureus, Clostridium perfringens,* and *Y. enterocolitica* to be major causative microorganisms. Meat and meat products, fish and shellfish, snacks, and foods prepared for immediate consumption such as in food service outlets, especially those catering to hospitals, old peoples' homes, cafeterias, and restaurants, were the primary carriers. Hemorrhagic colitis associated with *E. coli* has been frequently reported in nursing homes (Stavric and Speirs, 1989). Processed packaged meat products were found by Tiwari and Aldenrath (1990) to be contaminated by *L. monocytogenes.* Slade (1992) considered the presence of listeria to be ubiquitous based on an extensive review of food processing environments.

Keeping the stabilizing systems designed for food products safe and free from compromise is not as simple one might think. Interactive packaging and controlled atmosphere and modified atmosphere packaging (CA/MAP) control the growth of some spoilage microorganisms but certainly not all microorganisms. For instance, Brackett (1992) noted that spoilage was retarded in modified atmosphere–packaged produce, but *A. hydrophila* and *L. monocytogenes* were unaffected; visual spoilage was stopped, but toxic microorganisms were able to grow.

Idziak (1993) noted that many gas mixtures used to extend the shelf life of MA/CAP chilled products are the very ones used to culture anaerobic pathogens. Growth of anaerobes (if present) may proceed with no visual indication of microbial spoilage. Park and coworkers (1988) confirmed this possibility in a study of modified atmosphere–packaged commercially processed wet pasta stored at good (5°C) and poor (16°C) temperatures (the latter temperature is frequently encountered in open refrigerated shelves in retail outlets). At the end of the recommended shelf life of four weeks, sufficient staphylococcal enterotoxin had formed in a significant portion of those samples held at the poor refrigerated conditions to have caused illness in sensitive individuals.

Many products, for example, anchovies, unpasteurized acidified foods, and beef jerkies, are semiconserved. That is, they are safe, stable, wholesome products in their uncompromised stabilizing system — for anchovies this is their very high salt-on-water content. By themselves, anchovies can be eaten safely; it is a preserved product. When, however, they are used as ingredients in another product, for example, a pizza, a quiche, or a flan, the stabilizing system of the semiconserve is compromised; the potential for a food hazard to develop is very real. If the pizza or quiche or flan is held for a long period of time in a warming oven or in a poorly maintained refrigerator, the microorganisms in the anchovies, some of which are spore formers held in bacteriostasis by the high salt content, are no longer in such inimical conditions. Growth commences and spoilage or intoxication of the multi-component product can occur.

Another consideration for developers is the unusual behavior of sublethally injured bacteria during processing reported by Archer (1988) and Rowley (1984). Sublethal injury to microorganisms, whatever the nature of the injury, evokes an adaptive response. In some instances injured microorganisms become more resistant (Archer, 1988). The very action of pretreatment, that is, processing, of foods may alter the sensitivity of microorganisms to stresses.

This desire on the part of customers and consumers for less processed foods such as chilled foods opens up a Pandora's box of new considerations concerning safety and stability that technologists must deal with. These new safety concerns must be addressed by development technologists with the quality control and processing departments. Quality control individuals screen by detailing process requirements that will be necessary if products are to be safe and wholesome.

D. The Costs of Quality and Safety Design

Quality in new food products does have costs. These costs (overheads) can be crudely distinguished (Table 7.2) as indirect and direct costs of quality. Indirect costs are hidden in the development work of designing new products to minimize hazards and to design processes to manufacture products with these designed features.

TABLE 7.2
The Costs of Quality in New Food Product Development

Indirect Costs	Direct Costs
Quality design of product	Increased inspection; more bodies or more equipment are required
Product standards	Additional analyses are required for newly introduced product
Ingredient standards	Increased costs for:
Processing standards	Maintenance
Process design	Sanitation
HACCP programs	Hygiene
SPC* programs	Warehousing
Operator training for	Control systems
Maintenance staff	In-line instrumentation
Sanitation personnel	On-line instrumentation
Hygiene personnel	Off-line equipment
Equipment operators	Rework costs
Quality control line inspectors	
Laboratory analysts	

* SPC = statistical process control.

Modifications to plant equipment for the new product involves much costly work. Training of operators and technicians in the new procedures (sanitation, preventive maintenance, process and quality control, storage) novel processes and products may require contributes to indirect costs. Staff must be trained in new inspection routines to recognize and report hazards associated with something novel to them.

The direct costs include (a) salaries of additional inspectors and analysts to cover the increased need for grading, collection, and interpretation of data for incoming materials; (b) development and installation of in-line, on-line, or off-line process controls; (c) rework of failed (not meeting standards) material; (d) returned goods; and (e) lost customers. The more time, effort, and money spent on prevention in the design phase of new product development, the less the losses will be for failed products.

Every food company has a quality control policy. Large companies may have this in the form of a clear statement about the quality of its products. From this statement are derived procedures for quality in purchasing, processing, and ware-housing, indeed, for every aspect of the company's business. The end result is a manual of operations documenting all the product and process procedures that affect quality. Implementation of these procedures is the responsibility of the quality control department. All new products must conform to the company's policy.

Not to be forgotten is the impact that all new products may have on safety related systems. Supporting a plant's quality control systems and HACCP program are several interrelated programs that complement the company's quest for safety and quality. These are:

- Preventive maintenance
- Pest control
- Plant sanitation
- Statistical quality control procedures
- Worker-related programs
  - Worker safety and health (ergonomics)
  - Hygiene for food handlers
  - Training programs for food handlers
- Grounds maintenance
- Good manufacturing practices
- Environmental safety
  - Waste management
  - Water reclamation and effluent control
  - Odor reduction

Manufacture of new food products that require new raw materials and ingredients in any food plant introduces new hazards into that plant environment, and each of the above needs to be reviewed for its adequacy respecting these hazards. New raw materials bring new microflora into the plant that staff may not be familiar with. Plant personnel unfamiliar with new ingredients, strange raw materials, unusual products, or even new plant routines can introduce hazards. For example, purchasing departments may not have the expertise to purchase wisely on the commodity market, or the plant may not have the facilities to store the ingredients properly.

> My client, a developer of breaded coatings, had a moth infestation that had spread into other parts of the plant and, more damagingly, into other finished products. The cause was simply explained. The purchasing department had purchased a quantity of difficult-to-obtain crumb for a new product they had "bought long." They bought several months' supply; they stored these in an area of their warehouse made of porous cinder block. The crumb was infested; moths had ideal breeding spots in the cracks and crevices of the cinder blocks; they spread rapidly. A clean-up and disposal of the crumb had preceded my visit, but the problem had persisted. The plant manager was aghast during my inspection tour when I insisted an electrical control panel in the main factory be opened. It was alive with moths. A more thorough clean-up followed.

Anything new or unusual introduced into a controlled environment like a food plant introduces a new hazard. If the nature of the introduced hazard is not understood or anticipated, then the hazard cannot be controlled adequately.

E. HAZARD ANALYSIS CRITICAL CONTROL POINT (HACCP) PROGRAMS

The early history of the development of the concept of HACCP systems is described by Bauman (1990).

The developers must set out an HACCP program for every food product a plant produces and before any new product, including line extensions of already established products, is introduced into a plant for production. HACCP programs for established products may not be adequate even for line extensions of these same products. HACCP programs are, or should be, flexible and able to evolve continually to meet equally flexible, and constantly evolving, hazards of economic and public health significance. Archer (1990) cites the emerging microbial hazards such as *C. jejuni* and *L. monocytogenes* that can challenge an inflexible HACCP concept.

Development teams should work closely with their production members to prepare a flow chart of the processes and equipment to be used for the new products. It is here that feedback from technologists can help production staff assess the effects that the rigors of plant processes might have on the product and its sensitive characteristics. They identify points in the process where, if control were to be lost, a hazard might arise that may jeopardize the new product's manufacture. The experience of production people with their knowledge of the past performance of the process lines assists the other members of the development team. They will know the limits of process equipment; they can more readily identify hazards.

Once identified, hazards are either eliminated or minimized at that critical point in the process (or in the plant environment) or the product is protected from the hazard with remedial action. Combined teams of technologists, production staff, and quality control personnel can now elaborate how this control for elimination or reduction of identified hazards is to be carried out. Control limits for the guidance of production can be established with confidence.

The final step is the determination of process parameters or product-in-process characteristics to be used to monitor product integrity. A sound statistical process control program will then warn of deviations from accepted norms. Stevenson (1990) describes the steps and considerations that should be applied when introducing HACCP principles to foods.

F. Standards Necessary for Safety

Development teams have the responsibility to establish standards and specifications for ingredients used in new products; for packaging materials used in protecting the safety of new products; for the labels, label statements, and the placement of the labels on packages; and for a detailed description of the entire process with the various pieces of equipment used and the support systems for the new products including everything from the analytical procedures to be used to the detergents and sanitizers necessary to clean equipment.

First, specifications for the ingredients need to be clearly defined so purchasing and foremost departments have precise descriptions of ingredient characteristics for obtaining exactly the style, grade, cut, color, flavor, heat level, or whatever other criteria are required at the lowest cost or at a cost consistent with the product's original cost estimate. For this reason, product development teams must define as closely as possible all the essential characteristics of ingredients used in new products.

Only the essential characteristics of an ingredient in a new product formulation should be identified. If color is important, it should be so specified and the color identified; if particle size is important, the range of particle size should be identified. After the essential and attributes of each ingredient are identified and specified, a list of suppliers whose products meet these specifications as defined by the product development team can be prepared for the purchasing department.

A second requirement of specifications is that the quality control laboratory have analytical or functional tests that permit assessment to determine that ingredients, when received, meet the attribute standards. Included with this methodology should be sampling procedures suitable for the nature and form of needed ingredients. If attributes are specified for any ingredient or raw material, they must be checked frequently enough to ensure safety and frequently enough to keep suppliers honest. Even a good customer–supplier relationship needs to be verified; suppliers have been known to make mistakes, take liberties, or change formulations if vigilance is lax. It is the wise customer who subjects suppliers to the same degree of quality inspection rigor that they apply to their own products.

Where applicable for ingredients and more particularly for raw perishable materials, proper storage conditions need to be specified.

There should, of course, be finished product specifications. These are not the same as quality standards. Quality was designed into the product. These finished product specifications identify whether the product has met the specifications required in the trade of that product.

G. International Standards

Ever-expanding markets are a goal for every company introducing a new product. Ultimately this will mean the exportation of their product. There are no specific international standards for foods *per se* but two bodies, Codex Alimentarius and the International Standards Organization (ISO), provide guidelines for contracting companies, that is, between manufacturers of products and buyers.

Codex Alimentarius publishes such documents as *General Principles of Food Hygiene* and *Recommended International Code of Hygienic Practices for Canned Fruit and Vegetable Products*. It would be advisable for any company introducing new products to be guided by the general principles outlined in these documents. Walston (1992) discusses with reference to Codex Alimentarius many problems in international food trade. In the Uruguay Round of talks for the General Agreement on Tariffs and Trade, the suggestion that Codex be the standard for food safety sparked controversy. Problems arose because of perceived shortcomings in Codex. There are three criteria for rejection of food in Codex: quality, safety, and efficacy. However, many governments, especially those of Muslim countries, and many nongovernment organizations believe there should be cause for rejection on religious or ethical grounds. Such nonquality or nondefect related rejection of product could have serious implications for exporting companies. The product developer with an eye to exploring export markets for new product expansion should be aware of international standards and regulations.

The ISO documents, ISO 9000 to 9004, are much headier material. ISO 9000, *Quality Management and Quality Assurance Standards: Guidelines for Selection and Use*, purports "to clarify the distinctions and interrelationships among the principal quality concepts." It also provides guidelines for "the selection and use of a series of International Standards on quality systems that can be used for internal quality management purposes and for external quality assurance purposes." ISO 9001 describes requirements for a quality system in which a contract between two companies requires proof of the capability of the supplier to produce to the required level of quality.

ISO 9002 lays out the requirements in a contractual arrangement whereby the supplier must demonstrate the ability to control the process within specifications. ISO 9003 describes the necessary quality system requirements for end product inspection and detection. ISO 9004 is really a guideline for establishing total quality management (TQM) in a company. It provides the basis for establishing and maintaining a quality management system. TQM is described by Shapton and Shapton (1991) and by Taylor and Leith (1991). The latter reference is accompanied by descriptions of the application of TQM at several food plants (pp. 21–26).

Direct application of either the Codex Alimentarius or the ISO series of documents may not be pertinent in the self-manufacture of newly developed added-value products for domestic consumption. Where exporting is a major objective or where companies wish to contract out the manufacture of their products to copackers domestically or in other countries, the ISO documents may be very useful in negotiations. Many companies now require that their suppliers and copackers be ISO-certified. Possession of such certification may provide a powerful marketing edge. Ingredient manufacturers should be aware of the ISO documents and their possible impact on sales.

IV. SUMMARY

The foregoing description of the support roles of members of the development team shows that there are overlapping areas for the maintenance and confirmation of product integrity. Lawyers (Chapter 6) ascertain that the product is legally safe and that the company has the best advice and guidance in any contractual arrangements associated with the new product and arrangements for its manufacture, distribution, and sale. Quality control ensures that all the necessary actions in procurement, manufacture, and warehousing have been taken to maintain safety and quality at desired levels.

There are gray areas of responsibilities among the roles of engineers, food technologists, quality control personnel, and production personnel. It is the support groups' roles to cover all contingencies in maintaining the integrity of the product and the safety of the customers and consumers.

The point that is important in the management of the development process is that no opportunity should be missed to bring all the skills of the team to the process of screening and producing the best product that meets the needs of customers and consumers at a price the customer is willing to pay.

8 Going to Market: Success or Failure?

… with few exceptions, marketers generally stub their toes with new product introductions …

Gershman (1990)

I. THE FINAL SCREENING

Binkerd (1975) reviewed the variety of tests that typically would have been applied to the product so far. His list included the following: focus group interviews, concept screenings, blind product tests, concept tests, mini-market tests, and test markets. Table 1.5 also provides an overview of activities during development.

However one orders these schemes, the result is that development has progressed through several stages of screening. Often the product has changed physically and conceptually from what was initially presented because of technical limitations, continuing market research of consumers suggesting a need for change, limitations of production facilities, and requirements for the assurance of product integrity. The carefully prepared kitchen samples used to test the concept and the final factory run product are very different, and consumer tests on them are expected to be different. Data from screening tests between the initial kitchen sample and the final realization result in a changed product but one that is as close to the concept as possible.

II. THE TEST MARKET: WHAT IT IS

Test markets are introductions of new products into regions carefully selected for a variety of geographical, marketing, and company reasons. The product is introduced, and after a predetermined period of time for awareness, trial by consumers, and repeat sales (the so-called ATR period), results of the test are analyzed. These results dictate whether marketing is either continued for further data collection, or marketing is extended into other regions, or the product launch is dropped and the whole project reevaluated.

Test markets take many forms; indeed, it is often difficult to distinguish between extended use tests or mini-market tests. Lord (2000) describes three classes of market tests:

- Simulated test markets: a concept testing technique similar in many respects to a focus group.
- Controlled testing: similar to the traditional test market, but the entire test is farmed out to a market research company that manages the entire test from distribution through to promotions.
- Traditional sell-in test marketing.

However, as Lord points out, there are many variations of each and they are not mutually exclusive testing techniques.

A. EXAMPLES

Two examples illustrate the many different test markets that exist. Mazza (1979) test marketed native fruit jellies in a gourmet gift pack through one retailer's stores in two cities in the first year of introduction. In the second year of introduction of this seasonal product, three retail outlets were chosen and two additional cities included. Marketing and consumer evaluations were carried out through a questionnaire accompanying each gift pack sold in both introductions. Customers completed and returned the questionnaires. Response averaged approximately 20%; data obtained included why the product was purchased, what attracted the purchaser to the product, whether the purchaser would repeat purchase the product, and how the purchaser would rate the product. This was not a highly competitive marketing environment and demonstrates a very simple test market for the cottage trade.

Clausi (1974) describes how the General Foods Corporation moved to a test market. After making modifications arising from the results of in-home testing, the product was put into a test market in one or two cities. Then, depending on market data, a move to a larger section of the country was made to evaluate both consumer reaction to the product and awareness of the advertising and promotional campaigns.

Test market situations for small companies and large companies are very different. Development teams of large companies are much more conservative, cautious, and concerned with all the ramifications of making a mistake in the introduction. This awareness of consequences permeates down the chain of development command throughout large companies; more is at stake should a blooper have been made. People may have something to lose: their jobs. This is a very real concern of company personnel.

Small companies have more flexibility in their introductions; they can literally deliver and sell off the back of their company station wagons to small independent grocers or independent franchisees. Their introductions are also more hands-on operations, with all members of the team participating. They use sales at county fairs or local sporting events to introduce their products or have tasting sessions at church socials. My first introduction to retort pouch entrees was at a local county fair where a senior member of the research and development team introducing the product served me. McWatters et al. (1990) used a mobile kitchen traveling to local events to evaluate consumer response to akara, a snack product.

Small companies get immediate feedback of consumer reaction to their products' characteristics. They have the patience, the time, and the intimacy with and proximity to their customers to develop a market; this is something the big company does not have, particularly if they are driven by short-term gains. These smaller markets also do not require introductory fees (slotting fees) to be paid to smaller retailers.

III. THE TEST MARKET: ITS GOALS

A test market is the first, large-scale, controlled opportunity to evaluate how customers, consumers, retailers, and the competition will react to a new product. As such, it is a phase, the final phase in the development process prior to a more formal (even national) introduction. Nevertheless, many companies prefer to bypass a test market and introduce a new product directly into its intended markets. If so, it is now that the strategists, particularly the financial and marketing departments, assess the results of the work of the development team.

Advertising and promotional strategies based on the targeted consumer are in hand. The production department has filled the distribution channels, and the timing is right for a market launch into a test market. The ball is in the marketing department's court. The next and final stage in development is theirs.

There is a very high cost that can amount to many hundreds of thousands to millions of dollars involved in a test market, and it is all spent to find out everything about the new product:

- What are the targeted customers' and consumers' reactions to the product going to be? The test also evaluates the retailers' reactions to and acceptance of the product and the competition's retaliatory action.
- The technologists' skills in developing and stabilizing a safe, nutritious, attractive, and tasty product with a uniformly high-quality shelf life will be evaluated by the consuming public and the company's management.
- The production department's ability to manufacture consistently a uniformly high-quality product is being tested. The production capability of the plant is challenged to maintain regular product availability in a timely fashion and to produce the new product as well.
- The package designer's skill in creating a package that sells and protects the contents is challenged.
- The warehousing and shipping department's ability to store and distribute the product in top quality and on time is put to the test.
- The test market verifies (or not) the marketing department's skills with its advertising and promotional campaigns designed for the targeted customer.
- The skills of the sales department to use the advertising materials and to sell the product to retailers are tested.
- Management's strategic and tactical skills at countering competitive action, with the support of all the other members of the new product team, are tested.

Test markets are a significant part of the screening process. They provide unique opportunities for further study of and experimentation with the product, its package, and the reaction of customers, consumers, and retailers to the advertising message put forward. Developers with the support and analysis of market data are concerned with product maintenance, that is, looking forward to how the product can be improved, to what variations (line extensions) can be added to support it as its growth falls off (see life cycle curves, Chapter 1). It is a very complex experiment involving emotional, intellectual, political, and people issues.

The people element of new product introductions cannot be ignored; the personal careers of all members of the team, but especially the technical members, are under scrutiny and are often forfeit. It has been my experience that new products seldom fail technically in test market, but as Gershman (1990) notes, marketing often "stubs its toe." Nevertheless, the technologists face more directly the stigma that somehow it was their fault, that they could not duplicate the concept, that the concept was good but the research and development group could not match it. The other members of the team glide silently off to other positions within the company; the technologists are stuck in their laboratories and pilot plants.

Consumer research is as active during a test market as it was during earlier stages of development but is focused rather differently. The test market is used to answer many questions: How and when does the consumer use the product? Is the consumer misusing the product? Are preparation instructions clear? Is the product's message being misinterpreted? How is the customer reacting to the message? What is the competition's reaction? How are retailers reacting? Consumers' reactions to, and their usage of, the product may suggest new opportunities for repositioning a product or indicate possible line extensions. Test markets provide excellent learning opportunities for companies.

A. Some Cautions

The nature of test markets varies widely dependent on the type of product to be tested and the goals of companies doing the testing. (Types of new products and their characteristics are discussed in Chapter 1.)

Improved (reformulated) or repackaged (established) products for which new market niches are being explored or for which new marketing strategies are being tested, present unique test marketing situations to marketing personnel. These products are already established. Marketing departments are using the test markets to seek answers to such questions as: Will changes incorporated in established products be accepted by established consumers? Will they attract new consumers? Will new market niches be opened?

Introduction of line extensions into test markets may backfire on the overall marketing strategy. If already established products are valuable cash cows to companies, then new line extensions may cannibalize the company's existing bell-ringer products. For example, a newly introduced mildly hot sauce may take sales from an established spicy hot sauce. Rather than opening up a new market, the newly introduced product may cut into the sales of the established product. In such a situation, marketing personnel must carefully interpret consumer reaction to the new product but also to the remainder of the product line. The data from test markets must clarify what is going on in the marketplace.

TABLE 8.1
Advantages and Disadvantages of Test Markets

Advantages	Disadvantages
Information about the effectiveness of product, pricing, packaging, and marketing strategies is obtained.	They can be very costly ventures.
	They are time consuming.
Information about retail reaction is obtained.	The sales force is diverted to a new product launch possibly to the detriment of regular, bell-ringer products.
Information about competitive counteraction is seen and protocols can be developed to thwart the competition.	Test markets warn competition of company activity.
Development protocols are justified.	A successful test market does not foretell a successful full-scale launch.
	Loss of face occurs if the test fails. This could result in a possible poor trade reaction for other products.

In addition, where heavy advertising pushing line extension products to consumers occurs, this gives impetus to competitive me-too products from companies wanting to capitalize on the promotion. It carries all similar products including those of the competition.

1. The Costs: A Deterrent

Many companies challenge the value of a test market, although in truth, much of this may center around how companies define test markets. A successful test market is not a guarantee of a subsequent successful national or even wider regional launch. Consequently many companies try to accomplish their new product marketing research with alternative mini-market tests at a much smaller outlay of money than a traditional market test (Lord, 2000).

The expense of test markets can be horrendous. Management must weigh the marketing risks of foregoing a test market and the results it brings against the financial burden. If the risk of omitting the test is small and if there are existing financial constraints, it very well may be worth risking dispensing with the test (Kraushar, 1969). If getting into the market early without alerting the competition is important, dispensing with this expensive and time consuming exercise is a wise move. The advantages are that 6 or more months of lead time to build a dominant market share are gained, and expenses in excess of several hundreds of thousands or even millions of dollars are saved. Advantages and disadvantages of test markets are presented in Table 8.1.

B. CONSIDERATIONS FOR A SUCCESSFUL TRADITIONAL TEST MARKET

The where, when, and how of introduction are closely interwoven. Understanding the principal elements in what Lord (2000) refers to as the traditional test market enables a better understanding of the characteristics, concerns, shortcomings, advantages, and pitfalls of simulated market tests and controlled test markets. Separating these in this manner should be recognized for what it is: an explanatory device. The nature of the product being introduced will greatly influence the answers to these questions.

1. Where to Introduce

Marketing personnel want unbiased marketing information for making very important economic decisions. Therefore, the location of the test market should not introduce a bias into the data obtained. There is no area that represents a cross section of the population with all its ethnic, religious, cultural, and economic diversity. Therefore, any area chosen for the launch introduces a bias that market researchers must be aware of.

It is therefore necessary to be aware of what biases may be introduced. Some issues that require consideration are:

- Is the area chosen for the test market peculiar to the company? Launching a product into areas where the company and its products are well known may lead to distorted sales and marketing data. In areas where the company has not been a good corporate citizen or the company is in a labor dispute, an introduction may be influenced by the company's reputation.
- Is the area chosen for the introduction peculiar to a competitor company and its products? Introducing a new product into marketing areas that are heavily saturated by a major competitor's products is foolish unless that head-to-head confrontation with the competitor is deliberate. This is not usual practice. A test market is an expensive experiment; it is not the time or the place to challenge competitors. In addition, a market dominated by a competitor will be costly to penetrate. Heavy advertising, promotions, and trade allowances to get shelf space or significant market penetration will be a burden on profits. Indeed, one can expect the competition to disrupt the test market with their marketing tactics.
- The area chosen for introduction should be one where there is a competent sales force in position and a competent distribution system already established. The sales and distribution team should be representative of the company's skills. It is the strength of the product that is being evaluated, not a particularly skilled sales force in the chosen test market area.
- If the area chosen for the launch is dominated by large retailers or by a single large retail chain, it may not be possible to evaluate advertising, promotions, and sales efforts for the product. Dominance by large retailers or by a single retailer in the test area restricts the activities of the marketing personnel in planning promotions and advertising campaigns. Campaigns may not be conducted as a company may wish, but as retailers want.
- Is the targeted consumer in the chosen test market area? Introduction of products with a strong ethnic appeal in areas devoid of that ethnic group is remarkably stupid. Likewise, introducing products aimed at an elderly

population into a growing suburban area dominated by young families does not make much sense either.

- Does the product have a style, flavor, form, etc. foreign to food peculiarities in the chosen test area? Some geographic areas have distinct biases for flavors, colors, forms, or styles of products. For example, there are many different styles of "authentic" chili: meatless, *con carne* with ground meat or with chunk meat, and heat levels from mild to very hot, with each popular in different areas of the country. Likewise, pizza has many variations of styles of crust and shapes, for which there are regional preferences. Each area with its unique style considers that style to be the authentic one. Strong regional preferences based on the local variant (which is the criterion) will weigh against new products which do not conform.
- Test market areas should not be dominated by mono-economies; that is, the area should have a mixed economy. The economic health of an area influences the buying habits of consumers. In economically depressed areas, consumers may not be willing to try new products that are perceived as luxury items or may not purchase higher priced products in which the added value is not appreciated. Nevertheless, marketing departments should take cognizance of the economic mix of communities where new products are planned when interpreting market data.

A test market should be viewed as a carefully designed experiment based on a marketing plan to obtain as much marketing, sales, and consumer research data about the product as possible. The company must adhere to the plan as closely as possible. Some flexibility is required because the test market launch will alert competitors, who can be expected to retaliate in some manner that is planned to seriously bias the test market results.

Small companies rarely have any geographic area that is peculiar to them. They usually test market locally within range of their factories to keep costs low and to avoid stretching their distribution resources to the limit. Usually too, they use their own sales forces. They are rarely poor corporate citizens, and often their tactics in test markets, even those dominated by a single competitor, go unchallenged or even unnoticed. This is not so with large companies.

Small companies are at a distinct disadvantage in trying to introduce products where a large retailer dominates. The manner in which large retailers conduct their purchasing does not favor new introductions by unknown or one-product companies without heavy advertising support. For small companies, introductions are usually done with little fanfare in small independent grocery stores.

2. When to Introduce

The seasonality of products dictates when test markets are carried out. Promoting a hot soup in the summertime or promoting ice cream or frozen yogurt when the snow is flying outdoors are inappropriate times for the introduction of these products. It is not weather alone that determines seasonality. Products associated with national, ethnic, or religious holidays should be introduced at their appropriate times.

Promoting seafood products during, for example, the American Thanksgiving period is inappropriate. The Christmas period presents some interesting anomalies, and the development team had better be aware of these if their product is Christmas oriented. There are geographic, ethnic, and traditional variations (perhaps these could all be classed as traditional). For some communities, fish is traditional fare at Christmas; my daughter informed me of the scarcity of turkeys at Christmas time in the Hamilton area around the end of Lake Ontario where ham is the traditional fare. In other areas goose or turkey is the festive fare. It would be unwise to introduce ham-based products (ham rolls, smoked ham, etc.) in a fish-fare area or to introduce turkey rolls, smoked or otherwise, in a ham-preferring area.

The return to school, with the need to pack school lunches, is an ideal time to introduce nutritional snack products. Summer leisure activities are associated with foods such as salads, prepared meats, dips and the like, marinades, and barbecue items designed for outdoor activities or patio living. Winter outdoor activities bring in an entirely different range of products. The timing for the test market, the food itself, and the activity associated with it must fit as appropriately, for example, as hot and hearty soups, cheese fondues, and liqueur-flavored coffees or hot chocolate drink mixes for the *après*-ski crowd.

3. The Length of the Test Market Period

When to market is closely related to another question concerning time: how long should test markets be continued before an evaluation is made? The simple answer is that the test market should continue until reliable data has been obtained to evaluate sales volumes, the effectiveness of advertising and promotional strategies, and customer, consumer, and retailer response. Test markets must be long enough to measure the consumers' reactions to the product. This period must include sell-in to the trade, promotion, first purchase by consumers, and repeat purchases. Time is necessary to establish a pattern of purchasing both by the trade and customers. The nature of the product determines its usage rate and hence determines the frequency of repeat purchases to be made. Further time is required to analyze data and get information back to marketing personnel and the other members of the development team for any refinement of marketing strategy.

Too long in a limited test market without capitalizing on the advantages of early introduction serves no useful purpose either. Lead time is lost. Copycat products are introduced in other market areas by the competition; they get market share in these new areas and make further market penetration and expansion difficult. The team must remember that timing is very important and lengthy test markets are expensive.

4. Disruptive and Unexpected Elements in Test Markets

The best laid plans, however, can have a monkey wrench tossed in the works. Events in international trade, world agricultural pricing structures, or other events in the food industry can produce a short-term alteration in political events, in consumption patterns, or in the economics of an industry. Shortages can occur, with the resultant increase in raw material prices. Company management need to have their antennae out for unusual developments in commodities, government activities, and any unto-ward activities that could thwart markets in which they are introducing products. For example,

> the introductory test market launch of a precooked (microwaved) bacon product was seriously disrupted for a subsidiary of Imasco Foods Ltd. when the availability of pork bellies declined and prices rose. Estimated costs for the finished product went above a price that marketing felt would discourage purchases.

> In a similar occurrence, pricing for a newly introduced blended (sunflower seed oil with olive oil) salad and cooking oil was sent tumbling when Russia flooded the market with sunflower seed oil that had previously been high priced; we were left with expensive sunflower seed oil in stock when the price dropped. Countries can place trade embargoes on goods or buy up stocks, causing temporary shortages; consequently, prices rise. Then they flood the markets at the higher prices with the shorted commod-ities and reap the benefits of the higher prices; then prices are driven down.

Nature can play a role, but how to be prescient about nature is unknown.

> Natural disasters and, in less sensational fashion, weather events play a short-term economic role in raw material shortages. A highly successful hot sauce was seriously disrupted in its second year of production when rainy weather inundated the pepper-growing areas in California, and the hot peppers were unavailable.

Many such short-term events greatly influence the timing of test markets and can be very disruptive of tests already underway. When shortages of raw materials occur, extensive reformulation is required and pricing schedules are thrown off.

5. How to Introduce

How to get products introduced can be a problem for both small and large companies. Small and comparatively unknown companies find it difficult to get shelf space, sometimes difficult even to get an appointment with the purchasing agent of a large retail food chain.

Stores are becoming much more hard nosed about new product introductions. New products take away space from products with proven sales records. Stores want to eliminate slow moving items with poor margins. They are, therefore, reluctant to take on new products unless they are assured of good margins, rapid inventory turnover, and advertising and promotional support. A Catch-22 situation can result: products that stores want can only be developed by market testing, but stores are reluctant to give manufacturers the shelf space they need for market testing of new products.

New product introductions by large companies are accompanied by extensive advertising and promotions. There may be in-store demonstrations; couponing in magazines, newspapers, and door-to-door fliers; piggyback offers; and special pric-ing offers. Small companies cannot afford this.

Advertising and promotional activities must be measured for their impact on the introductions of products in any marketing area. Introductory promotions to consumers (and to the trade) are one-time events. Heavy introductory promotions cannot be carried out throughout the product's life cycle.

To interpret the volume of sales of the initial introductory period, the marketing department must understand the impact of in-store promotions, demonstrations, and couponing activity on sales; it must be aware of the price differential *vis-à-vis* the competition and know what the competition was, or was not, doing during this period.

6.   What Product to Market?

First impressions of products and people are always important. It can be very difficult to change these; this is especially true of first impressions in the marketplace.

All too often, a company will introduce a product in a specially designed package or even as a specially manufactured product. It is not the normal product consumers will see in repeat sales. Use of specially packaged product during an introduction can be a disaster. Consumers have been introduced to, have become accustomed to, or have come to expect, something specific; customers have been educated to a particular price, package, and product. Retailers have come to expect certain price deals and promotional support. A new package or modified product constitutes a new product. Data obtained on the sales of the product originally introduced may not be valid for this changed final product.

The test market must be carried out with the same product as the factory will run. The impact on retailers, consumers, and customers of pricing deals, couponing, special packaging, and other promotional gimmickry necessary in marketing introductions, but not permanent features that will continue throughout the life cycle of the product, must be clearly understood.

IV. EVALUATING THE RESULTS

At some stage during the test market period, there must inexorably come a time of reckoning, of evaluation while marketing of the product continues. Kraushar (1969) discusses problems with using test markets as predictive tools and makes apparent why many companies choose to skip them as too expensive for the information they provide; these companies may prefer to use other means to get predictive data.

A. THE MARKET: MISINTERPRETED AND MISUNDERSTOOD

The previous sections described many factors that can influence the results of test markets. Data can be easily misinterpreted. The natural desire to want to see the product succeed can color judgment. Interpretation is a task not to be taken lightly. Clausi (1971), detailing errors in interpreting marketing data during the introduction of a dry cereal with freeze-dried fruit, commented that "the strong initial purchase pattern coupled with overwhelming consumer acceptance of the concept tended to obscure the significance of the negative evidence." In short, they erred. Repeat sales were flat. Negative signs were misinterpreted.

... wisdom comes to us when it can no longer do any good.

Gabriel García Márquez, *Love in the Time of Cholera*

Enthusiasm for the project perhaps carries the team away.

1. Dynamism and Interrelationships in the Marketplace

Why are consumer research and test market data so easily misinterpreted? First, all the forces at play in the marketplace are difficult to research practically, to measure quantitatively, and to understand intellectually. There is the behavior of both the customer and the consumer, the receptiveness of the retailer, the activities of the manufacturer, the activity of the competition during the test market to determine, and advances in technology that affect all to understand and counter. All data obtained must be read against the backdrop of this complex behavior. In Figure 8.1 the major protagonists, the food manufacturer, the retailer, the customer, the consumer, and the all-surrounding competition, interacting within any marketplace and adding to the complexity of the marketplace (see also Fuller, 2001), are depicted.

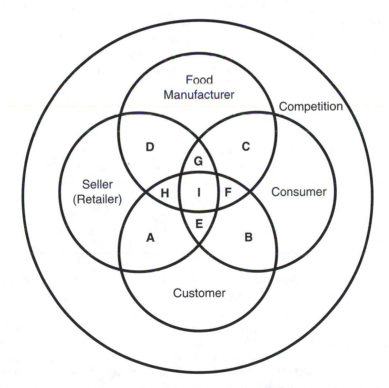

FIGURE 8.1 The major protagonists and their interactions within the various marketplaces. (Adapted from Fuller, G.W., *Food, Consumers, and the Food Industry: Catastrophe or Opportunity?*, CRC Press, Boca Raton, FL, 2001. With permission.)

In Figure 8.1 the complexity surrounding test markets that must be understood begins to emerge. Very distinct marketing arenas become evident. These are:

- (A) The seller–customer interface: There is direct interaction between the seller and customer. There can be several types of interactions here depending on who the seller and customer are and in what marketplace they are interacting.
- (B) Customer–consumer interrelations: The gatekeeper and user interactions that define usage, desire, cost, and other product characteristics.
- (C) The consumer–manufacturer interaction: This is the dominant interaction in new product development.
- (D) The manufacturer–seller interaction: It is here that selling policies are dictated, pricing and distribution defined, and in general, trade relations are established. These can be very complex depending on who the manufacturer and the seller are. Sellers, especially the large chain stores, may have supply chain management or order management policies in place with manufacturers.
- (E) The customer–consumer–seller interface: Most selling takes place here as typified by the mother with children in tow shopping in a supermarket. However, it can be much more complex, for example, the interrelationships between the engineer, his vice president of finance, and the technical sales representative of a food equipment manufacturer at an exhibition booth at a trade fair with their goal to negotiate the purchase of equipment for product development.
- (F) Customer–consumer–manufacturer interrelations: Most marketing research takes place here; promotional campaigns are developed here also.
- (G) The consumer–manufacturer–seller arena: Media and promotional campaigns are tested and used.
- (H) The manufacturer–seller–customer arena: Here there are mutual efforts by the manufacturer and the seller to attract the customer.
- (I) The main selling arena where all interactions are present.

Within this diagram all the interfaces can be seen that influence and hence must be reckoned with in any new product introduction, whether into a test market or into a major launch.

Against these interfaces and interfering with them as much as possible is the competition. How active or inactive (a rare event) is the competitor?

A colleague informed me of a test market conducted by one company when that company's competition was suffering through a strike. No competitive product was on the market; of course, sales were excellent and repeat sales good throughout the lengthy shutdown. This strike at the competitor's plant was ignored in the assessment of the new product launch — a disastrous miscalculation. If there is inactivity on the part of the competition, then why? Was the competitor shut down for any reason during the test market period?

Was the competitor buying up product either for chemical analysis, for their own test purposes (to get their own test information of consumer reaction), or simply to foul up the sales volume figures and so cause misinterpretation of sales data? Knowing what consumers are doing and how they are reacting to a new product is valuable information both to the company doing the test market and to the competition. It cannot be overemphasized: knowing what the competition is doing is vitally important.

2. Personal Opinions, Biases, and Self-Deception

Personal feelings and emotions can frequently blind the team to the reality of test market data. Product managers and marketing personnel, in particular, become emotionally attached to pet product development projects that perhaps they, or others, overpromoted to superiors during earlier phases of development. Justification for past actions may be read into their interpretation of the introductory results. Emotions must be kept out of the interpretation of data.

Even with all the data from the test market in place, the science of consumer research is still not precise enough to prevent misinterpretation of the information obtained from that data. The complexity of the test market, no matter how carefully designed the test is, is subject to all the interfaces seen in Figure 8.1. This, combined with the imprecision of consumer research science and the strategic goals of senior management, confound many test market results. In short, errors in interpreting test market results arise from a lack of objectivity by the interpreters of the data or from the highly subjective views and feelings for the project held by the interpreters of the data, from marketplace interactions that confound the data, and from the tools that are used to measure that data and their imprecision.

3. Criteria for Evaluation

Criteria for evaluating whether the launch was a success or failure must be established. Different criteria will be used according to the objectives of the company performing the test market launch.

Four measures can be used:

- Payback: When will there be profits? Is the company strong enough financially to wait for greater returns, or does a short-term payback mentality prevail in senior management? An unrealistically short time frame to achieve a satisfactory rate of return has been the downfall of many new product ventures. The interpretation of "unrealistically short" rests with the financial personnel in the company.
- Sales volume: Will sales volume goals or targeted percentage share of market or even significant market penetration be achieved? Will these criteria meet the company's objectives? Case movements must mean profit for the company and not disappearance in ill-advised promotions.

- Consumer reaction: Did consumers like the product? How can this be capitalized on (product maintenance)? Can simple strategies be applied to improve consumer reaction?
- Tactics: Being there in the market is the thing. Did this introduction preempt action by the competition, increase market share, provide market penetration, or satisfy the strategic marketing goals of the company?

The first two measures are very similar. One says it in money; the second uses case volume and share of market. The third is technical and asks whether product maintenance can initiate a family of products to capitalize on and support a new product line based on the introduction. The final measure is much harder to assess.

The second measure is perhaps a better measure of trade acceptance than the first. Nevertheless, it does require some caution in interpretation. If projections indicate the volume of units sold are related to enthusiastic consumer acceptance and repeat sales, then appreciable economies of production can be anticipated by scaling up manufacturing. Usually, on the evidence of learning curves, more units of a product can be made more economically than fewer units of the same thing (Malpas, 1977). This, in turn, will influence the rate of return of investment as manufacturing costs go down.

The volume of sales must be examined very carefully to determine precisely what it means. If these sales are consumer sales, this is a positive factor in interpreting the results of the test market. If they are merely case movements between warehouses or buy-up by the competition, sales volume could be deceptive for interpreting the results. The following incident describes one such event.

> I was on an acquisition study on the west coast. My object was a company that was test marketing a line of pouch-packed entree items. My company was interested in the product and the company. The product was being test marketed in three large super-markets in Vancouver. Prior to my meeting with the principals, I purchased two cases each of the four flavors for shipment to our laboratory in Montreal. Much to my surprise, during my meeting the next day with the president of the company, he regaled me with the tale of more than eight cases of product being sold in one store, so great was the demand for the product! What chaos would I have made of this company's sales statistics had I been a rival!

The third measure is an assessment of consumer reaction to the product. The test market is a commercial experiment in consumer studies. The reason why the consumer is buying the product and how the consumer is using the product, that is, the context of product use, is valuable. Can the context indicate strengths of the product to be capitalized upon? Are refinements needed or future line extensions justified? Is promotion and advertising directed properly? Should other market niches be explored? Weaknesses in the product need to be ferreted out by consumer research and eliminated before new marketing strategies are undertaken.

The last measure is more in the nature of a business tactic, a strategic tactic forced on the company. A product may be introduced or positioned to counter the activity of a competitor or to establish a position in a particular niche. Success would be measured by whatever small share of market could be gained as a foothold for strategic or tactical purposes at a later date. The company feels that it cannot relinquish a position in the marketplace; that is, the company must be seen and heard in this particular market niche.

4. Applying the Criteria

Small companies are more flexible about how they apply the criteria to measure success or failure. If each week more units are sold than the week before and the reports back from the marketplace say consumers like the product, then that product is usually considered a success. Small companies are generally less concerned with the element of market share that large companies use to define success; how much market share was obtained is not a question small companies ask themselves. Small companies demonstrate more patience and attempt more fine tuning of the product as sales and development proceed hand in hand.

Consequently, new product development costs in small companies cannot be determined with accuracy. Presidents of small companies are quite content to pay their bills and have an increasing bit left over each week. Financial criteria are not stringently applied because budgets for research and development in small companies are frequently not separated out as in large companies. These costs are bundled together with either quality control (usually the seat of research and development) or production expenses.

B. JUDGMENT DAY: THE EVALUATION

At the conclusion of any test market, all data, both subjective and objective, related to the product and the marketplace are gathered and analyzed. Questions related to the product itself, its protective package, the label, preparation or recipe instructions, pricing, and positioning of the product need to be asked. Salesmen must be interviewed. The need for changes in the product, the process, or the package must be reviewed to determine what changes, if any, should be made based on customer, consumer, and retailer reaction.

Whether the test market was a success or a failure and the reasons for its success or failure need to be examined. If reasons for the success of a product can be clearly determined, perhaps they could be applied in the future to other products.

If the product is a failure, the same canvassing of information is required. Now the question is, "Why? What went wrong?" What elements of consumer research, technical development, and analysis of data led to the failure? It is not a hunt for a scapegoat. These exercises should not be witch hunts. Too often, unfortunately, this backward look turns into a witch hunt, with everyone trying to obfuscate the facts, pointing fingers all too often at the research and development group, with the result that nothing is learned. The search should be for flawed systems or faulty information that led to incorrect decision making.

An apologia is required here. Often in a product failure, it is overtly or covertly suggested that the product could have been better somehow and research and development is blamed. The stigma of failure sits with the technologists. Others on the

TABLE 8.2
Elements in the Success or Failure of New Food Product Introductions According to Some Sources

Kraushar (1969)	Best's 4 plus 1 P's (1989)	Gershman's 12 P's (1990)	Morris (1993)	Wang (1999)
Product not appropriate.	Product.	Perception.	Inadequate market research: no market need or changing needs.	Lack of funding for long-term innovative research: short-term goals.
Product is faddish.	Place.	Pitch.		
Timing is wrong.	Price.	Packaging.		
Pricing is wrong. Too high or too low	Promotion.	Price.		
	Perspective.	Promotion.	Management: no commitment with budget and resources; no clear strategic focus.	Time pressures in days or weeks rather than months or years: emphasis on line extensions.
Product is wrong. Does not perform Not significantly different from competition		Promises.		
		Piggybacking.		
		Positioning.		
		Placement.		
		Premiums.	Risk aversion and short-term orientation: me-too products; line extensions.	Career risk in pursuing innovation.
Poor communication of suitable image.		Publicity.		Poor recognition of skills for development.
Lack of objectivity.		Perseverance.		
			Poor fit with in-house capabilities.	Poor management of development skills.
			No formal new product development process.	Cannibalization of existing products.
				Wrong research: established brands cloud new research thinking.

development team, who may equally have contributed to the failure of the product, usually have other avenues to pursue within the extensive framework of the large company; the failure does not blight their future. Not so the technologists; to pursue their chosen career paths they must stay within research and development. The stigma remains until it is washed off with a success; the pressure is on the technologists, but the reader should note Gershman's remarks at the start of this chapter.

V. FAILURES IN THE MARKETPLACE

Most frequently, much of the success or failure of a new product hinges on elements outside research and development. These elements are all in the realm of marketing. Kraushar (1969), Best (1989a), Gershman (1990), Morris (1993), and Wang (1999) have each described these (Table 8.2), with both Best and Gershman facetiously calling them the P's of marketing. A closer examination of all these authors' terms reveals they are all variations on a theme. Best takes the hallowed four P's (product, place, price, and promotion) to add perspective. Gershman splits the traditional four

— one can only with difficulty distinguish between promotion, publicity, piggyback-ing, and premiums. Kraushar's "lack of objectivity" deserves some further comment. Simply put, it is the inability to or unpopularity of saying "no." Saying "no" in the face of the development team's enthusiasm is hard, especially if some problems appear in the research (see earlier Clausi, 1971). Often, as Kraushar comments, these snags are minimized in the spirit of keenness of the moment.

Wang (1999) weighs in vaguely against management as a dominant factor in product failures. He provides a murky distinction between placing the blame at senior management's doorstep and at the product or brand management's level (the vagueness is due to the varied nomenclature of positions within companies).

Hollingsworth (1994) reported on the top four reasons for product failure as reported by Group EFO Ltd.:

- Strategic direction
- Product did not deliver promise (more P's)
- Positioning (still more P's)
- No competitive point of difference

From these opinions gathered from company executives, it would seem that nothing has changed since 1969 when Kraushar lashed out at these same reasons for failure. He could not understand how failures could happen in large companies with money and supposedly sophisticated and experienced marketing teams and could only lump the reasons together in his lack of objectivity category above (see "Personal Opinions, Biases, and Self-Deception" above).

Silver (2003a) lists the seven deadly sins of product development in a true biblical style with "thou shalt not" phrases. Her "laws," less biblically rephrased, are:

- Management commitment is essential. Proceeding without the full support of management is a signal of trouble.
- The formulator's dilemma in attempting to replicate cooking techniques should not be ignored. Scale-up from kitchen top samples to the pilot plant to the plant-scale production has difficulties.
- The sophisticated palates of consumers should not be belittled or ignored. They know quality.
- Cost-cutting methods should not be applied to testing. This is not the time or place to cut costs and take short cuts with necessary testing.
- Packaging should not be skimped on since it must sell the product. In addition, with packaging "form should follow function."
- Marketing should not be skimped on.
- Weak projects should be killed early.

Despite Silver's light-handed approach her points are not trivial.

Most new food products fail to survive their first year in the marketplace. This is a staggering loss of the time and effort of skilled personnel; of monies spent in consumer research, equipment, ingredients, and packaging material; and on monies spent on advertising and promotion. Careful consumer research, thorough collection of data and unbiased, detached, and unemotional interpretation of the data, coupled with sound product development work, should have made failure most unlikely. Hindsight, unfortunately, provides better vision than foresight. Hindsight allows one to make generalizations (or speculations) on what went wrong, and these observations need to be used to improve techniques for future development.

A. CAUSES OF FAILURE

An in-depth study of why a new product succeeded would have been a much more valuable contribution to an understanding of new food product development. One would then be equipped with guidelines to follow for future product development: a series of "if this, then that" conditions would simplify the process of development. Causes of failure merely provide developers with a series of "don'ts." However, predicting the success or failure of any product against the volatility of the consumer in a changing marketplace is still an art. As Clausi (1971) might have said, reading all the evidence correctly, including the negative evidence, will lead to success.

It is difficult to classify the reasons for a product's failure (Table 8.2) and harder still to pinpoint a particular product's failure and put these easily and neatly into pigeon holes, as Clausi (1971) found. Clausi, in describing one particular failure, could only suggest that the signals from the marketplace were misinterpreted. Signals are rarely objective and therefore require subjective interpretation with all the baggage this implies. How does this interpretation of failure fit among those listed in Table 8.2? An examination of specific product failures provides a very broad overview of probable causes, from which only generalizations arise. One cannot apply the generalizations at the start of the development process or at any other point up to and including the test market and say that this product or that product will fail because....

Simplistically, the causes for failure are broadly classified as those the company could not have done anything about and those they might have done something about. The former are reasons or causes beyond the control of the company and usually are external to it. (But should not the company have been aware of their weaknesses and been forewarned?) The latter are causes usually found within the company. These internal causes for product failure are not always manageable, for various company reasons.

Many small problems only partly assignable as either external or internal can trigger the failure of a product. Separating reasons into categories such as external and internal cannot always be done with clarity. For instance, too small a market (an external reason) can be a cause for a product's failure (see next section). That is a reason beyond the company's control. But if that were the case, should not marketing personnel have seen there was too small a market? Were the marketing capabilities and resources within the company incompetent or inadequate or both (internal reasons)? How else would marketing research have failed to determine the magnitude of the market beforehand?

Thus, if one states baldly that one reason is external one must, equally, understand that an internal reason may have contributed directly or indirectly to it.

1. External Reasons for Product Failure

After products have been introduced, marketing personnel may find that markets for them are too small. Growth potentials would be limited; possibilities of recovering development costs would be minimal. In certain markets, this knowledge may come unexpectedly. For example, changes in the purchasing policies of governments with respect to institutional buying for the military, for government-run correctional institutions or prisons, or for school meal programs may suddenly and abruptly be altered and the size of a market may change. Nevertheless, companies servicing such markets should keep themselves informed of pending government changes by close liaison (networking) with their government contacts.

Markets controlled by a dominant competitor are difficult to get footholds in. Companies introducing new products find themselves not just battling for customers but battling with competitors. For example, a dominant competitor is in a position to control, influence, or buy retailers so as to limit shelf space exposure for rival products. Consumer acceptance is too costly if advertising and promotional dollars have to counteract retaliatory action by the competitor.

Domination of markets by a single customer (e.g., the military, penal institutions, or large fast food chains) can present severe challenges to companies introducing new products into those markets. The cooperation of customers (retailers) is always essential, but when suggestions from customers become directions, then the situation can be fraught with stumbling blocks. After all, companies, not customers, have spent the development dollars and will risk most in a failure. But the dominant customer has the greatest say in pricing and marketing stategies in general. The tail, that is, the customer, wags the dog, the developer. The food service industry is one where this problem is apt to arise. The immense buying power of some retail chains has permitted them to dictate to producers what products and what development they want and at what price they want this for their marketing purposes.

There are product-related reasons for failure in the marketplace. With the introduction of a me-too product into a market that is saturated with similar products, consumers will refuse to buy another brand or variation if they cannot see a point of difference between it and already established products. The problem is perhaps beyond the control of the developing company; no one could have foretold the flooding of the market with copycat products, but close observation of the marketplace may have indicated some saturation. Markets do fragment and may provide a special marketing niche for new me-too type products.

Where technical novelty has been designed into a product and this technology is the dominant message to the consumer throughout introduction, the consumer can be forgiven for questioning, "So what?" "What advantages are there for me?" New forms of a product such as frozen for canned, tablets for powders, aerosols for liquids, and so on may fail disastrously if the consumer cannot see the advantage being offered or if the advantage (point of difference) over other similar products is insignificant. Flavored ketchups have not been successful for this reason, and all indications that I have received indicate that colored ketchups are not the success they were touted as, but this may be regional.

Products ahead of their time or for which consumers were not previously prepared have poor chances of getting market acceptance. They meet consumer resistance because consumers have not been adequately educated to their possibilities. Arguably, one might consider this an internal reason for a market failure, one within the control of the company; marketing personnel did not promote the product correctly. On the other hand, educating customers and consumers is costly since they can be quite quixotic. There is a danger here: why educate customers and consumers only to have a competitor reap the benefits with me-too products?

2. Internal Reasons for Product Failure

It is too glib to state plainly and simply that the intrinsic reasons a company fails to launch a new product successfully are a series of bad management decisions with the addition of a little bad communication. Failure can be said to occur because of the inability of management to recognize the strengths and weaknesses of their company. Consequently, management fails to understand what business the company is in. Hence clear company objectives are not available to direct growth and new product development — so the logical progression of thinking goes.

Marketing and research and development resources are not adequately developed unless there are clear company objectives. Obviously, if a company's marketing department is incapable of or incompetent at conducting reliable market and consumer research — a subjective evaluation a company must make prior to committing any research and development efforts to the project — new products are apt to fail. More damning, the company's marketing people may be unable to recognize their own shortcomings. This is unfortunate since many independent market research companies are available that could assist. However, one must question how much value would be gained if the internal marketing resources were so incompetent that they could not appreciate, understand, or communicate with the external resource.

Lack of production capacity can be a cause of product failure. If retailers cannot get product on time to stock their shelves because the plant cannot produce enough to satisfy demand, then customers and consumers cannot get product, and the impetus of the launch has been lost. Buyers, users, and retailers will lose interest. Again, one must question the lack of foresight in management not to have seen the likelihood of this and backed up production capacity with production contracted to copackers.

Unnatural adherence to and support for a project (unnatural in the face of negative evidence for its continuation) is, perhaps, more readily understood than the other causes for failure because of the human element entangled in it. People, whether in small or large companies, can become emotionally involved in their projects. It is their pet project, their baby. Their rather illogical reasoning goes something like this: too much money has been spent to stop the project now, so spend more monies to rescue the project. It is like the gambler who gambles more money to recover his losses. Costs need to be regularly evaluated to prevent their escalation.

A product can fail for technical reasons. It simply does not perform as promised or does not live up to the standards promised or as the customer and consumer had expected, that is, contrary to what promotions promised. This is missed communication. The cause for the poor performance could be inherent in the product itself.

That is, it was poorly designed. Or one of its ingredients was not correctly chosen or the packaging failed to give the proper protection.

It is an oversimplification to suggest that there are only two reasons for the failure of a new product. Nevertheless, there can be a great deal of truth in such a generalization. These two reasons are

Expecting too much
Not being lucky

The first, expecting too much, does not happen to companies whose objectives are based on a realistic assessment of their companies' strengths and realistic financial and marketing objectives. Good luck, or whatever one wishes to call it, comes more regularly, rather than randomly, to companies that utilize their resources well in order to research markets, consumers, and their products. At the very least those companies will reduce their margins of error.

3. Product Maintenance: Salvaging Failure

Kraushar (1969) makes an interesting point that should have been considered in product design. The growth curve for any product eventually displays a flat, no-growth phase that is usually followed by a dying away phase when no amount of promotion can liven up sales. The costs of promotion are prohibitive. He suggests that before any new product is launched, that is, during development, a product maintenance program should have been set up. Product maintenance programs are designed to make improvements in the product and are used, when sales stagnate, to pep the product up. They suggest new uses for the product, appeal to a new customer niche, or contribute unique added value. Product maintenance programs are an integral part of the development process and useful in retooling a failed product introduction.

9 Going Outside for New Product Development

Learn from the mistakes of others — you can never live long enough to make them all yourself.

Anonymous

Joint ventures are almost always bad. At worst, both parents neglect the stepchild in favour of their own.

R. Townsend, *Up the Organisation*

The sage in his attempt to distract the mind of the empire seeks urgently to muddle it. The people all have something to occupy their eyes and ears, and the sage treats them all like children.

Lao Tzu, Book 2, *Tao Te Ching*

I. A ROSE IS A ROSE IS A ROSE

Each of the above quotes contains advice for and warnings about going outside for help. It is very difficult to discuss going outside for assistance in development, however, without clearly understanding the following terms:

- Outsourcing
- Joint ventures
- Partnerships
- Hiring of consultants

They all describe working relationships, some without contractual obligations, with an outside resource, and all have similar features in practice. Closely allied with these are customer–supplier relationships (a form of joint venture) and copacker relationships. These too, have contractual arrangements and frequently play a role in new product development when the supplier and copacker provide their skill and expertise to the design and development of the process.

A. OUTSOURCING

Outsourcing is a comparatively new term but a very old business practice. Food companies have outsourced some of their activities for many years. For example, it has been normal practice to use trucking firms and customs brokers for distribution. A simple extension of this thinking was to use a third party warehouse (i.e., distribution center). This distribution center supplies much more than simple storage; it provides inventory control, invoicing of goods, and electronic data tracking of goods that allow improved distribution efficiencies, reduced capital costs for warehousing facilities (since these are shared with other companies), reduced staffing costs, and thereby lower overhead costs. Information technology and with it intelligence gathering are expanding so rapidly that they are commonly outsourced to companies more skilled in the operation of these technologies. By outsourcing, it is reasoned that the client has more time and resources to concentrate on its core business while experts take care of the outsourced work.

Thus, the nub of the argument for outsourcing is revealed. For distribution it means the task of logistics is left to the logistics professionals; information technology and telecommunications are farmed out to experts in these fields. Other services that are often outsourced are legal matters; accounting; public relations and advertising for radio, television, and newspapers; and more mundane activities such as general cleaning (grounds maintenance, landscaping, and office cleaning), plant security, and laundry services. The general area of outsourcing, then, might be interpreted as the use of expert services for specific tasks (consulting services).

The services that are peripheral to the company's core activities are contracted out to others who can do the services more adroitly. The company's core activities are defined as servicing its customers. One keeps the practices unique to the business, that is, those activities that make money for the client. Green (1996) suggests outsourcing can optimize return on investment capital. He does caution, however, that this will occur "in the right circumstances." Patterson and Haas (1999) warn that only those functions with clearly defined boundaries (they suggest nutrient analysis, for example) and that need little cooperative participation between the principles should be outsourced.

1. Cutting to the Core: Advantages and Disadvantages

Advantages seen by Green are presented in Table 9.1. It is important to note, though, that these advantages accrue in the right circumstances. The first three in the table are simply restatements of cost reductions that result from staff cuts; obviously if a department devoted to a particular activity is outsourced, there will be a saving because of reduced staff. The last two suggest that the company outsourcing some of its activities undergoes a degree of streamlining similar to that of the company providing the outsourcing activity and that this is advantageous. This is certainly not a logical extension of thought; it also might indicate an abrogation of managerial responsibility for those activities that have been outsourced by the company.

I encountered an example of this in a company I consulted for.

TABLE 9.1
Advantages of Outsourcing

Outsourced activity is trimmed (streamlined) or improved by experts in the outsourced fields.

There is consolidation of outsourced activities from a multitude of sites to one central location with staff cuts.

Information technology (for one) is consolidated and integrated onto a single platform (again with staff cuts).

There is a concomitant reengineering of business and information technology systems to reduce errors, duplication of work, and "non-value added activities" and to generally streamline tasks to increase speed and reduce costs.

There is a restructuring of internal support systems in order for them to be similar to those of a supplier to outsourced activities. That is, the company's internal support systems are streamlined to resemble the company to which it has outsourced some of its activities.

From Green, C., *Food in Canada,* 56, 8, 1996.

> They had outsourced plant sanitation and clean-up to a company specializing in this activity. No check-up on the efficacy of sanitation was performed. "It was in good hands." The sanitary cleaning company fell on bad times and cut back on its services, with the result that my client experienced spoilage problems. It is absolutely necessary to monitor how well outsourced activities are being carried out. My client was at fault for not maintaining managerial responsibility over plant sanitation.

The main resource the company has is people; if these are outsourced, what does the company possess; where do innovation and creativity come from but from people?

Cutting to the core business may have benefits, provided the management knows what its core business is. A "lean and mean" philosophy can result in an emasculated structure unable to achieve anything resembling active product development based on sound customer, consumer, market, and technological research.

B. OUTSOURCING, CONSULTING, PARTNERING, AND JOINT VENTURES

The similarities between outsourcing and consulting are obvious, so much so that one may consider one as a variant of the other. In outsourcing, a client depends on an individual or company to provide a necessary service that meets an agreed upon standard or level of service, that is, solution of a problem or performance of a service. Outsourced services are usually contracted for periods of a year or more. In some outsourced activities such as telecommunications and information technology these contracts may amount to many millions of dollars and continue for several years; the contracts can be extremely complex legal documents. For example, it is reported in *The Gazette, Montreal,* for December 3, 2003, that Canadian Pacific Railway Ltd. has outsourced its computer and technology section to IBM Canada Ltd. for $200 million in a 7-year contract.

In contrast, a consultant is seldom kept beyond the duration of the task at hand unless that consultant is paid on a retainer basis to provide a continuing service or to be available at a moment's notice. The relationships in outsourcing can take the form of contracted services, joint ventures, or partnerships. Which they are depends on the circumstances and the services. The service provider is responsible for delivering a product (service) of high quality and free from error that is both on time and within the standards described in the contract. The buyer of the service rids itself of both responsibility and accountability for performing the function; it is in the hands of experts.

Partnering is the joining together of two or more companies in a contractual agreement; the activities of all are complementary, and by working cooperatively they benefit one another. An example, provided and described by Kuhn (1998b), is a partnering of a flour milling and bakery operation, a processed meat company, and a pizza manufacturer with a combined operation centrally located; another example is a cocoa bean processing plant entering into a partnership to supply a confectioner with chocolate. Flavor houses often enter into partnerships with their clients in the development of a particular flavor for a new product the client is developing; supplier and customer collaborate and develop a trusting business relationship. None are rivals, and proximity benefits all parties.

Williams (2002) describes partnering between customer and vendor and building a relationship between customer and supplier, in this instance, between a customer (client) and a flavor house (consultant) for flavor development. Williams enumerates in detail the requirements for making such a partnering successful by knowing what both the client and the supplier want from such a relationship for product development. Conditions or situations are elaborated that can undo a working relationship, such as:

- Breach of trust between client and consultant supplier
- Poor communication
- Lack of benefit for one party (but see Harvey, 1977)
- Mixed messages
- Disincentive to add value

Poor communication and mixed messages stress the importance of clarity in these relationships. This latter point has consequences in long-term contracts, as the client's business goals may change, but the service provider does not deliver services to satisfy the new requirements necessary to the client. There will obviously be problems; the principals have not communicated the changed needs.

Joint ventures are somewhat different. They might very well involve rivals. They are often based on the principle of "let's stop fighting one another — there's enough pie for both of us." Such agreements rarely remove rivalries and rarely involve equals. There are always conflicting loyalties that confound joint cooperation. Joint ventures and partnerships do have some similarities; Townsend's comment heading this chapter describes the situation for both succinctly. Many managers feel that if a venture (e.g., construction of a factory for a new product development) is worth embarking on, it is worth doing alone, lest if it were done cooperatively it might be neglected for vested, self-interest projects.

Consultants are somewhat more complex entities. Shahin (1995) rather cynically described the consultant as follows:

> When you encounter him, you'll know him on sight. He'll glide into your meeting as radiant as confidence…. If he were a car, he'd be a Lexus. On cruise control.

Scott Adams, through the voice of his Dilbert cartoon character, told a time management consultant that he had become a consultant because he had been fired from every job he had ever had for wasting time. To which the consultant replied, "Welcome to the wonderful world of consulting." Putting cynicism aside, consultants are individuals or companies who are experts in unique areas of knowledge hired for guidance and advisory services in areas in which the company is ignorant.

C. A Classification of Consultants

For simplicity in characterizing and classifying consultants and consulting services, I have classified them as either professional or amateur (Table 9.2) following an analogy to amateurism and professionalism in sports; that is, the classification is based indirectly on income (Fuller, 1999). The term *amateur* is not used in a pejorative sense as a reflection on the competency or experience of a consultant. Very simply, the amateurs are consultants whose consultancy practice is not their main source of income. It is a hobby, a means of keeping busy, of keeping their hand in the business, of supplementing a pension, or of supplementing a regular income by moonlighting. It is a means of attracting research grants by universities, a ploy used by professors at universities to obtain monies for graduate students, or a stopgap tactic resorted to by many executives or research scientists who find themselves between jobs (often as the result of downsizing because of outsourcing by their previous company). They use consulting as a tactic to seek a new position.

The term *professional* is not a reflection on a consultant's competency, nor does it suggest that a consultant meets some standard of professionalism. To my knowledge there are no standards of professionalism in consulting. Professional consultants, pure and simple, make a business of consulting; it is their main source of income.

Both the amateurs and the professional consultants are excellent resources to assist clients in new product development. Nevertheless, the distinction between the two groups must be clearly understood by the client, as this distinction may have implications in future client–consultant relationships (see caveats in the next section).

Some university-affiliated consultant organizations (Table 9.3) are described in more detail by Giese (1999). Universities see a need (and a source for the generation of income) in making their expertise available to food manufacturers. Consequently, new centers of assistance and expertise are being formed constantly; the Institute of Food Technologists' *Food Online Newsletter* (August 20, 2003) announced that Texas A & M University's Institute of Food Science and Engineering had received a grant from USDA to establish a National Center for Electron Beam Food Research. Their November 5, 2003, newsletter described the high-pressure processing

TABLE 9.2
Outside Resources For New Product Development Classified on an Income Criterion

Class	Subclass
Amateurs	Executives-between-jobs exist as the result of takeovers, downsizing, and mergers. They are job hunting.
	Retirees: Early retirees often choose to work on aid programs in developing countries or to supplement incomes with consulting. (They often have restrictions on companies for whom they can consult.)
	Academics are allowed to consult with the prospect in mind that this activity will bring in industrial projects.
	Extension departments of universities provide consulting services for local, regional, state, or provincial industries.
	Student training programs: Students under the guidance of professors consult for small companies as part of their training.
	Research institutes: Groups of academics or departments at a university combine to form a research institute. This is a quest for financial support.
Professionals	Individuals working alone or networking with others to form larger consulting entities.
	Independent companies: More formalized than above with a broad range of consulting services.
	Private research institutes and associations are usually contract research groups much like independent companies. Companies subscribe by paying a membership fee and are then privy to research activities of the larger group.
	Government agencies: Governments have research groups that can be used by local industries or they provide individuals (often retirees) to guide fledgling companies.
	Trade associations often provide consulting services for their members.
	Specialized service providers provide unique services, for example, nonroutine laboratory facilities, information retrieval, forensic accounting, decontamination processes, retail sampling, and recall programs.

Adapted from Fuller, G.W., *Getting the Most Out of Your Consultant: A Guide to Selection Through Implementation,* CRC Press, Boca Raton, FL, 1999. With permission.

laboratory established at Virginia Polytechnic Institute and State University. Hollingsworth (2001) focuses on U.S. federal research and development programs funded by agencies or administrations and run cooperatively by university centers.

There are foreign universities and institutes that undertake contract research projects. Any follower of the Institute of Food Science and Technology's (U.K.) journal *Food Science and Technology Today* (now defunct) and the current *Food Science and Technology* will find any number of government, private, and university research facilities described, as the following arbitrary listing shows:

- Food Refrigeration and Process Engineering Research Centre (University of Bristol, U.K.) (March 1994, Vol. 8)
- Swedish Institute for Food Research (June 1994, Vol. 8)

TABLE 9.3
Listing of American University-Affiliated Research Centers Available For Assistance in Product and Process Development

Research Center and Location	Professed Specialty
Center for Advanced Food Technology (Rutgers University)	Problem solving for member companies Cooperative venture between industry, Rutgers University, and government
The New York State Food Venture Center (Cornell University, New York State Agricultural Experiment Station, Geneva)	Assistance in all aspects of new product development and introduction
The Food Processing Center (University of Nebraska)	Technical and business assistance to food industry in product development
The Northern Crops Institute (North Dakota State University)	Northern crop development and promotion
The Kansas State University Extrusion Center (Kansas State University)	Grain products and extrusion processing
The Spray Systems Technology Center (Carnegie Mellon University)	Spray systems and atomization
The Southeast Dairy Foods Research Center and Center for Aseptic Processing and Packaging Studies (North Carolina State University, Raleigh)	Integrated approach to product and process development from formulation to shelf life studies to scale-up and market research
The Food Innovation Center (Oregon State University, Corvallis)	Advice and technology for added-value products for Pacific Rim markets Lead organization for ValNET (Value Added Network of Export Technologies)
The Food Industry Institute (Michigan State University, East Lansing)	Research and outreach in food technology; workshops; equipment leasing
The Food Industries Center (Ohio State University, Columbus)	Pilot plant facilities, product development, and scale-up for fruits, vegetables, meat, and dairy products
The Center for Food Safety and Quality Enhancement (Griffin, Georgia)	Facilities for sponsored research in food safety and quality studies; consumer attitude and perception of quality studies
Institute for Food Safety and Security (Iowa State University, Ames)	Assistance to the food microcosm to combat food-borne infections, to prevent contamination of water and food, and to protect animals and plants from catastrophic diseases
The Institute of Food Science and Engineering (University of Arkansas, Fayetteville)	Value-added research for processing of agricultural products
The Food and Agricultural Products Research and Technology Center (Oklahoma State University, Stillwater)	Support and provision for basic research, training, and advisory services for food and agricultural processing in Oklahoma

From Giese, J., *Food Technology,* 53, 98, 1999.

- University of Milan, Department of Food Science and Microbiology (March 1995, Vol. 9)
- Campden & Chorleywood Food Research Association (September 1995, Vol. 9)
- British Industrial Biological Research Association (March 1997, Vol. 11)
- University of Nottingham, Division of Food Sciences; has four food groups: Flavour Technology, National Centre for Macromolecular Hydrodynamics, Food Microbiology and Safety, and Food Structure Research. The latter group is described in detail in *Food Science and Technology* Vol. 17, June 2003. The Interactif Club sponsored by the division consists of food companies with interests in the research of the food division.

To avoid confusion throughout this chapter, the company or any authorized individual within the company who buys, hires, partners with, or enters into a joint venture with an outside resource will be referred to as the client; all references to the consultant will describe any outside resource, or in the jargon of the day, any resource that is outsourced, whether this is an individual, a private research company, or a government- or university-based research institute hired by the client.

D. What Do Consultants Do?

Consultants perform numerous tasks for their clients. Specific tasks have already been mentioned. A tabulation of activities involving consultants, some of which I have been engaged in, are shown in Table 9.4. Some of these have a bearing on new product development. The first service, political or tactical use, has a direct connection with new product development. I have on two occasions been employed to conduct exploratory research in my own name. In one instance it was for a client wanting to study the effect of high pressure processing on their products, and the other was to contact and evaluate research houses for work on super-critical carbon dioxide extraction and to undertake some initial studies. Problem solving and advisory services are commonplace tasks for most consultants; some involve breakdowns

---

TABLE 9.4
General Classification of Services Provided by Consultants

Political or tactical use of consultants: The client uses the consultant to undertake some activity that management or the company prefers not to be seen to perform directly.

Problem solving: The client faces a crisis, for example, new product failure in the field, and requires crisis management skills.

Investigative research: Long-term basic research projects are placed with academic institutions. Projects may be contracted to outside resources with specialized pilot plant facilities, for example, extrusion processing, ultrahigh pressure processing, encapsulation technology, end-over-end can rotation thermal processing facilities, etc.

Specialized services: Training programs for staff, market research, product audits, forensic accounting, etc.

Advisory services: Support in application of novel technologies or strategies.

TABLE 9.5
General Reasons Outside Resources Might Be Consulted for New Product Development

Company lacks necessary skills in-house for customer, consumer, market, or technical research and development.

The skills are available but are already maximally deployed elsewhere in the company for maintenance of existing product lines.

An opportunity in a new applied technology or in a line of novel products has become available. The necessary skills are alien to the company. The company wishes to explore these without committing physical resources.

Economic factors: After due financial analysis, it may be discovered to be cheaper to contract with outside resources than to utilize existing manpower and facilities in-house or attempt to develop new skills and facilities in-house.

Time factor: An extended period of basic research and experimentation is required for the new project. The company does not want to tie up its own resources for long periods of time.

Secrecy: A company may feel more secure if the research and development is conducted elsewhere than on its premises.

in newly introduced products. A small food manufacturer asked me to act as liaison between it and the research institute with which it had placed a research project; they did not feel comfortable with their ability to discuss, ask questions, and digest the data and information they were being fed.

II. GOING OUTSIDE FOR PRODUCT DEVELOPMENT

A. THE NEED

Companies use outside resources to conduct some or all aspects of market research, new product development and research, associated engineering research, and marketing. The reasons for doing so are varied; some are shown in Table 9.5. Any one of these reasons may contribute to the client's need to use outside resources.

In essence, the client orders a product designed to its specifications based on its own market research or to specifications based on market research performed for it. If the client wishes, the work may at this time be passed over entirely to an external development company. If so, the client and consultant meet regularly thereafter to review progress and to determine whether the product in progress still fits the clients business and marketing plans. This regular review of progress is absolutely necessary because, with the passage of time, the client's needs may change or unforeseen difficulties in development arise. Both client and consultant may wish to avoid extensive research trials and therefore decide jointly to change direction with the project.

Alternatively, some consultants work closely with the client's technical staff, visiting regularly to work on the development of a product. The development work is done entirely on the client's premises with the client's staff under the guidance of the consultant.

Often a consultant will approach a prospective client with either a product — the result of new technology the consultant has — or with a product concept and some prototypical products. Client and consultant come to an agreement respecting ownership, licensing, patents, and any other legal matters.

The similarity to turnkey construction projects is obvious: the finished product, a newly constructed factory, is turned over to the client for an agreed upon fee. In new product development, the finished product is handed over to the client for a fee or for royalty payments.

B. Finding and Selecting the Appropriate Consultant

There are several ways to find consultants for product development. There are trade directories where consultants and their companies can be found along with brief descriptions of what *they* describe as their expertise. Consultants also advertise in trade and technical journals. The magazine *Food Technology,* published by the Institute of Food Technologists, has annually published a comprehensive listing of consultants and their services. The Institute of Food Science and Technology (U.K.) publishes a list of its members who consult along with a brief description of their areas of expertise. Consultants pay for listings in these two publications. Many trade associations will refer clients to members of their organizations who provide consulting or technical assistance service.

A more successful way to find consultants is through networking. Today, most business people network through the trade, professional, social, and philanthropic associations that they belong to. From their fellow members in these associations, clients looking for assistance hear about and get referrals to consultants. Here too, clients can get word of mouth verification of the caliber of work performed by these consultants.

Consultants are looking for business; they also network. They are constantly alert through business magazines and newspapers for situations where their expertise can be used. They solicit interested parties by postal or electronic mail. *They* are more likely to find the client needing their services than the client is to find them. They are available as speakers for lectures. They teach short courses; they advertise in-house training seminars for staff. They write articles in trade and technical journals. They have Web sites advertising their services.

If consultants are found from impersonal sources such as an e-mail contact, from a directory, or a referral from a professional or trade association, it is strongly recommended that the client request references and follow these up. Experienced and reputable consulting experts are wanted. Consultants referred through networking come with the personal comments of those who made the referral. If not, further enquiries can be made of others in the networking situation.

An interview with, or a presentation by, the consultant is also recommended for no other reason than for determining the compatibility of client, the development team, and the consultant and to familiarize the consultant with the client's requirements. After this consultation, the consultant should be required to present a proposal carefully detailing what work shall be carried out and within what time frames it will be performed and clearly outlining all the legal implication of the arrangement (e.g., who owns what intellectual property?). This proposal describing the work objectives should be clearly written without any hyperbole, jargon, bafflegab, or obfuscation of terminology; if necessary, the proposal should be reviewed by a lawyer. Clear communication between the client and the consultant should be such that both understand the problem and both know clearly what will be done and what both parties have responsibility for.

However, in over 30 years consulting in several countries, I have been interviewed only once and that by a company that was sold less than 3 weeks after the interview! They found me in a directory. All my business contacts have been by word of mouth; I have found this method of making contacts to be common for my colleagues as well. Thus the real world seems to favor networking to provide referrals for consultants.

The client's difficulty is usually not in finding a consultant; it is in finding the right consultant who can resolve the client's problems in new product development. What is important is what help the client wants in whatever phase of the development process difficulties have been encountered; that is, what is wanted is satisfaction of the client's needs, not what the consultant can offer. Clients must choose a consultant *not* on what the consultant can offer; what the consultant can offer is *neither* important *nor* even desired. It is the client's needs that must be resolved.

Both parties, the client and the consultant, need to know each other and respect each other and be clear on these objectives; the project should be a relationship based on mutual trust and confidence. This is no different from the relationship with any supplier.

The elements that are important before selecting a consultant are:

- The client must know what area of new product development it requires assistance in. What is the relationship for? What does the client want from a consultant? These needs should be written out by the client in a clear and concise statement describing what the client wants the consultant to do.
- The client must recognize the limitations (or extent) of its own knowledge and expertise. What advice, skills, and expertise does the consultant deliver that are not within the client's capabilities? Could the client do it alone?
- Closely allied with the above is the question of the consultant's capabilities. In short, does the client have more knowledge than the consultant? It is not unknown for consultants with their hyperbole to promise more than they can deliver; they are hucksters selling themselves. Often they subcontract work without the client realizing this. Security and confidentiality can be breached in this subcontracting. The circle of people familiar with the product development work grows larger.

These points can be illustrated by some actual occurrences. As vice president of technical services of Imasco Foods Ltd., I observed an instance of having more awareness of the technology than the consultant.

We wanted to measure the water activity of several products prepared in our various manufacturing plants. I had discussed the project, clearly I thought, in advance with a representative of a well-known midwestern research company. They had assured me of their ability to perform water activity measurements. (It was the early days of water activity studies.) Our plants sent in their products from various locations. In due time I received the results: the moisture contents of each of the products! An accompanying letter politely informed me that there was no such thing as water activity and what I obviously meant was water content. A flurry of telephone calls and faxes ensued until finally they admitted I meant water activity; they had erred; they could not measure water activity. Much time, money, and goodwill were lost on a simple development project. This research company was never consulted again.

An alert member of our research and development group caught an example of duplicity on the part of a consultant.

We had hired a New York–based consulting company to prepare some pour-over sauces for pasta. Samples were submitted along with their formulations for our evaluation. These were submitted to our manager of research and development of our manufacturing plant. He noted a strong similarity to formulations he had been working on using a supplier's ingredients and the supplier's accompanying book of demonstration recipes. We were expected to pay for what was basically public knowledge.

The next example occurred when I was at the Poultry Science Department at what was then the Ontario Agricultural College. We had quoted for a research project requested by a local turkey grower and processor.

We did not get the contract for this research project; it went to a large commercial research and development company. The owner of the growing and processing operation was a frequent visitor to our department, and during one visit I asked why we lost out. He told me we lost because he felt a bigger and more prestigious research and development company could handle the project better. Little did he know, and I never told him, that his bigger and better development company subcontracted a large portion of the work to us.

Buyer beware, indeed!

Giese (2001) discusses guidelines similar to those above for selecting laboratories for analytical services. He also provides a brief listing of testing laboratories and a useful checklist for working with outside laboratories. Of great importance, as Giese points out, is the need to determine the methodology that laboratories will use. It has been my personal experience, and my error, to find that very few outside laboratories use official or standard methods; they often have some in-house method that they invariably find "more accurate," "less time-consuming," or that "gives just as good results." Unfortunately, I had wanted comparative analyses of similar products from widely distant processing plants in the one instance, and in the other the characteristics of different plant varietals grown in different countries were being followed over an extended time period. It was necessary to work with different laboratories. Comparison of results between factories, products, or plant varieties were useless because despite my injunction to use official methods, different methodologies were used. It is absolutely necessary to spell out what methods are to be used, especially if the results are to be used for arbitration in legal work.

C. Caveats in Selecting and Working with Consultants

Patterson and Haas (1999) present a strongly reasoned argument for buying outside expert services to undertake tasks that the client considers to be not their core competencies. Their outsourcing is a broader interpretation of consulting than that which is generally understood. Their reasoning for choosing to outsource certain business activities closely parallels decisions a client must make in selecting a consultant to undertake new product development (see also Fuller, 1999).

One issue glossed over by Patterson and Haas is the extra cost required to monitor the outsourced activity. No matter how much the service provider and the buyer or client trust and respect each other, the client needs to monitor the outsourced activity with the same diligence they use to monitor their own activities. The client's production systems and products are constantly monitored by process and quality control activities; their copacked product is subjected to inspection. Therefore, outsourced activities require monitoring no matter how much confidence is placed in the consultant. The client has only themselves to blame if something goes wrong when they were not vigilant.

As in all purchases, the cautionary philosophy is *caveat emptor*, buyer beware. Similarly, in partnerships (or partnering) and in joint ventures, warnings still apply. Perhaps this is why one outsourcing contract was reported in the *Wall Street Journal* to be over 27,000 pages long. There are several caveats that clients seeking consultants need to be aware of. Patterson and Haas (1999) provide guidelines for the selection of partners in outsourcing.

1. Exposure

Exposure of sensitive business plans to outside parties is a very real danger since during discussions with a consultant, clients reveal the nature of the project, which in turn, reveals the direction of their research or business goals. Consultants, particularly those who are executives-between-jobs could very well at some later date be hired by a competitor. Consulting is an excellent way to job hunt.

Consultants on pensions from other companies understandably have strong ties of allegiance to their former employers; they have social and business ties with their old employers. A casual remark or an inadvertent comment can provide a good listener with hints of another company's activities. Where companies have in place an intelligence gathering training program for their employees, there could be serious leaks of confidential information.

I have been reluctant to place with universities any projects the nature of which I wished to be kept private largely because of the following:

> I experienced an egregious breach of security while on a prearranged consultation visit with an academic at an American midwestern university. The professor left to give a

lecture, and I was introduced to a doctoral student who conducted me on a tour of the laboratories during which work on a competitor's research project was in progress — openly displayed on the work bench. I had clearly identified myself and my company affiliation to both the professor and the student.

Many nonacademic people have access to university laboratories: staff and invited outside taste panelists or test subjects, students and their friends, professional staff and technicians, and equipment and instrument sales personnel, as well as representatives of industry. All are normal occupants of such facilities. They have eyes and ears and can talk.

Borrowing or renting pilot equipment from equipment manufacturers prior to purchase can have its confidentiality hazards.

One of our companies was interested in a thermally stabilized tray pack but required a special sealer. We called in the sales representative of a company manufacturing sealers for a discussion on the possibility of renting a unit for some trial runs. He gleefully told us we were lucky. A sealer was available across town in the premises of our competitor, who had just finished their trials with it.

The use of consultants always presents the possibility of a security breach respecting a client's research and development programs or business plans.

2. Loss of Client's Collective Learning Opportunity

By turning a product development project over to a consultant, the client loses both product and ingredient experience. The client's research and development team do not have the opportunity to work with the ingredients or to note the behavior of the product in different applications. The team loses the "feel" for a project and the knowledge and pride that accompanies the experience of bringing the product to fruition.

3. Employee Growth

Concomitant with the above, the client's technical employees lose an opportunity to learn "on the job" and grow. It is a lost training opportunity for the client's employees. Senior technical staff lose an opportunity to learn or demonstrate their skill in supervising a large project. Management loses an opportunity to evaluate staff, to reward them, and to select a cadre of future leaders.

4. Dissension

Both the client's technical and nontechnical staff often become disenchanted with the presence and interference of consultants in their (the client's) routines and their territories. They are alienated from the project because, having no input into the product development process, they do not see it as "their baby." There may be no interest in it. A deep resentment toward, jealousy of, and subtle uncooperativeness with the consultant results. An unfortunate example of this happened to me.

I had arranged several weeks in advance with the company president, the plant manager, and the research manager to run a small trial on plant-scale equipment for a condiment sauce. The research manager, who was in charge of this particular plant operation in addition to his research duties, made all the arrangements for raw material, staffing, and production time. He also helped design the research protocol. On the day of the test run, I found that the research manager had taken a day's holiday to attend a bridge tournament. The trial could not be postponed and Murphy's Law was fully proven: what could go wrong did.

The client's staff may be overcritical of the project in general or of the consultant's contribution. They certainly will not be overjoyed if their work schedule is interrupted to assist the consultant. The consultant will earn his fee, but those on the line may lose their production bonuses.

5.   Other Obligations and Responsibilities

The use of academics for consulting has some difficulties. Academics have other obligations and may not be able to work to the rigid schedules that prevail in industry. They teach. They advise graduate students in their research programs. They conduct their own research programs. They travel to present papers at conferences and to be guest lecturers at other universities. They have departmental business to attend to. They write up research grants and must network to find sponsors for research dollars. They are frequently not business savvy. Many lack any business, marketing, managerial, or practical industrial experience. Two experiences that I have encountered illustrate this point.

I lost a consulting academic hired to work on a project. He simply disappeared on sabbatical in the middle of the project without ever informing me. The project had to be shelved until he returned. He did not think it was important to inform me.

In the second incident:

We wished to improve the shelf life of a processed meat product and thought that lowering the water activity might provide the stability. We sought the help of an academic who was well published in this field. He advised the addition of glycerol. This, I informed him, was illegal. His answer was simply that nevertheless it would work and nobody could or would think to detect it.

I hasten to add that I have had some excellent assistance and advice from other academics.

D. ADVANTAGES AND DISADVANTAGES

As in all activities, there are advantages and disadvantages to the use of consultants in new food product development. Distinguishing between the two is often difficult since it depends largely on whose standards are used to measure the results and who stands to gain tangibly and intangibly from the results of the services offered.

1. Utilization of Resources

Some claim that a client's research and development dollars go further with the use of consultants (see, e.g., Patterson and Haas, 1999). Costs for consultants are a bottom line expense, a cost of doing business, and therefore they are deductible expenses. Where the research monies are placed with a university for long-term research, there are often taxation benefits and, coincidentally, goodwill is generated. By farming out research, the client is not saddled with the costs of hiring skilled staff, investing in expensive research and processing equipment, and devoting much company time and personnel to explore risky ventures that may not prove successful. The disagreeable task of ridding itself of the extra staff that were hired for the project is gone; no sophisticated equipment needs to be gotten rid of. Again, however, the question must be asked: is the company interested in short-term or long-term benefits such as growth and development of staff?

2. Flies in the Ointment

I have found that monies given to universities may lose 40 to 80% in value due to overheads (see Chapter 4 in Fuller, 1999 for an in-depth discussion of consultant fees). Thus, only 20 to 60 cents of each dollar given for research actually goes for research. However, I have also been informed by one food science department head that this overhead fee is highly negotiable.

Harvey (1977) suggests that joint ventures and partnerships may have flaws. One ought to include outsourcing; outsourcing as a buzz word had yet to be in the business lexicon in Harvey's time. In his experience joint ventures and partnerships should be approached with caution. He is particularly scathing of joint ventures meant to reduce competition, to share a risk ("risk is hardly divisible …"), or to share costs one company alone cannot afford. He discusses good and bad reasons for joint ventures.

Cost savings may be realized, it is true. But should there be changes required in the service as the client's needs change, the services must grow and develop as the demands alter. There will naturally be an escalation of costs to the client. There is a real danger (e.g., in information technology outsourcing or the consolidation of computer systems where long-term contracts for services have been signed) that the outside resource providers do not advance their services as the client grows; they are not pushed to deliver. Technologies change and a client's shifting business plans may change respecting product development, brand expansion, and the associated need for information technology.

Consultants are considered by clients to be both objective and unbiased. However, it must be realized that they have a self-interest to keep themselves employed. They want work. In over 8 years as the client hiring and working with consultants, I have yet to see a report that did not have a conclusion:

- That suggested more work should be done for a more definitive answer
- That suggested a follow-up review should be done to confirm that all was well
- That suggested the client would be advised to follow up certain "promising" avenues of research

All are tactics employed for soliciting further work and it will be noted that all are veiled attempts to put some doubt in the client's mind.

3. The Need to Monitor

One element that Patterson and Haas (1999) do not fully develop is the need for the client to monitor the consultant to assure that work is progressing toward the goals defined by the client. The same care, diligence, and monitoring the client gives to its own products and services must be applied to all its outsourced activities. Clients want a consultant who can work successfully, cooperatively, and without upsetting their internal operations too much; who can advise on improvements; and who can help interpret the client's ideas for development into new products. It is the wise client that maintains a tight rein on consultants.

Consultants with established reputations and experience in esoteric fields of technology bring distinct advantages to companies wishing to explore these newer technologies. Clients are not hobbled with struggling within their own companies to establish a foothold in these areas. They get a head start for product development.

There are some disadvantages to using consultants. Consultants can be disruptive to the orderly working of a manufacturing plant. The need for consultants to run trial production runs of samples is understandable, but for production staff who are perhaps on production quotas these stoppages are annoying.

III. COMMUNICATION: DOES THE CLIENT UNDERSTAND THE CONSULTANT'S LANGUAGE?

Obscurity is the refuge of incompetence.

Robert A. Heinlein

Another problem that presents a major disadvantage in all consulting activities is communication. Do both the consultant and the client speak the same language? This is somewhat tied in with the self-interest of consultants alluded to above.

A. "SPEAK CLEARLY, DAMMIT"

Consultantspeak, bafflegab, doublespeak, gobbledygook, even governmentspeak are all terms used to describe language that, when heard or read, sounds wonderfully pregnant with erudition at the moment, but later when one tries to remember what was said, one is left with a lot of buzz words "full of sound and fury signifying nothing." Margaret Thatcher put it succinctly: "You don't tell deliberate lies, but sometimes you have to be evasive."

Evasiveness can be accomplished with words, and words may be chosen to say much and mean little. I have told the following story before (Fuller, 1999):

> My first job upon leaving school was with the Canadian Food and Drug Directorate and my first assignment was to build a gas liquid chromatograph (there were no commercial units). This done I proceeded to research food flavors and citrus oils in

particular. A research director of a major food company asked my opinion of the method and its potential in food flavor work. I wrote favorably about the technique quoting my own public domain work as well as published work. All mail out of our department had to go through the department head for approval; he was very much the civil servant. We were after all THE GOVERNMENT. The department head severely edited my letter and filled it with government jargon. I was informed that should I be wrong in my opinions regarding gas liquid chromatography that this would reflect on him, how his department was run and ultimately on the government. I signed his letter but snitched a copy and took it home to my wife. Neither of us could tell whether I was for or against the method as a tool in food analysis.

Doublespeak is not rare. Words begin to have wonderful meanings: for example, restructure, downsize, destaff, right-size, unassign, all mean someone is going to be canned, fired, on the street. I have found all these in business and trade articles. Unfortunately, they seem to proliferate in articles written by consultants. I am not alone in this observation; some articles against this loose talk are "Professor Singles Out Double Talk" headlining a newspaper article and "Speak English, Dammit," a title for a business magazine article.

The following joke has made the rounds via the Internet. Its very presence is an indication of the prevalence of such language that is creeping into reports and communication:

1. Before your next meeting, seminar, or conference call, prepare your Baf-flegab Bingo card by drawing a square 5 cells across by 5 cells down, making 25 cells in total.
2. Write one of the following words or phrases in each cell:

synergy	benchmark	knowledge base
strategic fit	value-added	at the end of the day
core competencies	proactive	touch base
out of the box	win-win	mindset
bottom line	think outside the box	client focus(ed)
revisit	fast track	ballpark
take that off-line	results-driven	game plan
24/7	empower (or empowerment)	leverage
out of the loop		

3. Check off the appropriate cell when you hear one of those words or phrases during your meeting.
4. When you get five cells horizontally, vertically, or diagonally, stand up and shout "Bafflegab!" or "Nonsense!" as some prefer. Other words have been suggested.

I sent this to my daughter who is in food and nutrition development work with a strong marketing bent; she had already seen it and played it but without the shout. It had already made the rounds at her office. The game or ones similar to it are not new. Hammer (1994), a Food and Agriculture Organization consultant who analyzes project proposals and recommendations received by this body, produced a lexicon organized into four columns. By selecting a word from each column one could make up wonderfully forceful phrases meaning nothing. Prophetically, her paper is entitled "Why Projects Fail…." A few gems from her article are "centrally motivated organizational approach," "radically delegated technical initiative," and "strategically balanced development scheme."

These make perfectly meaningless phrases. My personal favorite for evasive doublespeak was reported in *New Scientist*, 1996. It described the Ariane space rocket, which blew up within 40 seconds of launch. The European Space Agency described it thus: "The first Ariane 5 flight did not result in validation of Europe's new launcher."

So much for the debasement of the language. Any client who spots bafflegab or evasive language in the reports, language, or discussions with his or her consultant should quickly show the consultant the door. Communication in a language that both the client and the consultant speak and understand is of utmost importance at all phases of the development process where outside resources are used.

> Beware the Jabberwock, my son!
> The jaws that bite, the claws that catch!
> Beware the Jubjub bird, and shun
> The frumious Bandersnatch!
>
> Lewis Carroll, "Jabberwocky" in *Through the Looking Glass*

The off-beat approach used in this section should serve to underline the importance of clarity and understanding in all communications.

IV. SUMMARY

Working with consultants can be very profitable for a company involved in new product development. When the consultant brings a new skill or knowledge, the client can be put into a more advanced stage of development, and the client's staff can get an opportunity to learn from the advice and guidance provided. There are, however, some pitfalls, and the client needs to be fully aware of these to ensure a good working relationship. The consultant is there to assist the client, not to supplant management's prerogative to provide sound management decisions. The client knows the client's business objectives and goals better than the consultant. All the consultant contributes are the tools, some arcane knowledge, and some ideas — all of which can be purchased for a price.

10 New Food Product Development in the Food Service Industry

Man does not ingest nutrients, he eats food.

Attributed to R. L. M. Synge

I. UNDERSTANDING THE FOOD SERVICE INDUSTRY

Two areas, the food service and the food ingredient industries, are major components of the food microcosm. They are sufficiently different from the mainstream food retail marketplaces to deserve separate treatments with respect to new product development.

More and more people are changing their eating habits. Today fewer meals are prepared from scratch and eaten at home; rarely are meals eaten as a family unit except for festive occasions. Meals are more likely to be put together from prepared menu components and eaten by the preparer who is "on the run." One result is an increase in meals-away-from-home, with a consequent increase in the number and variety of food service outlets.

II. THE FOOD SERVICE MARKETPLACES

One of the first challenges for the developer to face is the overwhelming range of food service establishments for which new products are required (Table 10.1). A more extensive listing is provided by Bolaffi and Lulay (1989). Each category in Table 10.1 is subdivisible. Fast food restaurants (now euphemistically referred to as quick serve restaurants) could be further separated into independent family-owned businesses, large multiunit chains either all or partly franchisee owned and operated, restaurants with sit-down service, and those with only counter or take-out service. Likewise, industrial worksite feeding could be an urban or suburban factory-run cafeteria offering a hot lunch, a cafeteria serving a 24-hour-a-day work force as in hospitals, or a food service organized for an isolated oil rig in the North Sea. Each category of food service establishment has its own peculiar needs to suit the menus, diners' needs, and the economic class of its diners.

TABLE 10.1
Variety of Outlet Types in the Food Service Marketplaces

Sector	Subclassification
Restaurants	Fast food (chain) restaurants: chain-owned and franchisee-owned
	Fast food family-owned restaurants (diners, cafes, bistros that may also be chain-owned)
	White tablecloth restaurants (stand alone)
	Hotel (again white tablecloth restaurants) and motel restaurants
	Tourist travel: cruise ship and barge boat meals
	Isolated work camps and industrial sites: floating oil rigs, mining camps
Food stores	Food stores in shopping malls
	In-store delicatessens with eat-in and take-out facilities
	Sandwich bars in food stores
Catering services	Hospital, convalescent home, and retirement home catering
	Visitors and staff
	Special diets for patients
	Bars, pubs, and nightclubs: foods for noshing (e.g., happy hours, light buffets)
	Transportation related: airline and train meals
	Charity
	Meals-on-wheels programs; soup kitchens
Institutional feeding	Penal institutions
	Sanatoriums and rehabilitation facilities
	School meal programs
Miscellaneous	Vending machines: ubiquitously sited (hot and cold food vending, beverages)
	Street vendors: finger foods
	Mobile canteens servicing urban work sites
Military feeding	Officers' messes
	Servicemen's messes
	Combat rations
	Field kitchens
Leisure feeding	Vacation camps
	Roadside eateries

Food service establishments can be partitioned into two sectors with different characteristics, which adds a further challenge to product development:

- The commercial sector: This sector runs the entire gamut of restaurants to the ubiquitous vending machine dispensers. When there is a free choice of food, that is, where consumers have the option of going to a variety of outlets if choices at one do not appeal to them, this option is characteristic of the free choice or commercial or noncaptive market sector. The commercial sector is profit motivated. There are commercial eating establishments that are operated marginally, but these are run as a service or a labor perk, and some person, company, or organization is subsidizing the operation. Examples of this type of operation that come to mind are a government- or university-supported hotel/restaurant training school's demonstration restaurants; employee restaurants underwritten by a company and part of a labor contract; and restaurants, often open to the general public, at private golf clubs and subsidized in part by members' fees.

- The noncommercial sector: In the noncommercial market (also called the institutional market), consumers have no freedom of choice in the menu or have only a very limited selection, i.e., the familiar "Will that be chicken or fish, Madam?" that used to be heard on airlines before they started charging for meals. Consumers in such situations are captive; they cannot go elsewhere to eat. Only drastic action on the part of paying travelers avoiding poor service carriers, prisoners rioting over the quality of food, or tales told by school children to their parents bring about changes in both service and quality of food (cf., the concern of parents over obesity in their children and the presence of soft drink vending machines and fast food franchise meals in school lunch programs). Travelers switching travel plans to more accommodating carriers, riots, hostage-takings, and demonstrations due to poor quality prison meals, and angry parents complaining to elected school officials about the quality of food in school lunch programs are powerful incentives for change. It is not profit driven, but neither is it run at a loss; the operator of the establishment determines both the level of service and quality of the food within constraints of budgets established by government, school or hospital boards, etc. Consumers are presented with limited or no-choice menus in the following situations:
  - Hospitals and convalescent homes, social services, institutionalized patients
  - Military feeding for troops on maneuvers (e.g., arctic survival training), on air and naval patrols, or on active enemy engagement
  - School meal programs in public schools as well as food service outlets on college and university campuses
  - Company cafeterias in isolated suburban industrial parks
  - Transportation meals served during air, bus, or train travel
  - Meals served in prisons and other rehabilitation centers
  - Relief feeding during famines and other natural disasters and in refugee camps

Bolaffi and Lulay (1989) separate out the military food service as a distinct entity for food developers by virtue of the product standards, labeling, and packaging requirements and the bidding and tendering requirements demanded by governments. However, in this discussion, military feeding will be included with the noncommercial sector.

A. Customers and Consumers in the Food Service Industry

There are customers whose needs are to be met, and consumers who dictate those needs to the customers. The food service industry is similar to the consumer food product retail industry in this respect. The concept of the customer as the gatekeeper is also valid but requires a stricter examination. There are two distinct customers in the food service industry:

- The customer is the individual who orders from the menu, or who answers "Chicken, please" to the steward, grabs a steamer from the street vendor, points to a steam table tray in a cafeteria lineup, or puts money into a coin slot at a vending machine. They either eat the products on the premises where they were purchased, munch on finger food on the street, dine in cafeterias at their work places, or pick up prepared food to eat at home. They are customer and consumer combined.
- The master chef of a restaurant, the owner of a restaurant or diner, a dietician manager of a hospital commissary, the food purchasing agent for a government institution, a food store owner with a deli bar, a business person who purchases a franchise of a fast food chain, and a military quartermaster are also customers. They purchase food ingredients, prepared meats, *sous vide* products, and preprepared raw produce from suppliers. They are also the consumers (users). They or their staff transform the basic food items into meals either through the artistry of chefs, by heating and plating *sous vide* products, by simply assembling prepared items, or merely displaying or dispensing items prepared by others. They have characteristics in some instances analogous to that of the gatekeeper.

This two-tiered customer and consumer element introduces many novel problems. Developers of products for the food service industry develop prepared or semiprepared products for highly trained and skilled people (chefs, cooks, etc.) and for people with rudimentary knowledge of food preparation and handling, nutrition, sanitation, and hygiene. These developers are preparing products that must directly satisfy the needs and expectations of their customers and consumers.

How does one screen a product such as a new ingredient when one does not know specifically how or in what finished product that ingredient will be used by a chef or cook or in what social setting those finished products will be served? How does one get market research, conduct a test market, or evaluate a test market?

III. CHARACTERISTICS OF THE FOOD SERVICE MARKET

Both the commercial and noncommercial sectors can be discussed as one for the sake of simplicity, and any distinguishing characteristics of either highlighted. Some of these characteristics are presented in Table 10.2. Products designed for the food service sector must accommodate the characteristics of the sector they are to serve, that is, simple to prepare and present to customers, not conducive to waste production in a close-quarters working area, and not energy intensive in their preparation.

TABLE 10.2
Characteristics of Typical Commercial and Noncommercial Food Service Outlets

There is small-scale preparation of a wide variety of menu items. Some outlets cycle menus every 2 or 3 weeks, while others have fixed menus regulated by the day of the week or regulated by availability of fresh produce.

Food preparation area and social setting for eating are combined or closely associated.

Uneven periods of food preparation and serving. Staff often required to stand by during lulls in activity.

Rarely are staff or outlet provided with a formal quality control and inspection procedure. Fast food chains do have procedures.

Cook or chef has full responsibility for products and ingredients. Specification for purchase subject to cook's judgment or seasonal availability.

Preparation is labor intensive with variable skill levels required.

Storage of raw and prepared products, food preparation, waste disposal, and utensil clean-up often share cramped facilities. Often all activities proceed simultaneously and in hot, humid conditions.

A. CLIENTELE

Consumers in high-class restaurants are looking for quality of taste, service, presentation, atmosphere, and relaxation. They are less likely to be concerned with price *but will demand quality and service for the price.* Consumers buying from street vendors are looking for cheap, good, quick, wholesome (safe) food while running errands on a lunch break or during breaks in play at a game, or at an outdoor event. Their only common characteristic is that neither the upscale diner nor the eater on the run are looking for a healthy nutritious meal; one is seeking gratification of the senses; the other eats quickly and of necessity to get on with some activity.

Expectations of consumers in the noncommercial sector are quite different but just as diverse:

- Safety and nutrition are important in prison feeding, hospital feeding, military feeding, and school meal programs. Food intoxication running through these populations can have devastating effects. Presence of allergens in foods are of major concern.
- Quality, at the very least, must be acceptable to the majority and certainly not so bad as to be rejected. Poor food has been known to cause prison riots. Transportation meals must not be so unacceptable that they are a factor in passenger's choice of carrier.
- Budgetary restrictions limit to a degree both quality and availability of choices.

Hospital care feeding presents a special case, which will be discussed later.

B. FOOD PREPARATION AND STORAGE FACILITIES

Five-star restaurants, family-style restaurants, ethnic restaurants, hotel kitchens catering to two or three classes of eating outlets, fast food franchises, mom and pop diners, or bars serving hot hors d'oeuvres need a wide variety of food preparation equipment. These range from warming units to steamers to friers to specialty wood stoves and smokers to tandoors to woks to hot stones to 50-gal steam jacketed kettles with mounted Lightning mixers to microwave ovens to potato peelers, grinders and slicers, to walk-in freezers and refrigerators and air conditioners and fume hoods for odor control. Some are well equipped. Some (street vendors and snack setups in bars at happy hour) have less equipment than a well-run family kitchen would have.

Many restaurant kitchens are run at one heating temperature for all menu items served. Preprepared products and ingredients must be designed or formulated that can maintain their high-quality attributes during preparation and holding at this operating temperature stress.

Storage is limited in food service establishments. Full service restaurants have adequate, but limited, storage facilities for supplies, food materials, and waste, while street vendors, for example, have minimally adequate storage and must rely largely on city maintenance facilities for waste disposal. In happy hour setups in bars, there are basically no storage facilities. Owners must rely on same day deliveries with the caterer bringing warmers and hot finger food and returning later to remove the dirty ware and warmers.

Food suppliers must design into their products stabilizing systems to withstand the challenges that these products might encounter in inadequate conditions of distribution, storage, preparation, or the vending environment. This concern of safety is great for products for food service operators supplying convenience stores, vending machines, and street vendors.

In the vending machine sector of food service, newer machines incorporate microwave ovens or deep-fat fryers making a wider selection of product offerings possible; combination plates, sandwich, salad, and fries or soup, and sandwich and dessert are now regular offerings (Williams, 1991). These new menu items are much more sensitive to time and temperature abuse.

If storage for food supplies is limited, then storage for waste is even more at a premium. Products for food service must be designed to produce as little waste as possible. Prepreparation is a key added-value feature. For take-away hot foods, fast food outlets face another problem: packaging. It must be environmentally friendly.

In general, food preparation and storage facilities in noncommercial establishments are better, or more adequate to their tasks, than those of commercial establishments; they were often built to specifications designed for the traffic. They are usually run by professionals trained in food service or are catered to by industrial caterers. While they are on limited budgets, they frequently can purchase in bulk and obtain the reduced prices that quantity buying permits.

Food preparation systems in use in U.S. hospitals employ one of five common preparation systems (Matthews, 1982):

- A cook–serve system whereby food is prepared on site, plated, and distributed on trays to patients.
- A cook–chill system in which the food can be prepared either on- or off-site ahead of use. After preparation, the food is rapidly cooled, held refrigerated for not more than a day (but see Mason et al., 1990), plated, and distributed to patients from specially designed heating carts.
- A cook–freeze system that is similar to the preceding with an additional step requiring the thawing of menu items.
- A thaw–heat–serve system where the main menu item is precooked and frozen either on- or off-site. On-site, the product is thawed, heated, and plated.
- A heat–serve system using thermally processed steam table tray containers. The main menu item can be heated and held hot before plating.

These systems are not unique to the health care food service industry. They are also used in military feeding, especially the heat–serve system, and in many commissaries servicing a variety of cafeterias and other dining establishments. Awareness of the rigors that such systems impose on food products and ingredients in them is necessary to guide developers in the design of products.

Mason and coworkers (1990) and Livingston (1990) discuss commercial food preparation systems used in hospitals and other food service outlets and evaluate these with respect to their influence on the quality of the food.

An interesting combination of commercial and noncommercial food service in one facility is demonstrated by some U.S. hospitals that have rented out part of their food service facilities to fast food chains and developed a separate source of income while accommodating all their food service needs.

C. Labor

Kitchen labor in both the commercial and noncommercial sectors is highly variable respecting skill and knowledge levels about food handling. To describe any one sector as characterized by unskilled labor simply invites a barrage of counter examples. Staff from chefs in upscale restaurants (where they are not owners) to temporary staff working their way through school are highly mobile or in transitional stages in their careers, supporting themselves through school, or temporarily unemployed. (I had no idea until recently that chefs and sous chefs had agents.) Their skills range from highly skilled and knowledgeable in food preparation and handling to rudimentary. Many of the latter are not careerists in food service; they want out at the earliest opportunity into other career paths. Training such candidates can be discouraging and fruitless for food service employers.

Therefore, design of food components used in preparation of finished products in many food service establishments:

- Must be kept practical, safe, and simple. Products for food service use must be as close to "error-proof" as possible in the hectic environment of a kitchen.

- Require the minimal amount of labor for their preparation. Ideally they should require only reheating or a finish cook.
- Produce the least amount of waste for handling during preparation, and their preparation should not contribute unduly to the microbial load in the kitchen.
- Must be such that storage of unused portions must present no complications for their safety or quality.

Quality restaurants have the skills of their chefs and sous chefs available. They can use premixes to prepare and present attractive, added-value products. They have the ability to take unused portions and reuse these, safely, in other quality dishes. Yesterday's unused poached salmon can be today's salmon mousse or salmon cakes. Such prowess reduces waste and thereby minimizes cost overruns. Nothing of chance can be left to personnel in fast food or more downscale fast-casual restaurants. Unused food or warmed and thawed portions must be disposed of.

The availability of skilled and trained labor is exacerbated by the rising costs of labor. A rather dated report by Pine and Ball (1987) found that wage bills (as a percentage of sales minus pretax net profits) ranged between 8 to 38% in the U.K. food service industry depending on the class of the establishment. Labor-cheap businesses were largely in the catering-only end of food service where labor consisted of putting together preprepared items or they were businesses lacking a personal service aspect (transportation catering). These figures were generally applicable in other countries.

The noncommercial sector faces the same labor problems as the commercial sector. Since many workers are often newly arrived immigrants, there can be communication problems if there are language barriers. In some penal and correctional institutions, unskilled inmates may serve as help.

There are, however, wide variations in skills in the noncommercial sector. Dieticians required in hospitals and nursing care facilities and military chefs are highly trained; school lunch programs are usually under the supervision of equally skilled personnel. Nevertheless, wages are at the minimum level for unskilled labor, and at the bottom end of the wage scale, job security is absent. These are not elements conducive to attracting skilled people. On-the-job training is difficult in this environment. The exception is the military, where training programs have proven very effective in raising the skill levels of food handlers and cooks.

D. PRICE, QUALITY, CONSISTENCY, AND SAFETY

Price, quality — more importantly consistency of quality — and safety from hazards of public health significance are equally necessary considerations in both the commercial and noncommercial food service sectors. Price — where it is an element in the noncommercial sector — is regulated by what the targeted consumer in these outlets (e.g., seniors' residences, long-care facilities) are willing to pay.

Price is also regulated by many other factors not determined by the targeted consumer but that may affect the consumer that the food service establishment wishes to target:

- Costs of raw materials and ingredients: Ethnic restaurants, upscale restaurants, or product-oriented restaurants (vegetarian or seafood) will each have unique product requirements that command premium prices. Control of costs is partly obtained by adherence to a strict portion control program. Plate waste, even if it has been paid for by the consumer, is nevertheless an indication of prepared product losses that could have served more customers. Portion control, therefore, becomes an important element of price and service provided by the supplier.
- Labor-related costs: The type of restaurant determines the skill level; for example, a Chinese restaurant will require a specialist cook/chef skilled in Chinese food preparation.
- Site-related costs: These are costs such as real estate taxes, local economy, and type of customer traffic around the area. The old real estate adage that "location, location, location" is a contributory factor to success brings many cost consequences.

Quality must be consistent with the price the consumer is willing to pay for that quality at each and every sitting, and service must not vary from restaurant to restaurant within a same chain. This is a problem that has not been mastered in some doughnut chains. I have found wide variations in the quality of doughnuts from the same chain in the same city, as well as variations in different states and provinces.

Food poisoning, intoxication, and allergic reactions of consumers (Gowland, 2002) are major hazards in the food service industry (Snyder, 1981; Snyder Jr., 1986). Protein salad foods presented the greatest safety concerns according to Solberg and colleagues (1990) in a study of meal items and food preparation facilities at Rutgers University's campus. These are foods generally associated with "summer sickness" caused by improperly prepared and stored egg, chicken, or turkey salads.

Suppliers of prepared menu items (e.g., *sous vide* items) can protect their new product introductions only by designing a suitable protective constraint to food poisoning microorganisms and by ensuring that products leave their premises in excellent and safe condition. Suppliers need to provide clear instructions on the handling, storage, and any further preparation of sensitive prepared items respecting time temperature tolerances, proper rotation of stock, and zone isolation for the preparation of sensitive components to prevent cross-contamination. A close supplier–client relationship can minimize chances of hazards occurring. The attendant media coverage of food poisoning outbreaks can destroy individual restaurants and seriously damage the reputation of fast food chains.

Consumers having reactions to potentially allergenic ingredients is a hazard for developers, especially for suppliers to the food service market (Gowland, 2002). The suffering and deaths of consumers as a result of allergic reactions to foods eaten in food service establishments are devastating incidents (Williams, 1992) since menus rarely have lists of ingredients to warn consumers. Rarely do these incidents get the same coverage in the news since they affect only an individual, whereas summer sickness at a company picnic affects many people and gets wide publicity. Allergic reactions are a major concern for suppliers to school lunch programs. The offending foods for children appear to be age-related (that is, sometimes children grow out of them): eggs and egg products, milk and milk-derived products, and nuts (especially peanuts) are often the source in the younger years. With age, seafood products frequently become problems. There is no protection for children who in a carefree, or careless, mood will exchange forbidden treats with classmates. Since only a few molecules of an allergen can trigger a reaction, developers must exert extreme caution to avoid allergenic ingredients in their menu components and provide truthful labeling statements.

Allergy specialists in Canada estimate there is at least one death a month due to allergic reactions. In the U.S., with a tenfold higher population, this suggests ten or more deaths a month due to eating prepared foods. Integrity of ingredients respecting the presence of allergens is a major problem for manufacturers; such integrity is very difficult to attain in multiproduct manufacturing plants.

To serve their customers (or shift responsibility from themselves), food service establishments, particularly fast food chains, have taken to posting allergy charts listing the ingredients they use in their menu items (Anon., 1988b). The consequences of this precaution forces manufacturers supplying food service establishments to ascertain that the composition of all ingredients they use in their products fit these charts. This requires vigilance on the part of manufacturers as they switch ingredient suppliers when looking for better sources and more effective ingredient cost control.

E. NUTRITION

Nutrition in the commercial sector is not a problem for developers. (Nevertheless, the litigious nature of consumers and the growing problem of obesity and diseases associated with obesity have caused a welter of class action suits against fast food chains; claimants accuse the chains of having caused their obesity-related ailments through the promotion of fatty foods.) People eat out for a social event, for comradeship, business (networking), a celebratory event, conversation with good friends, and pleasure. Nutrition is not high on their priority list; good, tasty, safe food is their priority.

By contrast, food service facilities within the noncommercial sector have highly variable requirements for nutrition and quality of taste and flavor that reflect the type of noncommercial food service outlet.

Meals served on an airliner are one-time occurrences; their nutritional content poses no risk to the well-being of the passengers, whether the service be a snack, a cold meal, or a hot meal. The long-term nutritional health and welfare of traveling passengers are not dependent on the food served; safety of the food, with taste and satiety close seconds, is always a major consideration.

Nutrition as a quality feature is two faceted. There are both customer needs and consumer needs, particularly respecting health care feeding. Customers (dieticians) in the health care field require unique products to satisfy special dietary requirements of postoperative patients and patients undergoing cancer therapy or drug therapy. They also require very specific nutritional information for consumers (patients or the elderly enjoying a meal service) in their care. They need much more detailed, indeed esoteric, nutritional data than that required by most nutritional labeling regulations.

Developers must be prepared to provide this information in order to assist dieticians in the preparation of the many different dietary regimens required for patients. Schmidl and colleagues (1988) provide an extensive review on parenteral and enteral feeding systems, a rather exotic branch of hospital care feeding.

Is there a problem of nutritional quality with the food service most used by the consuming public, the fast food sector? Fast foods have not received good publicity concerning the nutrition they offer consumers. For example, Hibler (1988) reported the results of a sampling of burgers, fries, and chicken, all of which are highly ranked favorites among frequenters of these outlets. The calories per gram of

- Burgers ranged from 1.9 to 2.9 cal/g
- Chicken entrees (nuggets or fingers) ranged from 2.4 to 3.1 cal/g
- Fries ranged from 2.7 to 3.5 cal/g

If the weight of entrees were constant (they were not), hamburgers would be the least calorific choice. The calories from fat were highest in the chicken entrees, ranging from 39 to 55% (average 49.5%) and almost identical in the burgers (36 to 50%, average 44%) and the fries (37 to 48%, average 42%). A typical burger and fries or chicken pieces and fries would put the percentage of calories from fat well over the recommended level of 30%. Ryley (1983) calculated that, even back in 1982, fast foods and snacks contributed over 16% of the daily fat intake per person in the U.K.

There is concern about nutritional quality in the feeding of the elderly at home. Turner and Glew (1982) studied the nutrient content (protein, energy, calcium, iron, and ascorbic acid) of meals delivered in Leeds (U.K.) and provided by six food service organizations. They found significant weight differences between the meals supplied by the different organizations as well as between the protein contents of the meals. Meals supplied between 20 and 48% of the recommended energy intake for elderly people, which represents under, to grossly over, the energy requirement for the main daily meal. Of particular importance to geriatric nutrition, it was found that the calcium content of the meals varied widely from inadequate to ample — dependent, primarily, on whether the dessert was milk-based or not. Iron content was found to be just adequate for the elderly, but ascorbic acid contents varied widely from providing more than 50% to less than 25% of the recommended daily intake. However, with ascorbic acid, significant losses were noted between the first and last meal deliveries, as might have been guessed from the lability of this vitamin. Keeping meals hot during transportation as well as the damage that the duration of hot transportation itself had on vitamin C were weak links in delivering nutrition to the elderly. The rigors that the delivery (serving) system imposes on meals must be considered by designers of products for such institutional feeding systems.

1. Standards

Some detail has been recounted in reviewing these studies on the nutrition of food service meals to point out a major quality problem for developers of products, whether they be *sous vide*, frozen or other chilled food entree items, or dry powder bases for soups, gravies, or desserts. What are the nutritional standards or guidelines for these products? There are no nutrient standards or even nutrient specifications for products meant for institutional or military or school meal programs. There certainly are no standards for the special diets (What's soft? How soft? How does one measure soft? By fiber content?) required by health care establishments. Developers could create added-value menu items with established nutrient content or closely defined standards for such vague terms as low ash diets, low calcium diets, liquid or semisolid diets, or semiprepared foods with fixed soluble/insoluble fiber content ratios for health care establishments if they knew what the standards were. These are excellent opportunities for industry to assist these establishments and find good, and profitable, market niches for their products.

A much greater concern is the natural desire for health professionals and food manufacturers to provide foods that have been developed with nutraceutical fortification to patients in hospitals, nursing homes, and institutionalized care facilities. Opportunities to provide nutraceutical fortification of foods might be expected that would complement or substitute for the orthodox or prescription medicines and mood-altering drugs patients use and thus reduce medical costs and reduce hospital stays. They would also benefit the general population following preventive diets.

There are hazards. Would nutraceuticals be incompatable or antagonistic with prescription medicines? There is concern — frankly there is as yet no firm knowledge — about whether nutraceuticals, if used in food systems, would react synergistically or antagonistically with other phytochemicals or other components in the food system. Would the concentrated phytochemical in capsule form prove effective? Which phytochemical of the many hundreds in natural products might prove effective? For example, in an extensive review of research work Giovannucci (1999) found that tomatoes, tomato-based products, and lycopene (a nutraceutical antioxidant in tomatoes) displayed an inhibitory effect on certain cancers. However, Boileau et al. (2003) found that tomato powder proved effective against prostate cancer in rats but that lycopene in concentrated form had no more effect on cancer than did the placebo. Both Giovannucci and Boileau and colleagues stress that tomatoes are very complex mixtures and attempting to single out one component as having the beneficial effect could be useless. The concern for all developers is that if varieties of tomatoes with greatly increased levels of lycopene, a major antioxidant, or products with added lycopene are developed, it may be found later that perhaps it was something else in this complex of chemicals called a tomato that was effective. Food legislators would be only too happy to curb false claims, and antibiotechnology groups would be quick to damn the plant geneticists. The adverse reactions would not benefit science, scientists, or developers.

Another hazard is that in the present state of knowledge there is little or no policing of the claims made about herbs and herbal extracts, particularly in North America. There is no informed knowledge of effective dosages of nutraceuticals. In practice, a slurry of the herbal preparation is sprayed onto snack items, but there is no information on what the dosage level is. Did the slurry contain those parts of the plant containing the active nutraceutical? Even a cursory glance through any modern herbal will reveal that leaves, flowers, roots, stems, or bark may contain the active phytochemical to the exclusion of the other parts. Developers of nutraceutically fortified foods for the food service sector of health care facilities and other areas of the food manufacturing sector should proceed with caution with new products, especially in the fortification of candies, drinks, and snack foods aimed at school meal programs.

Kuhn (1998a) and Neff (1998) discuss at length some cautionary comments of health professionals and manufacturers about nutraceuticals and provide a good overview of products and activities in this market although not directed specifically at the health care sector. Schwarcz (2002) urges standardization of herbal preparations and an accurate statement of the dosage in nutraceutical fortified foods, along with information on which part of the plant has been used in preparing the consumer products. He describes several interactions of herbal preparations and prescribed medicines: ginkgo biloba, an anticoagulant, can exaggerate the bloodthinning properties of aspirin and coumadin; St. John's wort, a mild antidepressant, is antagonistic to cyclosporin, an immunosuppressant; and ginseng, ephedra, and valerian can interact with anesthetics. (Severe warnings against the use of ephedra have been issued.) Kava kava has been banned as a liver toxin in many countries including Canada.

The use of these supplements either as herbs or extracts should be approached cautiously for development of products intended for children (snack foods) or for the health care food service sector. A minor caveat for developers is that many herbal preparations and extracts are not pleasant tasting, and flavor systems need to be developed to overcome their unpleasant taste.

2. Health Care Sector of the Institutional Market

This sector includes hospital feeding, convalescent care feeding, and feeding of the elderly in their own homes or in residences for the mobile elderly, or for those institutionalized in nursing and psychiatric homes. Most health care feeding presents two situations. It includes both the public cafeteria for hospital staff, students, outpatients, and visitors on a 24 hour, 7 days a week basis and the institutional sector where patients require special diets in accordance with their medical conditions as well as their personal, religious, or ethnic taboos.

Regional health boards frequently control or influence the buying power of hospitals in their districts. As a result, their purchasing power can vary widely. Meal items are often purchased from privately run food commissaries. Nevertheless, many hospitals do still prepare their own meals or prefer to prepare their own special dietary meals because of the lack of accepted industrial standards (Matthews, 1982; Burch and Sawyer, 1986).

Hospitals and long-term nursing care facilities, social agencies providing home delivered meals to the elderly or to the homeless in shelters or other humanitarian feeding programs, government-supported programs such as school lunches, and the military have a very special concern for the nutritional content as well as other dietary considerations (digestability, bulk, fiber content) of their meals. The nature of the meal and its nutritional adequacy have a direct bearing on the consumer's health. Patients recovering from the trauma of an operation, undergoing irradiation treatment, or who have medical conditions causing malabsorption of nutrients or are undergoing long-term convalescence require special diets affecting the nutrient content and texture of foods. Food must be attractive and flavorful in order to entice such patients to eat since they often have little appetite, and it must be nutrient dense to provide their nutrient requirements. Many of these consumers have weakened immune systems, and microbiological loads that would be tolerated with no difficulty by healthy individuals may fell these patients; safety of the food is paramount.

Confounding all the above is the lack of standards for special dietary foods. There are no standards for nutritionally designed foods or for any of the special types of foods dictated by convalescent diets. Consequently, developers wishing to put products into the special market niches have no guidelines. They must therefore work closely with dieticians in development.

3.   The Military Sector of the Institutional Market

It is important to any country to keep its soldiers healthy and fit. The dictum "an army marches on its stomach," attributed to Napoleon Bonaparte, was never truer. Food, good food, keeps an army going and its morale high. This requires good-quality, flavorful, capably prepared food. In military feeding, food safety is extremely important. An army cannot come down with a foodborne illness, which in emergency situations could incapacitate the troops.

Food products must be available and suitable for serving in highly variable conditions. Extreme conditions such as peacetime maneuvers in humid, tropical jungles, in dry desert conditions, or in arctic terrain or during hostile enemy activity in wartime make military feeding situations a difficult challenge for developers. Products must have a long, stable shelf life, must be easily portable (i.e., light in weight for it may have to be carried by the individual combatant), flavorful, and contain all the nutrients for an active stressful lifestyle. Variety of menus and ease of preparation with minimal equipment are also requirements. Mermelstein (2001) discusses some of the requirements for the basic combat ration, the Meal Ready-to-Eat (MRE):

- Shelf stable for a minimum of 3 years at 80°F and a minimum of 6 months at 100°F
- Able to be airdropped by parachute from 1280 ft
- Able to be dropped out of a helicopter from a height of 100 ft without the aid of a parachute
- Able to withstand environmental conditions of −60 to +120°F during storage, distribution, and handling
- Packaged to be resistant to wildlife

In addition, it must be acceptable organoleptically to the soldier. Mermelstein presents further details regarding requirements for military rations as well as for rations for relief feeding.

The military market is also highly fragmented. Food service facilities and operations range from officers' messes to noncommissioned officer messes to combat field kitchens to vending machines and snack bars to hospital care facilities. Most of these facilities must be operated 24 hours a day to accommodate the shift work involved.

When and where this food will be prepared and consumed introduces problems. Ease of preparation is important because food may be prepared or reconstituted in less than ideal conditions in combat zones, in cramped conditions in submarines or airplanes, and often by the unskilled personnel themselves. Packaging must be lightweight yet protect the food from all environmental conditions (which are unspecified) and from any treatment (including air drops) it might endure. Yet the package must be easily opened and the packaging readily disposed of lest obvious waste disposal dumps be seen by enemy aerial observers as a sign of a field kitchen nearby.

As is the case with all government procurement purchases, products must adhere to rigid standards and specifications for ingredients, processing, and packaging. These standards and specifications can be quite detailed. Suppliers are advised to obtain copies of them before attempting to manufacture products.

A possible cause for failure of new products, discussed earlier, could be their introduction into markets dominated by a single customer; the military represents such a market. On the other hand, the military market and the food service market in general do present unique opportunities to introduce new products within a select portion of the population, which may develop a liking for and a familiarization with (i.e., it is an educational opportunity) novel food products.

IV. DEVELOPING PRODUCTS FOR THE FOOD SERVICE SECTOR

Diversity is the key characteristic of this sector (Table 10.3). There is diversity of facilities and equipment in these outlets: diversity of labor skills and wages, from chefs in top-of-the-line gourmet restaurants to street vendors in their peddle carts selling sandwiches, hot dogs, and sausages and chili sauces; and diversity in the consumers and expectations of consumers.

A. THE PHYSICAL FACILITIES OF THE CUSTOMER

New products for the food service industry must be capable of being finish prepared with the equipment that the outlets have on their premises. Equipment can be primitive or very sophisticated, and the food product developer must recognize the restrictions in product design that the state of the preparation equipment may pose. While there is a restriction imposed by the available equipment, the finished product must be platable with the smallware available in the establishment. New product development must suit the chefs and cooks as well as the kitchen.

Settlemyer (1986) describes this situation perfectly with a description of the development of a buttermilk biscuit for a chain of restaurants. During development, it was determined that although ovens suitable to bake the biscuits were available in individual outlets, warming ovens to hold the biscuits were not. Thus, included in the development process was the need to research different brands of biscuit warmers. The needs of the kitchen have to be satisfied.

One fast food multioutlet company solved the problem of compatibility of products and equipment with a fully operational outlet at its development center, not merely a mock-up test kitchen. The sole purpose of this outlet was to provide

TABLE 10.3
General Problems in Food Service Establishments That Influence Product Development

Clientele: Runs the gamut of those seeking gratification of sensuous pleasure with cost no concern to those eating of necessity in order to get on with something else to those who have to be fed for humanitarian, military, or medical reasons.

Facilities: From well-equipped to barely adequate preparation equipment. Caterer often has to supply equipment. Many are "heat and hold" facilities.

Skill levels: Skill levels are highly variable within the commercial and noncommercial sectors of the food service industry; training levels are high in white tablecloth restaurants down to minimum in fast food chains.[a]

Labor and labor costs: In general, labor in the food service arena is mobile or transient or both; chefs and sous chefs are highly mobile with their skills; summer help is transient. Labor costs are high in such establishments. Labor in lower quality restaurants is usually lower paid and is often transient. Food commissaries have labor problems similar to those of any food manufacturing establishment; workers work in hot, dirty, noisy environments where often little formal education is required and basic language skills are not necessary (Fuller, 2001, p. 267).

Expensive real estate: Restaurants, company cafeterias, and fast food outlets are sited in high-traffic areas (downtown areas, office buildings, busy streets, or highways) and this results in high taxation and high property values. For these reasons, work area to income earning area is kept at a minimum.

Environmental areas of concern: Site locations make odor elimination and waste removal imperative; elimination and removal is expensive. Removal and elimination can irritate neighboring residents or establishments. Hours of operation (noise and traffic pollution) can result in local by-law regulations.

Energy costs: Restaurant operations are energy intensive.

Consistency, price, quality, and safety: Quality and price must vary with budget restrictions established for raw materials and ingredients. Consistency must be constant, and safety is paramount.

[a] The skill levels in the central commissaries of fast food chains are high, but levels in the actual serving establishment may be nonexistent and dependent on preparation and serving protocols laid down by the central commissary.

a one-store test to determine how operationally compatible products were before any test market was attempted (Peters, 1980). In both Settlemyer's and Peters's examples, the needs of customers (chef, owner, manager) were catered to first.

Dry goods storage space and refrigerater and freezer space is limited; new products must not require special storage conditions and must be adaptable to available storage space. Unused product must be easily resealable and capable of being safely stored.

B. ENERGY REQUIREMENTS

Types of energy used in food service facilities and their costs vary widely within any country and from country to country. Energy as a cost factor in food service operations is, as a consequence, highly variable. When prepared foods received from a central commissary are used, less energy is required for preparation and presentation. *De novo* preparation from raw ingredients requires more energy for food preparation and cooking. Energy is, along with labor, a considerable contributor to overhead expenses. Efforts to reduce the energy used in meal preparation would be greatly appreciated by food service operators. In military feeding, in particular, energy conservation would be very welcome since energy sources must be moved with the marching, sailing, or flying consumers.

To design products that conserve energy in their preparation, developers need to determine how and where energy is required in food preparation. If only the prepared food itself is considered, energy is absorbed by:

- The average temperature increases in foods as the masses of foods are brought to the necessary preparation and serving temperatures
- Phase changes in foods of mixed composition as solids are converted to liquids (dissolving and melting) or water to steam (concentrating) (Norwig and Thompson, 1984)

Norwig and Thompson describe the calculation of energy requirements using the frying of french fry–cut potatoes and the cooking of frozen hamburger patties as examples.

Snyder (1984) suggests that foods, particularly sauces and gravies, can be formulated with the correct amount of liquid incorporated initially to get the desired finished consistency. This reduces the need to boil the sauces down to evaporate water. No concentration step is required, and both time and energy can be saved. An additional benefit would be the improvement in quality since less heat damage to sensitive ingredients in the sauce would occur.

Many factors affect the energy absorption (and quality) of food products during heating: shape, thickness (McProud and Lund, 1983; Ohlsson, 1986), density (McProud and Lund, 1983), composition (Bengtsson, 1986; McProud and Lund, 1983), and method of heating: by microwaves (Decareau, 1986), by convection or radiant ovens (Skjöldebrand, 1986), by boiling (Ohlsson, 1986), by frying (Skjöldebrand, 1986), or by grilling (Bengtsson, 1986).

It should be pointed out that energy usage is not confined to food preparation in food service establishments. Energy is consumed in lighting, heating, air conditioning, and fans in both the dining and kitchen areas. In addition, refrigerators, freezers, serving cabinets, and warming ovens, all indirectly associated with food preparation, consume energy.

Designing products or ingredients to be energy efficient is an important consideration in products for the food service industry. Food service operators need to reduce their overheads; reducing energy consumption is a good way to do this.

C. Labor

Labor, its availability, its skill level, and its cost are problems for managers of any food service outlet. Preparation must be, therefore, simple.

Developers need to provide clear and explicit instructions for:

- Storage of the product and unused portions of the product
- Preparation of the product itself and preparation of any of its recipe variations (multiple uses of products or ingredients)
- Safe display of the product
- Serving the product

The product must minimize the labor used in preparation; it must be capable of being prepared in the rushed, hot, steamy, crowded, and temperamental atmosphere of the kitchen. Preparation may not always be by trained chefs or cooks but by young, minimally trained, part-time staff working at a job, not in a career as a cook.

In fast food restaurants, preparation must be simple and uncomplicated to produce products with uniform quality, and preparation time must be short. Periods of high-volume turnover demand rapidity of preparation. In addition, the outlet must maintain tight portion control. Pehanich (2003) discusses working relationships in food chains in particular.

Simplicity of preparation, versatility of usage, as well as rapidity of preparation are essential features of products for consideration by developers.

D. Waste Handling

Clean-up must be simple, and waste associated with the use or preparation of any new product must be curtailed. With storage already limited for supplies and food, storage for waste must be held to a minimum. New products or ingredients must produce as little waste as possible during preparation. Unused portions must be easily and safely stored without special storage requirements.

There is another consideration: sanitation. Any product introduced into the food service establishment must not introduce any unusual hazards or require unusual or extraordinary handling techniques with respect to hygiene, clean-up, and sanitation.

E. Customers and Consumers

So far only the needs of the customer (i.e., the outlet) have been considered. It can be seen that some of these needs and wants are very specific. A basic point about product development in the food service market must be made. It is no longer enough, if indeed it ever was, for a supplier who makes a conventional retail product to put that product into an institutional-sized container and believe it is truly serving this market. That is not product development in this unique market. Products designed for home use do not become products suited to the needs of food service managers merely by using larger-sized containers.

Catering is unique in that in addition to providing an eating occasion, with the emphasis on occasion, there is a sociability aspect attached to that provision:

> Catering systems that are centred wholly on the technical aspects and ignore the social aspects will fail.... production methodology or product formulation and presentation must recognise the social context within which the final product outcome is to be consumed (Glew, 1986).

Whether consumers "dine out" or "grab a bite," developers must recognize that these meals are being consumed in a social context. Food must please; food must entertain; food must satisfy; food must comfort. Product designers must fit their products into the social context of the food.

What does this conjure up for new product developers? Fun foods? Comfort foods? Entertainment foods? The context of food usage, the occasions on which it is consumed, must be understood as important factors in both development and marketing. An entertainment–sociability–warmth factor must be designed into these foods. Food retailers, for example, attempt to capitalize on this aspect of food with in-store bakeries wafting the aroma of freshly baked bread throughout their stores. The smell of freshly baked bread has a warmness to it and evokes memories. It is a smell that was most missed by soldiers in combat situations.

1. The Consumer: Nutrition

Today's consumer is concerned with nutrition; at least we are told that today's consumer is interested in foods containing nutraceuticals with disease-combating properties (but how to explain the growing problem of obesity?). How deeply this concern extends when the context of eating is social, for example, the candlelit dinner, is very difficult to assess. Nevertheless, customers (food service outlets) recognizing the consumers' desire for cutting down excess fat, getting plenty of good fiber in foods, and wanting minimally processed foods were driven to devise products that satisfy these needs. They came up with reduced-fat burgers and veggie burgers and provided vegetarian menu items and described menu items as "heart friendly." Yet quietly, reduced-fat burgers have disappeared from the marketplace, while veggie burgers have quietly gained wide acceptance. Is the entertainment–sociability–warmth factor (comfort foods) associated with eating occasions in consumers' minds greater than their nutritional concerns?

Many fast food chains have extended their menus to present healthful variants of standard products or to present new eating opportunities, for example, breakfast menus; soup and a variety of salads are being offered as calorically lighter fare. Extended menus require new ingredients and new food products. Where new products can serve a multitude of uses in a food service outlet to satisfy the consumer with a variety of tasty products and the customer by reducing the variety of ingredients carried, everybody wins.

Nutrition, price, taste appeal, convenience, and gratification of the gustatory senses influence acceptance of new products in food service markets. Nutrition is a factor in acceptance but not the only one, nor in my opinion, the most important one in eating out. The trend to healthy eating will remain, but this market must be allowed to develop at its own pace.

V. QUALITY IN THE FOOD SERVICE MARKET

There are two judges of quality: customers (chef/cook/owner) and consumers (diners/eaters/users). Quality is first determined by the management of the food service outlet, be it a commercial or a noncommercial outlet. If quality is to be understood as the satisfaction of needs, then product developers must first meet the needs of the kitchen staff. Here quality attributes must include price, convenience, minimal labor requirements, short preparation time, consistent quality, individually packed or portion-controlled items, and, if possible, multiple uses for the product. When these are satisfied, then perhaps, the product will be put to the consumer for final judging using the skills of the chef or cook. The ultimate judge of quality is the consumer. Criteria for assessing this judgment are the trash bin, not ordering, and loss of sales.

A. SAFETY

Concerns for safety from both hazards of public health significance and hazards of economic significance are paramount in food service. Programs (e.g., hazard analysis critical control point programs) used to resolve these concerns do not differ greatly from those discussed earlier for product development. No hazard of public health significance must be associated with new products or ingredients.

Extra consideration for the safety and stability of foods and ingredients must be given because of the special stresses that are normal in the food service industry. The limited storage area, the frenetic activity in the preparation area, and the need to display or hold product hot in serving areas present additional challenges to the quality characteristics of products or ingredients.

A further consideration regarding safety stems from the fact that in health care feeding, many patients have compromised immune systems, and special care must be paid to the microbiological safety of ingredients and products intended for their use.

VI. DEVELOPMENT OF PRODUCTS FOR THE FOOD SERVICE MARKET

Development for the food service market follows pathways similar to development for the food retail sector. This holds whether a company is attempting to introduce new products to operators of food service establishments or whether a food service company itself wishes to bring new items onto its menu (Peters, 1980; Settlemyer, 1986). Both should have first developed objectives that they wish to attain. From these there should arise strategies and then tactics to reach these objectives within the time and financial constraints management desires.

To enhance chances of being successful, product developers should speak to food service operators, discuss with them who their target market is and what their requirements for new products are, and be guided by these in new product development (Pehanich, 2003). Working closely with the customer permits the developer of new products or ingredients to better focus research to design products more suited to the physical and operational needs of customers. Unless food developers know what the ultimate preparation in the kitchen and presentation to the consumer will be, they will have difficulties designing suitable new products for this market.

To become successful suppliers to the food service industry, companies must work closely with food outlet operators or at least be very knowledgeable about the catering industry. This market is highly fragmented. Each fragment represents a marketing niche in itself. Development teams require:

- A clear, specific statement of what their new product is going to be part of, that is, dessert, main course, side dish, etc.
- Identification of which targeted consumer the product is meant for
- Information on which meal (breakfast, lunch, dinner, snack) the product is intended for
- The budgeted allowance of the product

Prices of menu items stick out like the proverbial sore thumb. On the menu, price is much more obviously comparable to the other items, soup to salad to entree. Price becomes a more important factor than in the retail food market. In a grocery purchase at a supermarket, any one item can be lost among all the others; its price does not stand out so noticeably as a proportion of the total purchase. On a restaurant menu items stand out; prices of items on a menu stand out. Prices of new products for use with menu items must fit into the price structure of a whole meal.

Companies that are serious about developing products for the food service market no longer rely on food technologists but now rely on chef-technologists. These are people who not only are skilled food scientists but are fully trained chefs. They combine those skills to be able to work knowledgeably with the food service outlet and the laboratory and the manufacturing plant through the routine of consumer research with focus groups, questionnaires, and interviews to get a clear and comprehensive reaction by the targeted consumer to the product concept. They understand the need to design meal occasions and menu items to meet the needs of the kitchen and the diner. Food service is a highly fragmented market to which developers must adapt with new products and menu modifications.

If competitive products are on the market, these chef-technologists can audit these products more competently than food technologists to provide some idea of the quality levels in the marketplace, the pricing structure, and expected consumer reaction to them. Criteria in screening are:

- Price and profitability: Do the products provide labor savings in preparation, rapidity of preparation, and serving time as well as portion control?
- Do the products answer to the constraints imposed by preparation and display equipment limitations?
- Are menu items of which the products are integral components well received by the kitchen staff and diners?
- Do the products provide additional or equivalent safety and quality characteristics to products they are components of against the stresses of that sector of the food service market for which they are intended?

Many products are prepared in a central commissary situated anywhere from several floors away in a hotel to a few miles away for health care feeding establishments to several hundreds of miles away for fast food chains. An obvious criterion of screening is whether the product provides better protection for both nutrition and safety against any abusive mishandling that distribution might impose. West (1994) reviews the similarity of good manufacturing practices for the food manufacturing and food service sectors. She emphasizes the need for designers of prepared foods and food components to know these.

Developers of food service items need to provide a clear list of instructions for the storage of the product, its preparation, whether it is capable of multiple use, preparation of all its variants, its display, the method of serving, and storage or treatment of unused portions. This requirement is less essential in retail food product development, but most food manufacturers do provide recipes, product information, and preparation advice on Web sites for good consumer relations.

At this stage of development, a consumer test can begin. This can be a small or large test of the product involving a few units in a local area to a dozen or more food service outlets spread widely. Marketing support can vary from simple table tents to TV, radio, and newspaper advertisements supported by coupons, in-store displays, and free samples.

VII. CRITERIA FOR EVALUATING A TEST MARKET

Reaction to new food service products can be found the following ways:

- Questionnaires (consumer intercepts) permit the developer to evaluate the consumers' reactions, that is, both the food preparer and the diner, to the new product and, if evaluation warrants, set about refining the product to better adapt to consumers' needs. The social context of the meal often precludes the use of consumer intercepts. The type of intercept depends largely on the product under study — is it one that only the food preparer would appreciate the presence or use of?
- Cash register data can provide information on the new product as a percent of sales. The impact of the new product on improving sales or the impact of its introduction on the cannibalization of some other menu item can be determined.
- More dramatically, analysis of the amount of new product found in the trash containers at test sites provides ample evidence of acceptance. A careful interpretation of all data will be necessary to provide information that may herald a successful introduction or prevent a commercial disaster.

There are three potential sources of waste in any food service establishment. Preparation, or kitchen, waste can broadly be classified as food purchased for kitchen use but discarded during preparation or spoiled in storage. Kitchen waste will be higher in establishments doing their own preparation work and not relying on preprepared foods. The second source of waste, service waste, is prepared food left in warmers or steam tables and not purchased or accepted by consumers. The amount of kitchen waste and service waste is largely, although not entirely, a measure of the management skills of the establishment. Finally, there is consumer waste: food purchased by the consumer but discarded. This is a measure of rejection.

How waste is to be measured and assessed presents some problems. Banks and Collison (1981) studied waste in 39 catering establishments in the U.K. and discussed the factors affecting waste, not the least of which is the size of the meal. Lack of attention by the establishment to portion control increased waste, but the amount of convenience food used by the establishment decreased it.

Kirk and Osner (1981) agree that consumer waste can be a sign of poor portion control:

> It may be thought that plate waste does not represent a financial loss to the establishment since the food has been paid for by the consumer. However, poor portion control can lead to more food being produced than is required or to a loss of potential sales.

There is another factor contributing to plate waste. People have varied attitudes toward the edibility of particular food items, for example, potato skins, the skins on other vegetables such as cucumbers or zucchini, and giblets.

Consumer research requires careful assessment. Do the data represent only the regional preferences of the test area selected for introduction, or can the data be extrapolated to wider market areas? Regional dishes that are accepted in one area of the country may not be equally well accepted in other areas. Impartial answers to these and other questions are required.

Nevertheless, analysis of consumer waste can be a useful tool in assessing consumer acceptance of menu items. Its interpretation must be used cautiously. Cash register receipts provide an indication of purchase, but the garbage bin audit can tell of the acceptance of the new product.

The introduction of any product, even one so seemingly simple as a different style of hamburger, into a fast food chain can involve several unexpected, interwoven variables that need to be assessed. One fast food chain "recently spent $1 million on thousands of taste tests to develop a better hamburger" (Anon., 1986). This chain experimented with nine buns, over three dozen sauces, three types of cuts of lettuce, two sizes of sliced tomato, and ten colors of four different boxes and some several hundred different names. It was even determined that the order of the condiments was important to consumers. All in all, this new product introduction represented a formidable task in market analysis!

11 Product Development for the Food Ingredient Industry

Progress in civilization has been accompanied by progress in cookery.

Fanny Farmer (quoted in Robbins, 1983)

I. THE ENVIRONMENT OF THE FOOD INGREDIENT DEVELOPER

A. CHARACTERISTICS OF THE FOOD INGREDIENT INDUSTRY

1. The Chain of Customers and Consumers: A Welter of Identities and Needs

A distinction between *customer* as one who buys a product to be used and ultimately consumed by a *consumer* has relevance in the ingredient industry. Customer and consumer may be one and the same, for example, the chef or cook in a restaurant purchases a product and then uses it (adds value to it; is, in effect, a consumer) to produce a new menu item that is then purchased by another customer (the diner). In the ingredient industry, the customer and consumer roles are often blurred, and the needs of each in the chain must be identified for a series of new products to be developed. The chain can be very long.

For example, in scotch whiskey manufacture, barley is sold to a malting company (the first customer and user) for soaking, malting, and drying over peat fires; the malting company then sells malt, smoked and dried to specifications, to a brewer (the second customer and user), who steeps and ferments the brew. The brewer sells the spent malt, called draff, to an ingredient manufacturing company or a feed manufacturer (either of which is the third customer and user). After suitable modifications, this spent malt may be sold to a bakery, where it is used in a baked finished product, which is sold to consumers, or it may be sold to other food and feed manufacturers. The chain is clear: each customer and consumer in this chain has needs that must be identified in order to be satisfied.

As another example, flavor extractors purchase spent seeds and skins expelled from the finishers from hot pepper sauce manufacturers for extraction of the color and heat principle for sale to confectionery manufacturers and pharmaceutical companies to be used in their products; this chaff may also be purchased for enhancing spaghetti sauces or dried and sold as a hot pepper sprinkle-on product. The lesson for ingredient suppliers is to know who their customers are, what their needs are, and how they use the product.

The general public seldom notes the ingredient developers' products. Some exceptions are diet products containing artificial sweeteners, fat substitutes, or a specific source of fiber where the ingredient's brand may be named for the particular cachet it carries. By-products are usually not identified by the original manufacturer's brand or trademark.

Some ingredient manufacturers have, in addition, a profitable retail market niche for food ingredients, for example baking powders, flavors, food colors, various types of flour (e.g., stone ground flour), and meat and vegetable hydrolysates. These markets are subject to all the pressures of the retail marketplace, and their development has been treated in the previous chapters.

2. Similarities and Dissimilarities to the Food Service Industry

New product development for the ingredient industry presents some very interesting differences from, as well as some similarities to, development in the retail food market and in the food service market. Similarities of product development in the food ingredient industry to that in the food service industry are very close, indeed startling. The similarities are threefold:

- Ingredient development tends to be reactive or crisis oriented; the demand originates primarily with consumer product manufacturers (customers) who want an ingredient for a certain function in a food. These consumer product manufacturers believe their customers want, for example, nutrified products, and they demand the unique ingredients with which to make these. In effect, ingredient developers are problem solving, reacting as suppliers to customers' needs.
- A corollary of the preceding is that product concepts for ingredients are obtained by a thorough knowledge and awareness on the part of ingredient suppliers of their customers' markets. Ingredient concepts originate with customers with whom the ingredient supplier works closely, often in a partnership arrangement. The focus group used to determine customers' needs is replaced by a consumer product manufacturer's specific requirement. Ingredient developers, then, work those requirements into a concept statement. Food service providers must follow the tastes of their customers for various cuisines (nouvelle cuisine, ethnic cuisine, fusion cuisine) that suit these changing styles.
- Ingredient suppliers serve at least two masters: the immediate customer who purchases ingredients for use in a product and the succeeding customers down the chain who use that enhanced product and rely upon its unique properties. Each customer in the chain puts in added value; each customer has different needs that have to be catered to.

There is one difference, a difference of degree perhaps, rather than a difference of substance. Development of products for the food service market is focused toward the customer or consumer or both; that is, products must contribute to the overall aesthetic characteristics of a product without being obtrusive but also must provide a service, that is, contribute convenience in the work environment of a busy kitchen at a price the customer is willing to pay.

The situation in food ingredient development is slightly different. Here, the efforts of manufacturers are still largely concentrated on developing ingredients with properties of texture, viscosity, mouth feel, spreadability, or flavor delivery to satisfy the technical demands of the product rather than the aesthetic tastes of the end consumer. That is, the ingredient supplier is concerned that the ingredient, if a color, remains stable throughout processing, that a desired texture does not break down during freezing, or that a flavor is not cooked or baked out until the consumer bites into it. Important too is that the ingredient not contribute to any undesirable changes or undue increase in product cost.

Ingredient developers are caught up in technology. For example, they have shown great skill and versatility in making fat replacements from so many different substances at such a rapid rate that the casual observer gets the impression that there is a contest to see how many different substances can be developed into fat replacements. Developers have produced products that substitute for oil and have most of the characteristics of oil except its nutritive properties.

To use an analogy, a shotgun approach has been used, and the shot used has produced a wide scatter. They have attempted to hit everything within the target zone, with the result that nothing has been satisfactorily targeted. A narrow scatter shot, to continue the analogy, might have been more effective in focusing on fewer targets but with more telling effect.

3. The Ever-Present Government

The likelihood of government intervention is high in the developmental activities of ingredient suppliers. This interest takes the form of safety testing and approvals for the many novel ingredients developed from new nonconventional food sources (e.g., phytochemicals from nonconventional plant sources), from genetically altered animals and microorganisms, and from chemically modified conventional food sources. Indeed, the whole question of calling such ingredients natural or even calling them environmentally friendly will be challenged not only by governments but by consumers as well.

Introduction of novel and imitative products faces hurdles. Righelato (1987) put this succinctly:

> Regulations exist primarily to protect the consumer, but they are necessarily concerned
> with existing products and hence serve to maintain the status quo. In doing so they
> protect the existing producer, who, in fact, probably helped frame the regulations.

In short, the introduction of novel ingredients faces challenges imposed by legislation and by the suppliers who provided opinions and hence advised the government on the formation of the legislation in the first place.

4. The Proliferation of New Ingredients

New ingredients and ingredient technology have grown at an amazing pace. A classic example of this growth is in the dairy industry; many products can be derived from milk, each with its own unique flavor or functional property that it contributes to foods in which it is used. A quarter of a century or so ago, a dairy product tree based on milk would have numbered only a handful of products:

ice cream	evaporated milks	cultured milk
market milk	cream	butter (salted and unsalted)
cheeses	dried milks	

Today milk constituents have been prepared into a wide variety of ingredients (Kirkpatrick and Fenwick, 1987):

- Products based on whole milk: pasteurized milk and powdered milk, sterilized milk and flavored UHT milk as well as milk from other species (goats and sheep)
- Products derived from compositionally altered milk: fat-reduced milk, protein-enriched milk, lactose-reduced milk, sweetened milk, reduced-mineral milk for infant foods
- Milk powder products from modified milk with heat stability properties or high dispersibility
- Products based on milk fat ranging from the common cream, butter, and anhydrous milk fat to compositionally modified milk fats with better spreadability or altered fatty acid composition to fractionated milk fat with controlled and defined melting ranges
- Products based on the proteins isolated from milk: whole protein coprecipitates, components of proteins such as casein rennet, whey protein derivatives (whey protein concentrates), lactalbumin, milk proteins combined with nonmilk proteins, and modified protein fractions
- Whole cheeses, cheese powders, sprinkle-on cheeses, reduced fat cheeses, processed cheeses, flavored cheeses, modified cheeses
- Products based on lactose where this sugar's low sweetness can be utilized and on the other end of the sweetness scale where its enzymatic conversion to glucose and galactose can produce sweeter products
- Biologically active materials that can be obtained from milk

These dairy ingredients find uses in dietetic foods (McDermott, 1987), in meat and poultry products as calcium-reduced binding and emulsifying agents (van den Hoven, 1987), in confectionery products (Campbell and Pavlasek, 1987), and in bakery products (Cocup and Sanderson, 1987).

Plant materials such as sea weeds, fruits, herbs, and spices are being similarly purified, fractionated, and blended to produce functionally important fiber ingredients, viscosity-adjusting agents, and antioxidants. Underexploited plants and underutilized fish caught as a by-product are finding uses as new ingredients (fish protein concentrates) or foods.

B. Focusing on the Customer, Who Is Also the Consumer

All product development gets its impetus from the customer or consumer or both; they have some perceived need that can be profitably satisfied. Ingredient developers challenge themselves by asking, "How can we convince potential customers that our products will improve their products?" or "How can my ingredients be modified to enhance the high-quality characteristics of my customer's product?" The possibilities for the development of close supplier–client relationships are apparent (which also contributes to the large numbers and diversity of ingredients).

The technological ability of an ingredient developer to transform a raw material, be it rice bran or a protein into ingredients by micronization to simulate the properties of fats, is unimpressive unless the use of these altered products has been fine-tuned in some desirable added-value product that the consumer product manufacturer can sell to other consumers. Ingredient suppliers must adapt the properties of their ingredients to their customers' food systems and demonstrate the value of their ingredients to their customers. Ingredients are designed to complement the quality and stability the customer has designed into their finished product.

1. Customer Research

The ingredient company's strength is the ability of their technical sales force to articulate the needs and desires of customers back to technologists in their research and development department. The technical sales force is also their primary route to customer and consumer market research.

Focus groups made up of an ingredient company's clients are clearly out of the question. These customers are actively competing with one another in the marketplace and are hardly likely to sit down together to discuss the supplier's product concepts or products. Likewise, questionnaires for use in individual, mail, or telephone surveys to gather information about customers are not likely to be successful. These are intrusions into business office routines and enquiries into what may be confidential areas of their clients. Of the various surveying methods, only the Delphi technique (see Chapter 2) of querying company executives appears to be of any general help to ingredient developers; it looks ahead too far to be useful for innovation.

Demographic and psychographic data about industrial customers are nonexistent. There are trade, business, or commodity associations such as the American Association of Meat Processors, the American Spice Trade Association, the Chocolate Manufacturers Association of the USA, the Milk Industry Foundation, and the International Ice Cream Association where general information of ingredient trends and needs can be obtained. Ingredient manufacturers have a very high profile in such organizations. These associations, with their sponsored exhibitions and conferences, are places where excellent contacts can be developed. *Prepared Foods* as well as several other national and foreign food trade magazines issue an annual index of trade associations and exhibits (e.g., Anon., 1991).

After a contact is made, the ingredient supplier's technical sales personnel will interview their potential customers' technical staff to communicate back the problems encountered by customers. Customer research is very much a hands-on business in ingredient development. That is, the ingredient supplier works one-on-one with its client. While an ingredient may serve several clients, each customer's needs are different and each customer's product is a different food matrix. Ingredients must provide industrial customers with finished products that have distinct points of difference. Ingredient users cannot always rely on off-the-shelf ingredients to obtain this desired point of distinction.

a. Partnerships

Often customers will contact ingredient suppliers, and partnerships may develop as they work together toward a common goal. Each customer of the ingredient supplier is unique. The distinctiveness of an ingredient must belong to that customer alone.

Flavor houses have developed this art of focusing on the needs of the customer to a high degree. They work closely with customers to produce any flavor sensation their customers want. They can blend from natural flavors, create flavors imitative of natural ones, or create unique flavors not found in nature. Ingredients and the technical service supporting the use of the ingredients are designed to satisfy the customer's needs.

The ingredient developer's goal is to provide a quality service with a product distinctively designed to meet the perceived needs of the added-value manufacturer. To develop ingredients, the supplier works backward with the customer, so to speak, saying, "What does this manufacturer need and how can I satisfy that need competitively?" Development is directed to these needs and their gratification. Ingredient suppliers do sell ingredients but they are just as likely to sell services.

An ingredient developer cannot create a family of ingredients, each with slightly varying properties, and then approach consumer product manufacturers with these samples saying, "Try these. One of them should work." The ingredient supplier must know the ingredient works before it is passed on to the customer. Perhaps it will not work to the full satisfaction of the customer, but it must work well enough to encourage a partnership relationship between supplier and customer. Hence the need for a close working relationship between supplier and customer through technical sales personnel.

There is an analogy with the food service industry: suppliers to the food service industry must adapt their products to conditions in the kitchen, to the skills of the kitchen's labor force, and to the style of the food outlet and its clientele. Similarly, developers of food ingredients must adapt their processes and products to satisfying the equivalent requirements of their customers.

Development of ingredients for the retail food ingredient business is similar to development for other retail food products. Standard consumer research techniques provide the necessary information that permits selected targeting of customers and consumers or the development of specific niche markets. Heavy promotion through cooking schools and cooking demonstrations in schools, church basements, carnivals, and agricultural fairs, plus recipe booklets and free samplings usually accompany retail sales. Feedback from these promotional tools provides its own consumer research information.

2. "Consumer" Research: Yes and No Possibilities

Is there an opportunity for consumer research in the food ingredient field? The answer is an ambiguous "yes and no." The use of quotation marks around the word *consumer* in the heading emphasizes that there are two consumers that the ingredient manufacturer deals with: first, there is the client customer-consumer with whom the supplier works to develop an ingredient specific for this client's needs but with whom no actual traditional consumer research is done; then there is a general population that the ingredient manufacturer can survey to see what interests there may be in low-carbohydrate food, high-energy performance snacks, high-calcium foods, and so on for which suitable ingredients can be developed.

a. The Yes Side

For the yes side: ingredient developers can make themselves aware of the consumers' activities in the marketplace. They should know about the health concerns of the consumer, for example, that low-calorie, low-fat, high-fiber, low- or no cholesterol, no salt foods are in favor, and develop suitable products for different food systems. Nutritious and diet foods, once relegated to the slow moving section of the supermarket, are now mainstream and prominently displayed. Green is in, and food manufacturers are attempting to draw attention to the green changes in their products. Natural ingredients can be prominently displayed in the list of ingredients for their image of purity and wholesomeness. The consumers' desire to self-medicate and a need for foods to combat disease has caused suppliers to rush to develop nutraceuticals (functional ingredients) for manufacturers of foods. All the foregoing provide consumer research for ingredient suppliers.

So, yes, ingredient manufacturers can research consumers, determine consumer trends, and fabricate ingredients that incorporate these desirable characteristics into their products. In a sense they leapfrog their customers, the consumer product manufacturers. This leapfrogging can be used for what Lee (1991) describes as "pro-active product development." In pro-active product development, ingredient manufacturers bring to fruition product concepts that if adopted by consumers as products in the marketplace would result in heavy usage of the manufacturer's newly developed ingredient. The ingredient makes possible products (for which, it is understood, there was an undiscovered marketplace need) that food manufacturers could neither produce previously without this ingredient, nor produce at a reasonable cost, nor produce at an acceptable quality.

Textured vegetable proteins, surimi-based products, and the mycoprotein product, Quorn, developed by Rank Hovis McDougall are typical examples of such ingredients. All find wide use as analogues in various engineered consumer products, for example, surimi (Duxbury, 1987; Brooker and Nordstrom, 1987) and Quorn (Godfrey, 1988; Best, 1989b; Bond, 1992; Wilson, 2001). A consumer desire for new texture is identified, which manufacturers can satiate with new products using the textures these new ingredients provide.

b.   The No Side

For the no side: ingredient developers find it necessary in such pro-active product development to create or capitalize on a hitherto unrealized ingredient and develop a need for it by demonstrating new and exciting innovative products; this is expensive. They must have an accurate and intimate knowledge of the general population of consumers and their needs; for example, if an ingredient is "in" for any reason (cf., Oatrim™ for fat replacement), then there is a scurry to make a variety of fat replacements from a multitude of sources of fiber. Such products require extensive development resources (surimi, an exception, had a long history of development and application and was readily adapted; Quorn is much slower to be adapted) and require the education of consumers to accept the product or to learn how to adapt it to local food traditions. Then, ingredient developers must convince food manufacturers of this opportunity by demonstrating a novel application and rely upon manufacturers to develop, market, and promote the finished products.

C. THE DEVELOPMENT PROCESS

Development of new ingredients differs from the development processes discussed previously; it is complicated (a) by the need to establish usage levels for different types of product and processes and (b) by the possible need to determine the safety of new ingredients such as concentrated extracts of non-food-source phytochemicals.

1. The Development Process and Food Legislation

Legislated standards for food products and regulations for permitted usage levels in products are more significant factors in ingredient development than they are in the development of products for either retail food outlets or the food service industry. There is currently no rationalization of food legislation among nations; consequently, manufacturers are constantly having to reformulate products if ingredients are proscribed or if permitted usage levels differ in an importing country. There are attempts at rationalizing trade through free trade agreements in several areas of the world that will eventually harmonize food legislation among all the signatory countries to these agreements.

Ingredient developers need to stay aware of food legislative activities in their governments in order to anticipate possible changes that harmonization of legislation may bring. Changes in food legislation have a devastating effect respecting the acceptability, or not, of ingredients, which foods the ingredients are permitted in, and at what levels additions are permitted; changes in trade barriers may affect the costs of ingredients and the availability of the raw materials from which they are made. Again, time and money in partnership with customers spent problem solving their product development must not be wasted by experimenting with ingredients for products that do not conform to local, national, or international laws or do not meet or respect religious customs.

2. What Criteria for Screening?

Screening in the development process includes safety of the ingredient at the recommended levels of usage, acceptability in the clients' products, and functionality, that is, does the ingredient do what the client wants at a price the client is willing to pay? Ingredient suppliers need both basic research teams and applied research teams to do this competently: the basic team to modify and study the physical and chemical properties of the ingredient modifying the base ingredient and the applied team to study specific food formulation reactions with the ingredient.

a. Financial Constraints

Financial criteria for new ingredient development have different time horizons. Return on investment can be accepted over a longer period of time, measured in years rather than in months; ingredient developers do not expect a payback in 3 to 6 months as might be expected in the retail food market. Standard ingredients are sold to many diverse clients such that profits are maintained as long as demand continues. Ingredients do not require the same level of advertising and other promotional gimmickry that consumer food products do.

> An ingredient which has been well researched and developed should be filling a market need and will sell itself to some extent, whereas it is often necessary to create a market for a new consumer product by intensive advertising (Lee, 1991).

Needless to say, ingredients must be profitable, but there is no easy way to predict or evaluate potential financial returns for a new ingredient. The financial criteria used as a screening tool for retail food product development cannot be applied to assess the financial success of ingredient development. Difficulties arise when trying to assess financial criteria in a partnership arrangement between the ingredient supplier and the consumer product manufacturer. Here there is only one customer since such arrangements often involve exclusivity agreements; the two parties must enter into a legal agreement.

A novel criterion similar to that encountered in the food service industry is introduced. Costs of ingredient development must be balanced against the customer's financial criteria. After all, the cost of the ingredient in relation to its usage level must not force users to price their consumer products out of competitive ranges. High ingredient costs may force exclusive users to seek alternatives. For example, marketing boards may set artificially high commodity prices in order to establish good prices for farmers. These prices may compel consumer product manufacturers using these commodities to find cheaper alternatives. This sharing of costs is a hazard of partnership agreements that must be worked out.

b. Foreign Market Constraints

Ingredient suppliers are able to distribute their products more widely than consumer food product manufacturers since ingredients, generally speaking, carry no ethnic, cultural, or nationalistic biases, although religious concerns do arise regarding whether ingredients are of animal origin or are kosher, pareve, or halal. Consequently, ingredient suppliers usually have large export markets since their ingredients carry no cultural or traditional taboos. Three major criteria for screening are unrelated to the specific ingredient itself. These are:

- Can ingredient suppliers adequately service foreign markets and provide the technical sales support, research and development, and marketing activity required to support these market niches?
- Do ingredient manufacturers understand the foreign market and its local food customs sufficiently well to identify potential consumer product manufacturers and to establish a rapport while recognizing the difficulties that language barriers and different business customs bring?
- Are suppliers aware of national, regional, and local food regulations that must be observed?

Obvious solutions are the use of local agents familiar with local conditions in the foreign markets or the establishment of satellite operations in the foreign country. Both solutions have shortcomings:

- The use of agents interposes one more hierarchical level between customer and supplier through which communication must be filtered. That is, ingredient suppliers must rely on others, their agents, for both reliable market information and for providing competent technical support to clients.
- Satellite operations, unless they have all the research facilities of the parent company, must send samples from the foreign customers back and forth to the head office for experimentation. Delays and inconveniences for customers result.

Both avenues represent added costs: agents want fees and satellite facilities are costly to maintain with the double teaming of technical and marketing staff that they require.

c. The Ultimate Criterion: Test Market

There is no formal test market for ingredients. Ingredient manufacturers cannot select a geographic area representative of targeted consumer product manufacturers and proceed to launch new ingredients supported by advertising and promotions.

What is more likely to happen is that ingredient manufacturers, after extensive business and customer research, will target potential high-volume users of their newly developed product. They will conduct carefully rehearsed and well-researched individual presentations with demonstrations to show each candidate the value of using their new ingredient.

New ingredient launches are usually heralded by announcements in trade magazines and technical journals or demonstrated in the carnival atmosphere of food ingredient trade fairs. A common routine for any ingredient supplier at a trade show is to hand out free samples of the ingredient or demonstrate it in a prepared food handout in which the new ingredient has been used. Admittedly, this is primarily a gesture of goodwill, but it is frequently hazardous. Products are presented under less than ideal conditions and are not examples of "best foot forward" presentations of what an ingredient can do.

d. Applying the Customer's Criteria for Screening

Potential customers evaluate ingredients on the basis of advantages that accrue from their use:

- Does its use reduce ingredient costs?
- Does its use reduce labor costs, simplify production, or increase production efficiencies?
- Does its use increase the high-quality shelf life of the product? Is there a satisfactory quality vs. cost ratio that justifies using the new ingredient?
- Are there added nutritional or safety benefits for products using the ingredient that are valued by the consumer? Will the use permit the manufacture of more environmentally friendly products or avoid the use of chemical preservatives? Will its use provide a clean label?
- Can the consumer (ultimate user) perceive the advantages that use of the ingredient provides?
- Does this ingredient permit the development of products previously impossible to manufacture?

There must be affirmative answers to some of these questions if the newly introduced product is to be successful.

D. THE FUTURE OF INGREDIENT DEVELOPMENT

New ingredients and biotechnology appear to go together. Best and O'Donnell (1992) reviewed a number of future new ingredients, many of which are of a biotechnological origin:

- Bacteriocins of microbial origin
- Transgenically altered milk production to produce natural preservatives, pharmaceuticals, and altered milk-fat fatty acid profile or to remove a milk allergen producing off-flavors in UHT milk
- Genetic manipulation of poultry to improve texture and water-binding properties of the meat

The authors advise that the promise of the new ingredients be tempered with several cautions that are just as apt today as they were in 1992. Their first caution, noted previously, is the certain intervention of government in the safety of and permission to use genetically (biotechnologically) derived ingredients, especially plants with enhanced phytochemical concentrations. Certainly, there is disagreement among nations regarding the need to label transgenic organisms in foods.

The second caution concerns the economic impact such novel techniques may have upon certain commodity industries, for example, the dairy industry. Already, this particular industry faces two conflicting government interventionist policies. On the other hand, in many countries including the U.S., the industry is encouraged to produce milk with a high milk-fat content (yet children and teenagers are the prime consumers of milk and milk products), but government health policies exhort consumers to eat less fat. Other industries also have concerns about the new technologies.

Finally, the desire for new ingredients with new functional properties and health or disease prevention properties will put enormous pressures on ingredient manufacturers to produce, separate, and purify the products to the degree that the consumer products industry will want. This will be expensive for ingredient suppliers.

The desire of consumers for natural products and for products manufactured using natural ingredients will pressure ingredient suppliers to explore the use of ingredients derived from natural sources and, concomitantly, the use of novel, minimally processed (i.e., modified) derivatives from these natural ingredients. Manufacturers of consumer products will want to present to customers and consumers a label with an ingredient list that appears less chemically and synthetically derived and more natural appearing.

Natural ingredients will be "in," but whether this will mean transgenically modified or biotechnologically derived ingredients will also be considered natural and therefore "in" by customers and consumers is a moot point. With the consumers' fears of big science so prevalent, a massive educational program will be required to allay these anxieties. Ingredient manufacturers must use caution.

This intense interest in naturalness in colors, preservatives, antioxidants, indeed anything natural that can replace chemicals, is attested to by the abundance of research papers and reviews of literature in these areas: natural colors (Engel, 1979; Francis, 1981; Gabriel, 1989; Shi et al., 1992a,b), antimicrobial agents (Shelef et al., 1980; Zaika and Kissinger, 1981; Baxter et al., 1983; Zaika et al., 1983; Beuchat and Golden, 1989; Daeschel, 1989; Barnby-Smith, 1992), and antioxidants (Kläui, 1973; Pokorný, 1991).

Ingredient suppliers, and that includes chemical and pharmaceutical companies, have established intensive research programs for new sources of ingredients from plants that may be useful as natural ingredients and to identify the active components. The need for natural ingredients with functional properties such as preservation, thickening, emulsifying, coloring, flavoring, taste modifying, etc., which also might have nutritional and pharmaceutical properties, will drive future new ingredient development.

A major caution to the use of all new ingredients in general, and ingredients of biotechnological origins in particular, will be their acceptance by customers and consumers and their legalization by governments. A major requirement for that acceptance is that the use of ingredients in a food serve a purpose; that is, they attract and please the customer and consumer. It is not enough that the use of an ingredient merely satisfy a consumer product manufacturer's need; it must have the means of creating something desirable in the product.

E. MEETING THE CHALLENGE: NEW INGREDIENTS

In this text, no clear distinction will be made between an ingredient and an additive (for a discussion of additives, see Fuller, 2003). Readers should be cautioned, however, that in the legislation of many countries such a distinction is made. At present, there is no uniformity from country to country respecting the classification of substances as being either an ingredient or an additive, as a natural substance or nature identical, or as organic or nonorganic. In this text, anything deliberately added to a food will be considered, rather simplistically, as an ingredient.

Regulations respecting the use of ingredients vary from country to country: acceptance in one country does not guarantee that the ingredient will be accepted in another. Therefore, mention of ingredients in this text does not imply they are permitted but merely illustrates their potential applications as ingredients. Even when the ingredient is permitted, there are restrictions regarding usage levels and types of foods in which the ingredient may be used. If food products have standards of identity — in the U.S. more than 300 foods do — the addition or substitution of ingredients may be restricted.

1. Marketing's Impact on the Direction of Research and Development

Marketing personnel have an influence on what kinds of ingredients are developed; they have promotional plans based on whatever market research they can do, and this determines the direction of research. Their requirements must be discussed early in development with the food technologists. If marketing personnel want to pursue a program of development of ingredients with a health promotional benefit, for example, lite or low calorie, low or no fat, high fiber, or with anticancer properties, then the research teams of the ingredient supplier have a direction. Food technologists must use sources of ingredients that provide the promotional characteristic that marketing personnel want. Where there is a client–supplier relationship, obviously the direction is dictated by the client's needs. Nevertheless, marketing's research either directly or indirectly points the direction for development.

If export markets are eventual goals for marketing personnel, they must consider that labeling laws for nutritional and health claims are different in many countries (e.g., between Canada and the U.S.). Consumers' interest in green labels (their desire for natural ingredients) and marketing personnel's obvious desire to please consumers will limit the ingredients that technologists can choose in formulations. Development teams need to consider early and carefully all the implications that the use of any ingredient may have for labeling, for promotional claims, for export markets, and for environmentally concerned consumers.

a. Fat Ingredients

The nutritional guidelines of many countries recommend that their citizens reduce the amount of calories derived from fat. Thus, today's consumers are spurred on to the desirability for reduced-calorie foods with their government's endorsement, with their own awareness that obesity is related to many disease conditions, and with the strong suggestion reported in the media that high-fat diets and breast cancer may be directly associated. Weight for weight, fat contributes more than twice as many calories as protein or carbohydrate. The easiest route to lower calories, then, would be to eliminate or lower the fat content of food. The only techniques are:

- Reduction or removal of the fat by trimming, skimming, pressing, rendering, or solvent extraction
- Reduction or replacement of the fat in the food by substituting, wholly or in part, some less calorically dense material (fat extender). Fat extenders still provide calories, but weight for weight they provide fewer calories than the fat they replace
- Reduction or replacement of the fat with some nonabsorbable substance (fat mimetic or replacement) with fat-like properties

The first solution is impractical. The resultant food product would be greatly altered in flavor, palatability, and appearance; fat provides many desirable organoleptic properties to food. The second two solutions require a substance, a fat-extender or mimetic, that provides all the sensory properties of fat and is nontoxic. It is in this area that developers of new food ingredients have concentrated their efforts.

Fat replacements or mimetics have been derived from a number of products. They can be roughly grouped into three classifications:

- Replacements based on physically or chemically modifying a polysaccharide fraction such as rice bran, pea fiber, oat fiber, corn starch, or tapioca dextrin.
- Replacements based on altering the physical character of a natural protein. Simplesse™, a trademark of The NutraSweet Company, was originally made from milk solids, then from milk and egg solids (Bertin, 1991). It is a suspension of protein microparticles that has the mouth feel characteristics of an oil.
- Replacements synthesized from natural foods such as sucrose or other sugars and esterified with fatty acids; these have fat-like properties. Silicones and long-chain members of alkane hydrocarbons with characteristics resembling fats have been used as fat substitutes (LaBarge, 1988) and can be included in this category.

Synthetic or engineered fats (Singhal et al., 1991) are somewhat different alternatives; they are derived from modifications to the basic skeleton of a fat molecule itself. This is done by:

- Replacement of the glycerol backbone with other polyhydroxy compounds; sucrose, raffinose, and amylose have been esterified with natural fatty acids. Propylene oxide can serve as the backbone. Sucrose polyester (Toma et al., 1988), one such fat replacer, is nonabsorbable and also able to reduce the cholesterol level in the body. Other replacements provide fewer calories per gram than do fats and qualify more as extenders,
- Substitution of acids, such as branched carboxylic acids (sterically hindered fatty acids), dicarboxylic acids (fumaric, succinic), or shorter-chained fatty acids for some or all of the long-chain fatty acids of the triglyceride structure. Medium-chain fatty acid triglycerides are specialty fats prepared by the hydrolysis of vegetable oils, the subsequent fraction-

ation of the fatty acids, and reesterification of these fatty acids to a glycerol backbone (Megremis, 1991).

- Substituting a polycarboxylic acid for the glycerol backbone and then esterifying suitable long-chain alcohols to the acid. (This has been referred to as reversing the ester linkage.) Common acids used for the backbone are citric acid or tricarballylic acid, the tricarboxy acid of glycerol. LaBarge (1988) discusses these and feeding trials using them.
- Reducing the ester linkages of the triglyceride to ether linkages and thus changing the properties of the fat. LaBarge (1988) describes these ingredients as much more slowly susceptible to hydrolysis than the esters but more prone to oxidation.

(Food product developers are cautioned that mention of these ingredients here should not be interpreted as meaning that all, or any, countries have sanctioned the use of these products. Developers must verify that the ingredients they use are permitted in the class of foods they want to use them in and in the amounts they wish to use.)

Medium-chain triglycerides have limited value as fat replacements where only caloric reduction is the major requirement. However, they also serve as processing aids by acting as flavor carriers and are used in confections, where their low viscosity can prevent sticking and provide gloss (Megremis, 1991). Medium-chain triglycerides are absorbed and are metabolized in the liver as rapidly as is glucose (LaBarge, 1988). Their biggest value is their usefulness in special diets to provide a rapid and concentrated source of energy to people with intestinal malabsorption problems (Crohn's disease, colitis), a potential niche market (Babayan and Rosenau, 1991; Kennedy, 1991). Babayan and Rosenau (1991) also describe the use of a medium-chain triglyceride oil (chiefly caprylic and capric acids for the triglycerides) in cheddar- and fontinella-type cheeses.

Some long-chain fatty acid triglycerides are needed to meet the essential fatty acid requirements of the body. Therefore structured lipids (interesterified lipids formed from medium-chain triglycerides and long-chain triglycerides) provide unique benefits. Kennedy (1991) summarizes these, in part, as:

- Improving immune function
- Lessening cancer risk
- Lowering cholesterol

Jojoba oil, a natural oil from the seeds of a hardy bush native to the southwestern U.S. (NAS, 1975), has also been explored as a possible substitute for fat (Hamm, 1984). Its oil is a liquid wax comprised of esters of fatty acids and alcohols, and it solidifies at refrigerator temperatures, which may limit its use in some foods. It has been used in such alternate medicines as aromatherapy and as a cosmetic. It is also believed to be suitable as a diesel fuel.

Gums also serve as aids in fat-reduced foods by stabilizing emulsions and providing viscosity, thus simulating some properties of oils and permitting the reduction of oil in the formulation. Dziezak (1991) describes the properties of gums and reviews their applications in emulsification.

There are problems with the use of fat replacements and extenders in foods. Product developers should carefully assess the properties of each before using any in products. Some, like the alkoxy citrates, are not thermally stable (Singhal et al., 1991); some may reduce the absorption of other macro- and micronutrients (LaBarge, 1988); and some may result in anal leakage when consumed (LaBarge, 1988). The major problems presented to food technologists are threefold: to determine which substitute will be stable in the particular food system, which will be permitted in this system, and which will be most suitable in the system in which it is to replace fat.

Dziezak (1989) provides a very readable capsule review of some of the properties of natural fats and oils for technologists unfamiliar with fat chemistry. Her accompanying treatment on fat substitutes is now dated. However, as she points out, this area of ingredient development is undergoing tremendous change.

Product developers rapidly reformulated products containing saturated tropical oils to remove these ingredients from products. This occurred when the prevailing wisdom declared that these saturated vegetable fats behaved as saturated animal fats in the body. Berger (1989) describes concisely the use of tropical oils in the U.S., presents a cogent argument for their continued use, and disputes attacks against their bad nutritional value.

b. Sugars, Sweeteners, and Other Carbohydrate Ingredients

Sherman (1916), in his classic *Food Products*, describes only the manufacture of cane sugar, beet sugar, molasses, refiner's syrup, maple syrup, open kettle cane syrup, and honey. Wiley (1917) included sorghum syrup. In the past, these were the sugars in common use; sucrose from sugar cane or sugar beets was, by far, the main sugar in food processing.

There are many reasons for the popularity of sucrose in manufacturing. It was readily available and, therefore, comparatively cheap. It served many functions in foods in addition merely to sweetening them. Along with the advantages, the use of cane sugar also brought disadvantages. Sucrose is cariogenic. Consumers have become more concerned with health problems associated with high caloric intake from the refined sugars typically found in North American diets. This awareness has led to the search for substitutes.

Alternative sweeteners to reduce or eliminate the need for sucrose in food products fall into two categories:

- Those that are intense sweeteners in their own right but contribute no or very few calories to the food. These can totally replace a sugar *where sweetness is the only function demanded of the sugar.*
- Sweeteners that are caloric alternatives, that is, on a weight basis they are more intense sweeteners than sucrose. They also contribute other properties, including calories, associated with sugars. These are frequently referred to as the bulk sweeteners (Giese, 1993).

A partial listing of the low-calorie group includes:

cyclamates	saccharin
sucralose	aspartame and other peptides
glycyrrhizin	miraculin
monellin	thaumatins
stevioside	rebaudiosides
neohesperidin	dihydrochalcones
acesulfame K	L-sugars

Caution must be taken in choosing intense sweeteners since some alternative sweeteners are unstable in certain food systems; some produce a slow onset of sweet sensation that peaks late in the mouth. Others have bitter or undesirable aftertastes. The sweetener chosen must complement the food system where it will be employed.

The alternative group, by far the more important for developers, comprises crystalline fructose, high fructose corn syrup, mannitol, sorbitol, xylitol, hydrogenated sugars (e.g., maltitol), hydrogenated starch hydrolysates, and isomalt. Members of this group possess, in addition to sweetening, many of the functional properties of sugar:

- They provide mouth feel (viscosity) in some beverages and control texture in solid foods such as baked goods.
- They are used to regulate a food's water activity and hence its stability. They provide humectant properties in jams and jellies.
- They serve as bulking agents in hard candies and other confections and can be used to control graining in soft-centered candies; they assist along with pectins in the control of gel texture for jams, jellies, and marmalades.
- They stabilize color in some food systems.

The uniqueness of sugars, sucrose in particular, is best exemplified by a simple demonstration described by Chinachoti (1993). In what she called her "magic trick," she mixed corn starch and water to produce a wet, unpourable powder. Next she added another dry powder, ground crystalline sucrose. Mixing produced not a drier mixture but eventually a flowable slurry. The corn starch–water mixture, despite its dry appearance, has a high water activity (approximately 0.98). With the addition of the sucrose, the dry appearance disappears after mixing. The seeming solid becomes a liquid with a water activity that has dropped to 0.93. As magical as this may appear, Chinachoti went a step further. One would expect the corn starch–water mixture to spoil faster microbiologically than the corn starch–water–sucrose mixture if one's prediction is based solely on the two water activities. Chinachoti demonstrated that this was not so. The mixture with the lower water activity spoiled more rapidly. The addition of sucrose had indeed lowered the water activity, but she demonstrated it had also increased the amount of mobile water. As Chinachoti explained, "Sucrose undergoes a phase change from crystalline to dissolved sucrose upon hydration.... solvation of sucrose promotes mobility and helps facilitate the spoilage."

The lesson for product developers and spoilage modelers is this: sucrose and other crystalloids lower water activity, but stability and predictions of stability should not be based on water activity. Other factors such as water mobility and phase transitions may play a more important role than water activity alone in product stability.

The importance of phase transitions and the physical state of food components, with some references to carbohydrates, are treated in depth by Slade and Levine (1991), Roos and Karel (1991a,b), Noel and coworkers (1990), and Best (1992). Noel et al. (1990) and MacDonald and Lanier (1991) discuss the importance of phase transitions and the glassy state in the storage of frozen foods and freeze dehydration and as a cryoprotectant for foods. If the aqueous system of a food can be formulated to maintain the matrix in the glassy state, then water crystallization and damage caused by crystal growth is minimized. The drying of pasta products is greatly improved by close control of humidity, temperature, and drying rate to control glass transition; this maintains pasta in the glassy state (Noel et al., 1990).

The quality of extrusion-puffed snack products owes much to the control of phase transitions. Roos and Karel (1991a,b) suggest that glass transition temperatures may be a factor in browning. Grittiness of lactose in ice cream, caking in milk powders, and stickiness of hard candies are all related to the behavior of ingredients in the glassy state (Noel et al., 1990).

c. Fiber Ingredients

Dietary fiber is an all-encompassing term for a group of poorly defined, natural plant components that includes the unavailable, polymeric carbohydrates, cellulose, hemi-celluloses, pectins, gums, and a noncarbohydrate component, lignin. None of these are themselves compounds of fixed composition; hence, fiber often has other plant materials with it. Historically, dietary fiber has been referred to as roughage or bran, regardless of the source. This lack of uniformity is a disadvantage that is overcome by purification, "chopping" of the polymers to desired lengths, and blending to produce fibers of specific, uniform properties. Cellulose is a linear polymer of varying lengths of glucose units. Hemicelluloses are a complex mixture of linear and branched polysaccharides with side chains composed (most often) of xylose, galactose, and other hexose and pentose sugars. Pectins are equally complex mixtures of polygalacturonic acid units, as are many of the plant gums and mucilages. The seed gum of mesquite is a polymeric galactomannan (Figueiredo, 1990). Lignin is a polymer of units of phenyl-propane.

Dietary fibers come from many different readily available plant sources, often as by-products of processing the raw plant: prunes; corn; oranges (Hannigan, 1982); soy beans; peas; oats; rice; spent barley; sugar beets; tofu processing; various vegetable gums such as gum arabic, guar gum, locust bean gum, and mesquite (Figueiredo, 1990); wood; potato peelings; carrageenan; and psyllium grain husks (Anon., 1990a). Chitin from the shells of crabs and from microbial sources and chitosan prepared by the deacetylation of chitin are biopolymers with promise as nonplant sources of fiber (Knorr, 1991).

Each source produces fibers with unique properties, and when used in combinations, many interesting applications in food systems are possible. A particular fiber can frequently be used synergistically with other fibers or with starches. Arum root, the source of konnyaku, a vegetable jelly for many centuries a mainstay of Japanese cuisine, is the source of konjac flour, a glucomannan (Downer, 1986). Tye (1991) reviews the synergism of konjac flour with kappa carrageenan and starches in maintaining the structural integrity of shaped foods during thermal processing and to simulate the texture of fat and connective tissue in sausage-like products.

The use of any particular source of fiber poses technical considerations in addition to any health benefits to be pondered when choosing a fiber or combination of fibers to serve some functional purpose in a food. Fiber will affect color, flavor, oil and water retention (hence the water activity of foods), rheological properties such as colloidal emulsion stability, texture, gel forming properties, and other viscometric properties such as thickening and mouth feel of products (Penny, 1992). They are used to control crystal formation in sugars and to act as cryoprotectants in freeze–thaw food systems. For example, Ang and Miller (1991) list some of the many products in which one fiber, cellulose powder, is used:

bread	rusks
cakes	cookies
pasta	cheese
soups and sauces	yellow fat spreads
comminuted meats	meat and fish pastes
slimming foods	dietetic products

Cellulose powder has also been used in stabilizing frozen surimi analogues. The many applications of fiber in foods are detailed in Ang and Miller (1991) and Penny (1992).

Heightened awareness of the importance of fiber in the diet dates back to the mid-1980s when the National Cancer Institute (U.S.) promoted the potential cancer prevention benefits of a high-fiber diet (NCI, 1984). This campaign alone gave an estimated 30% boost in sales to one well-known ready-to-eat bran breakfast cereal (Anon., 1988c). However, the beneficial use of high-fiber diets in aiding bowel regularity has been promoted for many years, and its laxative effects have been recognized for many more years. For developers wishing to obtain background information on fiber as an ingredient in diet, the following articles are pertinent: Rusoff (1984) and Schneeman (1987) on physiological responses of ingested soluble vs. insoluble fibers, Wood (1991) on the physicochemical properties of oat beta-glucan, and Ripsin and Keenan (1992) on oat products and their value in reducing blood cholesterol levels.

Claims that certain sources of fiber, such as oat, rice (Normand et al., 1987), pectin (Reiser, 1987), and guar, can clear cholesterol from the circulatory system suggest there is a role for fiber in lowering the risk of cardiovascular disease. Furthermore, fiber may play an as-yet-undetermined role in the prevention of some cancers, particularly colorectal cancer. However, Fuchs et al. (1999) found no evidence of any preventive effect of dietary fiber against adenomas or colorectal cancer in several thousand women between 34 and 59 years of age over a 16-year period.

Nevertheless, demand for high-fiber products has grown as food companies try to satisfy this increasing awareness by consumers that food, and in particular certain constituents in food, and health go together. Fiber is now advertised in a wide variety of foods from breads, muffins, and cookies to pastas, beverages, and processed meat products.

The function that fiber is to serve in a food system determines which fiber, with which desirable characteristics, to select. The two most important characteristics are particle size and shape for texture considerations and soluble-to-insoluble fiber content ratio. (See Olson and coworkers, 1987, for a discussion on analytical procedures for soluble dietary fiber and information on total dietary fiber and insoluble dietary fiber.) Soluble fiber has excellent water-binding properties and as such has a pronounced influence on the texture (moistness) of baked goods, on the stability of products, and on the viscosity of beverages. In general, soluble fiber is highest in vegetable gums and lowest in cellulosic fractions of plant materials. From the natural sources available, high to low ratios of soluble to insoluble fiber may be obtained to suit whatever purposes developers have in mind.

Crude fiber should not be equated with dietary fiber; crude fiber consists largely of cellulose and lignin remaining after the treatment of a food material with sulfuric acid, sodium hydroxide, water, alcohol, and ether.

d. Proteins as Ingredients

Proteins as ingredients have suffered from a lack of attention, as most interest has centered on fats because of their high caloric density and carbohydrates because of their contribution to sweetness and other functional properties in foods. Substitutes for both fats and sugars lead to low-calorie products.

Vegetable proteins, textured to simulate the fibrous structure of meat, have been available for several years. They were popular when meat prices soared in the U.S. during the late 1970s and textured vegetable proteins were used to extend meat products. When beef became more readily available, many of these products failed, partly because they did not meet consumers' expectations sufficiently to justify a long acceptance period, and consumer demand proved to be transitory. Quality has improved considerably, and textured proteins are now widely used in many products, especially dried soups and vegetarian dishes.

How strong is the growth in the vegetarian food sector is a moot point; suffice it to say it is there in the marketplace and is gaining prominence. Textured vegetable proteins have had a reasonable success with main dish items for vegetarians. A mycoprotein developed by Rank Hovis McDougall has been successfully texturized and used by the Sainsbury company (J. Sainsbury plc, U.K.) in a series of meat-flavored products. In test market locations these have been well received (Godfrey, 1988). These products have a dual advantage over meat, which marketing personnel can promote: they contain no animal fat and they contain fiber. The mycoprotein has been tested by Sainsbury for vegetarian (i.e., non-look-alike meat) dishes.

Surimi, a textured fish protein, is freshwater-leached fish muscle. It is hardly a new development — indeed it is a very old technology. Lee (1984) has described its processing in detail and its application in some products. Lanier (1986) discusses the functional properties of surimi, particularly its excellent gelling properties.

Surimi has been used successfully with crab meat, shrimp meat, and smoked salmon in fabricating substitutes for, respectively, crab legs, butterfly shrimp, and lox. Sausages made with surimi, very popular in Japan, have not been successfully received in North America.

Soy beans have become a popular ingredient with a wide range of applications in ethnic cuisine. Their popularity is supported not only by the taste sensation of umami that they contribute in their various forms but also by their value for the cancer preventive properties it is claimed that they have. Soy milk has been used with various flavors as a noncarbonated drink. Tofu has been used as an ingredient in many vegetarian dishes. Japanese miso and Chinese chiang are both fermented soy bean curd products used as flavoring. Tempeh, mold-fermented soy beans of Indonesian origin, has been used as a meat substitute. These products enable developers to create a wide range of added-value products, especially ethnic dishes, with unique textures and forms.

II. INGREDIENTS AND THE NEW NUTRITION

A. Opportunities Provided by the New Nutrition

The nutritional sciences have changed dramatically, and this change has opened up many opportunities for ingredient developers. Initially, the concern had been with diseases caused by the lack of certain nutrients, and great efforts had been made to optimize nutrient intake (fortifying foods and calculating required daily allowances) to prevent deficiency diseases. Then there came a move to remove things from foods, for example in low-fat, low-salt, low-sugar, low-calorie foods. Today the major thrust of ingredient development is to develop new ingredients from food and from the components they contain: to prevent specific health problems, to improve effects of stressful situations, to improve physical prowess, even to improve or recover failing cognitive and memory abilities. The nutritional sciences have become more concerned with how diet and specific food components operate at the physiological and genetic levels to influence health and disease (Young, 1996).

1. Biologically Active Nonnutrients

Health professionals are still learning (and unlearning) that some foods contain biologically active nonnutrients that have as yet a poorly understood connection at a physiological and genetic level to mechanisms of disease prevention. This understanding is growing through an accumulation of anecdotal and solidly grounded scientific evidence. The published data, whether from information on medicinal lore found in herbal books or research data in technical literature, has provoked food manufacturers to develop new foods exploiting this disease prevention connection.

a. Functional Foods aka Nutraceuticals

Throughout this text these nonnutrients referred to as nutraceuticals (*pharmafoods* and *medifoods* are terms devised by the media that have also been used). It has been suggested that the term *functional food* be used only for foods containing prebiotics or probiotics that provide a health benefit; the International Food Information Council defines functional foods as "foods that provide health benefits beyond basic nutrition" (Deis, 2003). *Nutraceuticals* is the popular term found in newspapers and food trade magazines and is used extensively in *Food Technology*; the term *nutraceutical* has been used by many scientists; it is descriptive of the value that a nutraceutical is purported to have in nutrition without the need of quantifiers or qualifiers. More on the definition of functional foods and problems with procedural rules for food producers can be had in Culhane (1999), Eagle (1999), and Deis (2003). This text uses the most descriptive term, *nutraceuticals*; functional foods serve a function in foods as colorants, sweeteners, emulsifiers, enzymes, stabilizers, preservatives, and so on.

b.   Prebiotics and Probiotics

Prebiotics and *probiotics* are terms used to characterize nutraceuticals. Both are nonnutritive elements that provide health benefits to consumers of these products, but here the similarity ends. Prebiotics are an uncharacterizable grouping of chemical entities found in a wide variety of foods; those of plant origin are often called phytochemicals, but the reader should note that some prefer that this term be applied to nonnutrients present in plants in very small amounts, thereby excluding soluble and insoluble fiber.

Probiotics (Table 11.1) are active microbial cultures or foods containing these, for example, yogurts and krauts that act on the intestinal microflora in a beneficial manner (Salminen et al., 1999). Microorganisms suspected of involvement are described by Hoover (1993), Lee and Salminen (1995), and Knorr (1998).

Prebiotics run the gamut of chemical structures that can be guessed at from the nomenclature in Table 11.1. The potential of plant parts with curative properties has been known for centuries. Books on biologically active plants range from ancient texts to current self-medication books:

- John Gerard, *Historie of Plants* (1597) [*Gerard's Herbal*, edited by Woodward, M., Senate, Twickenham, U.K., 1998].
- N. Culpeper, *Culpeper's Complete Herbal, and English Physician*, J. Gleave and Son, Deansgate, Manchester, 1826 [reproduced from an original edition 1981, copyright of Harvey Sales].
- A. Boxer and P. Back, *The Herb Book*, Octopus Books Ltd., London, 1983.
- V. E. Tyler, *The Honest Herbal: A Sensible Guide to the Use of Herbs and Related Remedies*, Pharmaceutical Products Press, Haworth Press, Inc., New York, 3rd edition, 1993.
- L. Bremness, *Herbs*, Stoddart Publishing Company, Toronto, 1996.
- And many articles on biologically active nonnutrients: Caragay (1992); Ramarathnam and Osawa (1996); Kardinaal et al. (1997); Garcia, D. J. (1998); Katz (1998); Ohshima (1998); Zind (1998; 1999); Anon. (1998); Sanders (1999); a series of columns in *Food Technology* by L. M. Ohr, particularly Ohr (2003) on fats; and Clifford (2003) on dietary phenols. Even the much abused caffeine has a protective effect against irradiation, at least for mice (George et al., 1999).

TABLE 11.1
Some Biologically Active Nonnutrient Factors Determined to Have or Believed to Have Beneficial Effects against Disease Conditions When Consumed

Classification	Category and Food Sources
Probiotics	Bifidobacteria • Fermented milks, yogurt *Lactobacillus* species • Fermented milks (acidophilus milk), yogurts *Streptococcus* species • Fermented milks, yogurts
Prebiotics (phytochemicals)	Fatty acids • α-Linoleic acid (canola and flaxseed oils) • Conjugated linoleic acid (safflower, sunflower, and soybean oils) • γ-Linoleic acid (evening primrose oil) • ω-3 Fatty acids (various unsaturated oils; e.g., salmon and tuna oils) Lecithins • Phospholipids (various oils, especially soybean oil), phosphatidyl serine, phosphatidyl choline Unsaponifiables of oils • Phytosterols (canola and soybean oils) • γ-Oryzanol and ferulic acid (rice bran oil) Organosulfur compounds, especially plants of the Cruciferous family (broccoli, cabbage, and cauliflower) and Allium (garlic, onion, and leek) family • Isothiocyanates (mustard oils of Cruciferous vegetables) • Sulfides (e.g., diallyl disulfide) and oxides (allicin) from garlic and onions Terpenes • Monoterpenes (limonene, perillyl alcohol) • Tetraterpenes (lycopene, β-carotene) Polyphenols (including flavonoids and catechins) • Anthocyanins (blueberries, cranberries, tomatoes, red wine, tea, onions, kale) • Various other phenolic compounds Nonsteroidal phytoestrogens (isoflavones) • Genistein, daidzein, biochanin, and formononetin (soy bean, whole grains, berries, flaxseed, licorice, red clover) Fiber (and associated material) • β-Glucans (oats, barley, wheat, rice) • Lignans (flax) • Other soluble and insoluble fibers (fructooligosaccharides, e.g., inulin in Jerusalem artichoke) Saponins (derivatives of pentacyclic triterpenes) • Ginseng, soybeans, grains Herbal components • See, for example, Tyler, 1993; Anon., 1998d

Adapted from Fuller, G.W., *Food, Consumers, and the Food Industry: Catastrophe or Opportunity?*, CRC Press, Boca Raton, FL 2001. With permission.

Archeological and biblical records show that plant materials have long been used both to flavor foods and to cure various ailments. Each culture has its store of traditional medicines based on extracts or brews of different parts of plants. The herbal medicinal practices of native cultures, folk literature and oral traditions describing the use of plant materials, old herbals, and ancient medical records have been treasure troves for pharmaceutical, chemical, and ingredient companies for phytochemicals to use as the basis for new product ideas (Wrick, 2003).

The "self-care shopper," as Hollingsworth (2003) calls these customers, has become an obvious target for consumer products containing nutraceutical ingredients. This target is one that should be aimed at with caution, not only for safety concerns but also for presenting customers and consumers with a confusing brand image: food or medicine?

c. Prebiotics and Phytochemicals

Prebiotic and *phytochemical* are terms that are often kept distinct by some scientists; they differ in the observation that phytochemicals are found in smaller amounts in plants than are prebiotics (insoluble and soluble fibers, gums, and mucilages), and phytochemicals act differently in the body. In this text no distinction will be made.

Prebiotics (Table 11.1) include the plant-derived vitamins and a variety of phytochemicals from flavors, colorants, antioxidants, noncaloric sweeteners, narcotics, texturizing agents (pectins, starches, and other polysaccharides), and so on — many of which have beneficial physiological effects in the human body (Zind, 1998). Food scientists and food manufacturers are interested in those with desirable physiological effects such as:

- Antioxidants: vitamins C and E have been used as such in foods; β-carotene, long used as a food colorant, also has antioxidant properties (Astorg, 1997; Ahmad, 1996; Elliott, 1999). Foods high in antioxidants are prunes, raisins, blueberries, kale, strawberries, and blackberries.
- γ-Oryzanol and tocotrienols: plant sterols found in rice bran oil (McCaskill and Zhang, 1999); tocotrienols are also good antioxidants.
- Lycopene and carotene: The carotenoids are found in a variety of plant sources, in particular tomatoes (Astorg, 1997).
- Isoflavones: These are a class of nonsteroidal phytoestrogens; genistein, daidzein, and glycitein are found in soy-based foods (tofu, soy milk, and miso) (Zhou and Lee, 1998); genistein, daidzein, biochanin, and formononetin have been extracted from red clover. All have beneficial physiological properties in human beings. Protein extracts containing these phytoestrogens have been prepared from these sources.
- ω-3 fatty acids and the related ω-6 type found in the fats and oils of both plant and animal sources (Ahmad, 1996). Examples are palmitoleic, oleic, linoleic, and arachidonic acids.

Drug, food ingredient, and consumer product manufacturing companies are very interested in these products: see Block (1985) and Petesch and Sumiyoshi (1999) on garlic and onions; Kinsella et al. (1993) on antioxidants in wines and plant foods;

Gehm et al. (1997) on resveratrol in grapes and wine; Katz (1999) on European trends in nutraceutical products. Potato and corn chips, corn puffs, candies, and other snack foods have been laced with ginkgo biloba, St. John's wort, kava kava, and many other herbs, spices, and essences of flowers in the hopes of relieving tiredness, calming mood swings, improving memory, and combating depression.

As is ever the situation, contradictory evidence is usually found. Fuchs et al. (1999) could not find evidence of any protective effect of dietary fiber, a prebiotic, against colorectal cancers or adenomas in a study of 88,757 women between 34 and 59 years of age over a 16-year period. The safety of kava kava (in Canada the government has recalled all products with kava kava) and St. John's wort have both been questioned, and even has the efficacy of tea and ginkgo biloba. So even with the nutraceuticals there is confusion about their benefits in the diet. Developers are warned to proceed with due safety and legislative caution.

d. Prebiotics: Red Flags

The warning cannot be overemphasized: product developers in the consumer products and ingredient industries should use caution in their development programs with nutraceuticals:

- The safety of all prebiotics, especially as isolated pure compounds, has not been established, particularly at the doses shown to be efficacious *in vitro*.
- There is still no reliable data on effective dose levels for all consumers, especially the very young and the elderly. Clifford (2003) presents challenging dietary data for polyphenols.
- Reliable information is lacking on interaction of prebiotics with other prescribed medication that consumers may be taking.

At present, research is needed for developing methodologies for accurately detecting and quantifying the presence of prebiotics in foods. Two questions that need to be answered are where and in what form do the prebiotics go in the body to produce the preventive effects they have? Proper dosage levels and mechanisms of action in the body must be understood. Once this is accomplished, then isolated fractions containing the active components can be used as medicines, their presence in foods can be enhanced through conventional or unconventional breeding techniques, for example, carrots with more carotene (Zind, 1998), and concentrates of isolated prebiotics can be used as ingredients in consumer products.

With the availability of nutraceutically enriched foods and concentrated extracts of phytochemicals, self-medicating consumers run the risk of overmedicating to harmful levels, an occurrence that may provoke the introduction of strict legislation.

e. Probiotics in Health

Both Fuller (1994) and Knorr (1998) reviewed the history of live microorganisms used by many cultures in foods, either for the preservative action they confer or their ability to convert foods into something more desirable. Only fairly recently has it been learned that there is a beneficial health effect that these added microorganisms provide for individuals. Probiotics are added adventitiously with the fermented foods as the vehicle or they are added directly to foods as live or dead cultures.

Typical food sources of probiotics are yogurt and cultured milk products (Speck et al., 1993), some fermented meat and cheese products, sauerkraut and other fermented vegetables, and selected kimchi, which is considered to have both probiotic and prebiotic elements (Ryu and West, 2000). A major problem for developers of consumer products is not so much how to create new products but how to popularize the enormous number of traditional fermented vegetable, meat, and dairy products available in other cultures, that is, to adapt these to new products suitable for a wider variety of cultural traditions, particularly North American traditions.

The extensive product development occurring in Japan since the 1950s is discussed by Ishibashi and Shimamura (1993). Problems in the design of novel products for both healthy and sick consumers are discussed by Brassart and Schiffrin (1997).

A growing popular interest in alternative medicines leads to their use by consumers and to research by consumer product manufacturers into herbal preparations that are used in folk medicines or leads to the use of extracts, tinctures, and tisanes prepared from herbs with purported benefits. This will spur the development of both nonprescription medicines, food supplements, and ingredients as well as new foods.

There are now opportunities to develop food products directed to specific health problems (anticarcinogenic foods, foods to improve cognitive ability) or for use in particular stressful situations (space flight, athletics). There are already, for example, sports drinks designed to rehydrate the body rapidly, to provide energy, or to enhance energy metabolism during strenuous activity (Brouns and Kovacs, 1997).

Prepared Foods (Anon., 1990b) reported that a significant percentage of research and development executives "expected their companies to increase efforts to reformulate ingredients with 'additive' label connotations out of existing product lines." This because the so-called unnatural ingredient list on many products has become a consumer concern, and executives wish to promote their products with as green an image as possible, that is, as natural or made from natural ingredients.

2. Other Ingredients: Some with and Some without Nutritive Properties

a. Antioxidants

Historically, food preparers skilled in the arts of cooking and food preparation have known that certain herbs, spices, and foods could extend the shelf life of foods and make them taste better. Later, as the science of food preservation grew, scientists found that some food components that occur naturally have antimicrobial and antioxidant properties. Thus these natural additives in foods gave rise to the hope that they could be used to give foods an extended shelf life. They can, but as might be expected, the very strong flavoring that herbs and spices contribute to a food limit their use.

Kläui (1973) provides the following list of herbs, spices, and other food materials with antioxidant properties:

oregano	rosemary
coffee beans	wine
nutmeg	mace
tea leaves	onions
sage	turmeric
citrus fruits	buckwheat
allspice	cloves
tomatoes	carrots
marjoram	thyme
beer malt	heated gelatin and casein
alfalfa	oats
rice bran oil	rice germ oil
pea flour	

However, the efficacies of these foods or their extracts as antioxidants depend on the fat system in which a particular material was tested. This is similar to the findings of Davies et al. (1979). They use:

cocoa butter	cocoa powder	cocoa shell
oatmeal	rosemary	defatted cocoa powder
sage		

or extracts of them in different fat systems. All have antioxidant activity, but again Davies et al. conclude that "the antioxidant activity of a particular material and its extract appears to depend upon the nature of the oil or fat used for testing and the type of solvent used for extraction of the antioxidant components." Extraction might not have extracted all of the active antioxidants.

Pokorný (1991) reviews several sources of natural antioxidants: tocopherols from sesame oil; phospholipids from olive oil; phenolic compounds from cereals; flavonoids from herbs, spices, and algae; carotenoids; polysubstituted organic acids; proteins, peptides, and amino acids; and Maillard products. He suggests in conclusion, however, that the best approach for protecting foods against oxidation is to avoid antioxidants altogether by removing oxygen from the food, avoiding or eliminating the oxygen-sensitive substrate, or decreasing the oxidation rate by low temperature, low light storage.

In an undated report, the Specialty Food Division of Ingredient Technology Corporation (Woodbridge, New Jersey) presented evidence of natural antioxidants in molasses and attempted to characterize them.

In many instances the source of the antioxidant activity has been identified and incorporated into commercial antioxidants. The active component in cloves is eugenol. Unfortunately eugenol's strong clove odor limits its use to baking products. Its isomer, iso-eugenol, has a more acceptable woody and spicy odor (Heath, 1978) that is also found naturally in nutmeg, basil, and other oils, according to Kläui (1973). Gordon (1989) has also reviewed the natural antioxidants and attempted a brief outline of some of the mechanisms involved that can assist developers' understanding of their use.

A wide variety of natural substances have antioxidant activity, but not all can be used because of other characteristics that they may bring to the food.

b. Antimicrobial Agents

Some spices and herbs, as well as other natural materials, have been known through traditional food uses to have antimicrobial activity against a wide spectrum of microorganisms. Some microorganisms are surprisingly sensitive to these natural components, as Beuchat (1976) found with *Vibrio parahaemolyticus* and oregano. Shelef et al. (1980) screened over 40 microorganisms associated with foods for sensitivity to sage, rosemary, and allspice; allspice proved the least effective as an antimicrobial agent, while both rosemary and sage demonstrated antimicrobial activity. Generally the spices were more effective against Gram positive microorganisms. They ascribe the antimicrobial action to cyclic terpenes in the spices.

Zaika and Kissinger (1981) demonstrated both inhibitory and stimulatory factors (which they were able to separate) in the behavior of oregano against *Lactobacillus plantarum* and *Pediococcus cerevisiae* alone and in mixed culture. At low concentrations of oregano, there was stimulation for both microorganisms. As the concentration rose, an inhibition of both acid production and viability was observed.

Zaika and coworkers (1983) extended their study to include oregano, rosemary, sage, and thyme against lactic acid bacteria. Again, microbial growth and acid production were retarded as concentrations of these herbs were increased. They did observe one possible hazard for developers. Microbial resistance to a particular spice could be induced. When microbial resistance to a spice occurred, that microorganism in which resistance was induced was also resistant to the other herbs. This suggests that the mechanism that induces resistance in a microorganism is a broad ranging one. Antimicrobials in foods have been reviewed by Beuchat and Golden (1989) and the phytoalexins by Mertens and Knorr (1992).

The components in these spices and herbs that explain the antimicrobal activity are substituted phenols. Oregano and thyme both contain oils high in thymol and carvacrol; sage and rosemary both contain cineole in large concentrations; allspice is high in eugenol (Heath, 1978). Sage, rosemary, allspice, oregano, thyme, and nutmeg seem to be the herbs and spices of greatest applicability as antimicrobial agents for foods.

The growing popularity of ethnic foods, which are often highly spiced, combined with the trend toward low-salt foods and vegetarian foods, both of which benefit from the addition of flavorful ingredients, are making highly flavored foods more acceptable. The high flavor impact from spices and herbs along with other flavorants such as vinegar, lemon, onions, shallots, and garlic can mask the lack of salt and the blandness of many products. Developers can take advantage of the preservative action of more flavorful natural ingredients to prolong the acceptable shelf life of products.

i.   Bacteriocins

Bacteriocins are nitrogen-containing substances with potent antimicrobial activity; as such they are emerging as a new class of antimicrobial substances for possible use as natural food preservatives.

Fermentation, particularly lactic and acetic fermentation, has been well established as a stabilizing process for foods, as has the addition of acids such as lactic or acetic acids. The lactobacilli, in addition to producing lactic acid with a resultant pH change, produce several substances that have antimicrobial activity against various microorganisms; they produce hydrogen peroxide, which is toxic for some organisms, as well as diacetyl, which has antimicrobial activity. Daeschel (1989) reviewed the antimicrobial substances from lactobacilli (nisin is one such substance) and proprionibacteria. Some show promise against many foodborne pathogens including *Listeria monocytogenes.*

Another review (Stiles and Hastings, 1991) published just 2 years later demonstrates the amount of research taking place in this field to discover new preservatives. Stiles and Hastings classify the bacteriocins as follows:

- Those produced by *Lactococcus* spp.
- Those produced by *Lactobacillus* spp.
- Those produced by *Carnobacterium* spp.
- Those produced by *Leuconostoc* spp.
- Those produced by *Pediococcus* spp.

As Stiles and Hastings point out, bacteriocins have little in common except their ability to inhibit microorganisms and the fact that they are all proteins. They vary widely in everything else: the spectrum of microorganisms they are effective against, their molecular weight, how they perform, as well as their biochemical properties (see, e.g., Juven and coworkers, 1992; Barnby-Smith, 1992).

c. Colorants

The gradual removal of many synthetic colorants from permitted lists because of questions about their safety is known to all product developers. Equally well known is the importance of color to the appeal of foods. The hunt by suppliers for replacements for the labile natural colorants spurred much research to find those that remain stable under the rigorous conditions frequently found in food manufacture, for example, in the manufacture of hard-boiled sweets.

Engel (1979) examined factors (heat, light, oxidation, presence of metal ions, reactions with other food components, pH effects, etc.) that influence the stability of three classes of natural colorants:

- Anthocyanins and anthocyanidins: Anthocyanins are ionic and their color is greatly influenced by the pH of food. They are not stable to heat, oxygen, or in the presence of sugar and sugar browning products. Sulfur dioxide bleaches them.
- Carotenoids: There are over 300 carotenoids. Many are fat soluble only, but some, bixin (from annatto, which is used in coloring cheese and baked goods), astaxanthin, and crocin (from crocus blossoms), are water soluble. Their highly unsaturated, conjugated double bond structure provides a range of red to yellow hues; blending increases this range of hues. Their unsaturated nature, however, makes them susceptible to oxidation and light sensitivity.

- Betalains are ionic, water-soluble quaternary ammonium compounds and as such are subject to changes in color with changes in pH. They are temperature, oxidation, and light sensitive.

Francis (1981) also reviews the anthocyanins, betacyanins (betalains), and the yellow-orange pigments (carotenoids and turmeric, which contains the colorant curcumin). Some interesting miscellaneous pigments, Francis notes, are cochineal and laccainic acid, which are of insect origin; a red colorant produced by the mold *Monascus purpureus,* which has a long history of use in the Orient; and chlorophyll, used in some specialty pasta products.

Gabriel (1989) and Shi and coworkers (1992a) worked on the separation and identification of anthocyanins from sweet potatoes. Shi and coworkers (1992b) studied the stability of anthocyanins in model food systems.

A major problem with natural colorants, in addition to their instability in many food systems, is the question of their safety; this is far from established for all of them. In their natural, *in situ* state, safety concerns are perhaps negligible. When they are extracted from their natural source and concentrated, an assumption of safety is not justified. Another factor against the use of natural colorants is their cost; many natural colorants are scarce, and their extraction and concentration is expensive. As Francis (1981) points out, 75,000 handpicked crocus blossoms are required to obtain a pound of crocin.

Francis (1992) reviewed a new group of anthocyanins substituted on the B-ring. Acylated B-ring substituted anthocyanins show promise as a new class of stable colorants because they have greater pH stability and produce brighter colors. They are, however, all from nonfood sources and would require extensive safety testing.

3. A Cautionary Note

Ingredients used in the formulation of food products exist in a delicate balance. Each contributes some characteristic to the product. Great care must be taken in understanding what contributions each makes in the product, either singly or in combinations where synergism of actions may play a role. Altering, removing, or substituting any one component for another upsets the balance in the formulation. There is a real risk that substitutions with other, perhaps cheaper, ingredients may occur later. These can have disastrous results on the safety, stability, or quality of a product.

> I well remember the terrible taste in fried peppers when our supplier substituted a so-called new and improved antioxidant into our fractionated peanut frying oil without informing us. The change had been meant as an improvement.

B. CHALLENGES FOR THE NEW NUTRITION

This new medical nutrition, that is, one directed to disease prevention, opens up a Pandora's box of problems. Food products directed to specific health problems become possible, but questions regarding their safety in concentrated form, dosage levels for effects, standards of identity, and advertising ethics also have arisen. The questions that arise from these new opportunities will continue to be a concern for both food manufacturers and for legislators.

1. Problems Presented by Enriched Foods

A great impetus was given to the food microcosm to capitalize on the public's quest for a long and healthy life free from debilitating diseases by developing products enhanced by or fortified with nutraceuticals. The scientific community and food manufacturers are interested in identifying these nonnutritive, disease-combating food entities and in developing products that contain them for the money-making potential they represent.

Manufacturers will want to inform both the customer and the consumer of the benefits of their fortified products. Yet labeling and advertising regulations must protect the customer and the consumer from the hyperbole and misrepresentation of health claims that may accompany the label statements and advertising claims. The Institute of Food Technologists' Web site at one time in the first half of 2000 displayed these announcements: "add more than value to your food and beverages … add life, with Polyphenols" to describe products put out by Templar Food Products; and "herbally-active, protein enriched frozen juice bars" contained chromium, manganese, 100% of the daily requirements of vitamins A, C, and E, as well as protein. These products were from Cold Fusion Foods.

With the wide availability of nutraceutically, phytochemically, or herbally enhanced products, there also came a warning from health professionals of the dangers some nutraceuticals might pose for people on prescribed medications and young children eating enriched candies and snacks.

Manufacturers want to inform the public of the benefits of a new product in a market niche where, metaphorically, shifting sands prevent the establishment of a firm product identity. Are these enriched products medicines or snacks? Is the company a pharmaceutical company or a consumer food product manufacturer? Management must decide its direction.

a. Delivery Systems

Delivery systems of these nonnutritive substances have taken many forms, and the ingredient companies have had to be prepared with neutraceutical ingredients (extracts, concentrates of phytochemicals) that suit these food systems. The most used vehicles for delivery appear to be candy, snack food, and beverage products. Zind (1998) reported that one candy manufacturer was adding phytochemicals to the candies it manufactured, which were purported to prevent cancer, bolster the immune system, and reduce cholesterol. Fiber and calcium compounds have been added to fruit juice beverages. Iced tea preparations have been fortified with polyphenols.

Potato chips, corn chips, corn puffs, and other snack foods have had added to them herbals and plant extracts claimed to have beneficial properties (Abu-nasr, 1998); some examples are ginseng (a long life promoter), St. John's wort (a depression preventative), ginkgo biloba (a memory improver), and kava kava (an aid to relaxation). Each of the claims above should be preceded by "said to be" or "said to have."

Baked goods have also served as vehicles, with ingredient suppliers developing muffin mixes with flax seed and blueberries. Some breads are fortified with ω-3 fatty acids such as docosahexenoic acid. Ice cream has also become a vehicle for supplementation. It has been flavored with phytochemicals from green tea (a source of catechins), ginger, avocado, sesame, wasabi, and capsaicins (hot pepper principle) to become nutraceutically fortified as well as flavored.

Old products have been given a new role by making them nutraceuticals with added calcium to enhance calcium intake, added caffeine to combat drowsiness, creatinine as an aid for body builders, added fiber to assist in lowering cholesterol, and so on. Soft drinks with calcium, protein, vitamins, or concentrated herbal extracts (or all of these) have moved from a refreshment role to a functional one as a nutrified beverage.

C. SUMMARY

Phytochemicals are biologically active materials. As such, pharmaceutical companies have pursued them and used them in the preparation of medicines or used the base molecule to prepare more efficacious medicines. They are being extracted from their natural sources and concentrated by suppliers and sold as extracts to the general public and to consumer product manufacturers for addition to prepared foods. Their safety in prepared foods at higher concentrations than are naturally found in a source food may not be established for all segments of the populations. Indeed, it is not apparent that the responsible active component or complex of components that an extraction produces and that is used to fortify a food is the correct one. Their presence in a multitude of common nonsource foods, that is, foods in which they are not normally found but have been added as an enhancement, may be equally as harmful to some populations who eat a lot of the enhanced foods. For example, snack foods targeted for children or teenagers who are voracious snack food consumers may not be appropriate vehicles for supplementation with some phytochemicals.

Kardinaal et al. (1997), Chung et al. (1998), and Hasler (1998) all have expressed concerns for the safety of consumer foods enhanced with phytochemicals and proclaimed a need for further research. These concerns are worth repeating here for developers:

- How these phytochemicals work in the body to prevent disease is not clearly understood. Do concentrates in foods have any adverse side effects? What are the nature of these side effects?
- Does the nature of the diet or the presence of other nutrients in the diet influence the activity of added phytochemicals? Do phytochemicals in enhanced concentrations influence the absorption or metabolism of other nutrients in the diet?
- There is no toxicity data on many of these substances and very little verifiable medical evidence of their effectiveness. There is a large body of anecdotal information, evidence based on epidemiological studies (e.g., groups eating large quantities of soy products have lesser incidences of certain cancers), and observations (again unverified) derived from tradi-

tional medicine. The tannins, for example, display hepatotoxic activity, antinutritional activity by forming complexes with digestive enzymes and proteins, and from epidemiological studies have demonstrated high levels of mouth, throat, and oesophageal cancers in people using betel nuts, drinking the herb tea *mate*, or in people whose dietary staple is sorghum (Chung et al., 1998). Chung et al. also report on antimutagenic, anticarcinogenic, antimicrobial, and antiviral beneficial effects (e.g., of tea drinking). Information on what acceptable dose levels might be for all segments of the population is lacking.

- Physiological interactions of phytochemicals with prescribed medications that consumers might be taking at the same time are a concern. Do they enhance or depress the activity of these medications?

Hasler, in particular, sees the need to balance benefits and risks of the use of foods containing physiologically active substances.

For herbs and herbal preparations, a different cautionary note must be sounded. Their long history in medical folklore as cures for many maladies has gained for them a veneer of respectability such that they are readily available in both health food stores and mainstream supermarkets where they are sold as preparations for teas or tisanes. Tyler (1993) and others have suggested several dangers in the sale of herbs and herbal preparations that ingredient suppliers (and customers) should be aware of:

- Are the raw materials what they are claimed to be? Not all parts of herbs are effective: roots, leaves, seeds, stems, flowers, bark, etc. may contain no or high concentrations of the phytochemical. Ingredient suppliers need to be certain of their sources of raw materials.
- Herbal preparations are not standardized with respect to the active ingredient(s) that they contain. Do they contain the active phytochemical in the concentration claimed, and are these safe concentrations?

Ingredient suppliers must realize they are, or may be, using unstandardized materials to prepare their ingredients for food products.

Tyler (1993) does not give all the herbal preparations a clean bill of health. Ginseng has a very low level of risk even with excessive use. St. John's wort also is relatively safe but may cause photosensitivity in some people at high dose levels. Ginkgo biloba extracts do promote vasodilation with improved blood flow, but very large doses can have unpleasant side effects and may exacerbate the effect of aspirin. Already warnings have appeared from scientists and the medical profession about the overuse of products containing nutraceuticals, especially when these are used in combination with or in place of conventional drug therapies.

The issues about the use of nutraceuticals in new products have been emphasized rather than the uses of these valuable substances in new products. Many have thundered onto the market only to be slowly having their efficacy or their safety questioned not by radicals but by competent medical professionals. Even as recently as December 2003 it was reported in *Journal of the American Medical Association* and elsewhere that echinacea proved of no value in reducing the number of colds in children up to the age of 11 vs. a control group not given the echinacea. Kava kava, gingko biloba, St. John's wort, green tea, fiber, and even ginseng have all received some negative publicity within the past 2 years suggesting they are not as efficacious as some of the hype had promised. Caution is the byword for developers in this new and promising field.

12 What I Have Learned So Far

Life is the art of drawing sufficient conclusions from insufficient premises.

Samuel Butler

I. LOOKING FORWARD AND BACKWARD

A. BEING SURE OF THE CONCEPT

This chapter's title is adapted from Jacob Cohen's (1990b) well-known article. His paper about the application, interpretation, and philosophy of statistics is very much worth reading, especially if one is not a statistician or is familiar with statistics only by plugging data into some exotic statistical computer software program that then whirrs, giggles, and spews out a decision on whether the results are significant or not. Many industrial food technologists, marketing people, and financial people are classed in the latter group, the pluggers-in of data. They often do not understand what assumptions the very competent software designers used in establishing their programs or what limitations there are to the programs; they just know data in, results out. Ergo, conclusion!

The path to new product development is being flooded with software programs that do much the same thing as these statistical software programs; they model the development process, even the test market. Data is plugged in and decisions come out often with the pluggers-in not understanding the machinations of the programs they use; they do not understand what assumptions the software programmers used in devising the program. They realize only that somehow the laborious calculations were made much easier and decisions regarding significance or irrelevance were made for them.

By no means do I decry the use of these new product development programs, whether they be for statistics or modeling. They serve very useful purposes and provide great assistance in seeing subtle nuances in the data being digested; they reduce the amount of work necessary to reach conclusions.

I do decry the blind use of these without some understanding of their pitfalls and limitations. I decry the unthinking acceptance of the conclusions these reach without questioning and examining these conclusions and assuring oneself they are backed up with some sound logical reasoning. The programs and the algorithms upon which statistical, decision-making, and modeling software are based must be understood for what many of them are: models, pretend playthings to be used to assist careful thought, not to replace it. Too often, machine- and software-generated results are accepted as Gospel without a challenge to their validity or to the assumptions that were made (by the originators of the software) in reaching these decisions.

The makers of the programs do not know the user's precise situation: the results and conclusions reached using them might not be pertinent to the user's real world situation. There must therefore be an evaluation of whether the results are valid for the users situation.

When the blind lead the blind, they will both fall into the water.

Chinese proverb

1. The Value of the Earlier Literature

Despite the age of Cohen's article — it is more than 10 years old, as are many of the references in this book — it still presents very sound thinking that extends beyond the statistics in it. I mention the article's age because recently a prominent, well-published academic expounded to me at an Institute of Food Technologists' annual meeting and conference how he refused to allow his graduate students to cite any papers more than 5 years old. Such papers were in error because of newer technology according to him!

Technology, that is, the bells and whistles of the scientists, may change, but sound thinking stands on its own merits — *plus ça change, plus c'est la même chose*. I would have had no literature citations with such an attitude toward the older literature when I worked on capsaicin for my master's thesis. The only literature meeting the 5-year limit was a study done by a master's degree student from our department and a follow-up research problem worked on by a fourth-year student who frankly admitted to me that he couldn't repeat the master's student's work! *Beilstein* became my bible for information.

This disparagement of the earlier literature cannot be an uncommon attitude. In a search for material on another subject, I asked a librarian in a technical library (his official position was designated as information technology officer) for some books. He informed me that the books I wanted were 10 years old or more and hence they were in the archives. I asked permission to access these references but was told such material was kept only a short time and then destroyed! There has probably been a great loss of much valuable earlier work, and many research workers have had to "reinvent the wheel" by not looking backward in the literature.

Some clarity is required on sorting out what the early food literature is:

- The early journals contain scientific articles, reviews, and letters. In my own search for early articles, I have found that many journals have changed their names, often several times, and that the publishers or societies that originated the journals do not themselves have copies of their earlier publications. Indeed, one society asked me for copies of the articles I had so that they could build up their collection.

- Cookbooks often provide capsule social histories recording the changes in the popularity of particular foods, food customs, availability and kinds of food ingredients, the cooking appliances, and how preparation and cooking was carried out. A sociologist has recently been awarded a substantial grant to study cookbooks. When these are read against the backdrop of the political history of the time and the history of science and technology, one understands how people used food and how scientific developments and controversies as well as political and social upheavals influenced their daily lives. The pamphlets and brochures supplied by ingredient and appliance manufacturers provide added insights to the food scene. Parallels to today's circumstances lead astute developers into exciting ideas.
- Textbooks on food processing and nutritional sciences, dietary guides, popular commentaries (often written by those now considered quacks but who at the time were revered authorities) and such are like frozen capsules of the state of collective food knowledge.

a.   Displaying Some Humility

The value to be found in the older scientific literature should not be overlooked. First, the early literature is an excellent teaching tool; one sees how scientists worked and planned without all the trappings of the modern laboratory; they were not black box scientists. One can learn how ideas and technologies developed. We should be humble enough to realize that the modern equipment in today's laboratory will be deemed historic pieces of junk 50 years from now, if indeed they last that long.

Second, we often laugh at the beliefs, customs, and understanding of former times, but we should realize that similar fallacies (cf. how diets and nutritional science theories wax and wane) are rampant today, and future generations will laugh at us for our naiveté.

Cookbooks and nutrition texts show how previous generations applied and coped with the discoveries and theories of nutrition then being advanced and how they tried to adapt these. Later developments have shown that sometimes these earlier theories were wrong. It teaches one humility to read the early literature and see how wrong scientific experts can be. One advantage I have found in my own collection of old cookbooks is the assistance they have provided for new product development in finding recipes for old dishes and cookies with the cachet of "just like grandma's."

Third, the literature is much more cogently presented by people with some knowledge of, and skill in, writing — they were, perhaps, never exposed to technical or report writing classes.

Finally, many of the earlier scientists and researchers were scientists and researchers because of an aptitude for and a love of science. Today many choose science because we are a technology-driven society; there is a demand for scientists and hence the promise of good jobs. In anticipation of future prospects, many students finish their education with a graduate degree in business administration — betraying that they do not plan to remain scientists. Their quest for MBAs always reminds me of a remark attributed to Napoleon Bonaparte:

Every French soldier carries a marshal's baton in his knapsack.

The MBA has become a substitute for the marshal's baton.

Among today's mass-produced scientists there are many gifted individuals. Unfortunately, there are many times more individuals of mediocre or even poor abilities whose ambition is not to make any contribution to science but to climb the administrative ladder of success and, with their indifferently learned science and their MBAs, ultimately direct scientists who are more capable than they are.

2. What Customers and Consumers Want or What Retailers Want?

Is what customers and consumers want different from what retailers want? Or do customers, consumers, and retailers all want the same thing? And very simply what is that thing?

Customers want to be able to obtain food items they desire at a price and a quality they can afford and with the knowledge that these food items are safe. Consumers want to have the food items they had asked their customers (the gatekeepers) to get for them. Retailers (or sellers, to use a broader term) want to attract customers and consumers, make shopping (or dining) a joyful experience to encourage more shopping (or return dining experiences) and make a profit on the food items they sell. No right-minded business person can argue with that.

Retailers want to sell products that are most in demand and that are most profitable per unit of shelf exposure. Does this mean most customers' needs have been met? These most profitable products are given prominent placement by the retailer — and incidentally their positioning is often paid for by the manufacturers. Thus they become those products most prominently available and therefore situated to the advantage of the retailer.

To answer the questions presented at the beginning of this section, I suggest there has been a subtle, but a very real, discontinuity between customer, consumer, and retailer needs. They are not necessarily the same. (We are not discussing the food service arena, where the customer and consumer have choice and dominate the restaurant owner.)

Market research on consumer buying has shown that customers and consumers can be broken down into a myriad of market niches. The retailer is now thinking in terms of profits, that is, stocking those items that sell, and not necessarily in terms of satisfying all the customers and consumers identified in the many market niches. Let it be very clear that I do not suggest that the seller stock items that do not sell or sell only infrequently. I only suggest that data describing which products sell and therefore that typify customer buying habits may bear a strong bias for retailers and dominate, or certainly greatly influence, their business operations.

3. The Changing Scene in the Food Microcosm

I have suggested elsewhere (Fuller, 2001) that four major elements exist within the food microcosm: the primary producer and gatherer, the manufacturer, the retailer

TABLE 12.1

Shifts in the Power and Influence that Drive the Food Microcosm

Past Times	Modern Times	New Times
No. 1	No. 1	No. 1
Primary producer	Customer and consumer	Retailer or seller
No. 2	No. 2	No. 2 (tied)
Manufacturer	Retailer or seller	Manufacturer
No. 3	No. 3	No. 2 (tied)
Retailer or seller	Manufacturer	Customer and consumer
No. 4	No. 4	No. 3
Customer and consumer	Primary producer and gatherer	Biotechnology company (seed, feed, and animal genetics companies)
		No. 4
		Primary producer and gatherer

or seller, and the customer and consumer. All else in the food microcosm are derivative of these. All have shifted their positions of dominance with time and therefore have influenced the driving direction of research and innovation. This fundamental change in power structure and influence is, I suggest, a contributing factor bedeviling the interpretation of much market research.

My points are best understood through an examination of Table 12.1. This table is divided arbitrarily into three time periods: past or historical times, near past and modern times, and future times. No dates have been committed to these times; in some geographic areas, these power shifts are not as advanced as in others.

a. Past Times

In historical times the primary producer or gatherer dominated or drove developmental momentum in the food microcosm. The farmer was often all things: manufacturer doing slaughtering and butchering; manufacturer of sausages; and miller, brewer, and baker. Fishermen dried and salted their catches. Olive growers pressed their olives for oil; grapes and dates were dried or fermented by the producer. These primary producers often sold their produce and finished goods themselves in their farm markets. It is true some specialization occurred. Even today many small towns have throwbacks to the tradition of market days when farmers brought their products to central marketplaces, but nowadays these are largely organic farmers, boutique food crafts people, or antique dealers.

With time, specialized crafts developed. Brewing and wine making matured with specialists appearing; millers and bakers along with confectioners emerged as craftspeople. These craftspeople were the sellers, selling from their own establishments. Retailing had not developed into the science it was to become.

The customer and consumer still had roles that duplicated those of the food manufacturer. They often baked their own bread and brewed their own beers. They "did down," that is, they preserved many of their own fruits and vegetables in season for over the winter months. They did not have wide choices of available fruits and vegetables, and most were locally available produce; nor were customers and consumers researched to any great degree to determine what they wanted. No efforts were made to make shopping a pleasure or a convenience.

b.   Recent Times and the Present

During the last years of the second millennium, there was a major shift of power and influence in the food microcosm. The customer and consumer dominated (Table 12.1) as the retailing and selling sectors and the manufacturing sector tried to understand customer and consumer habits and capitalize on this understanding. It was the infancy of consumerism.

Food manufacturers responded to demands of customers and consumers; primary producers and gatherers plugged along as best they could, responding to the demands of the manufacturers. Primary producers were often also at the mercy of retailers; for example, a major retailer demanded of a producer with whom they had contracted that they wanted X number of chickens weighing between 4 and 6 lb for sale at $Y/lb for a promotion, and advertising monies were expected from the producer. Manufacturers, too, contracted with producers for raw material of the size, shape, maturity, and condition that producers had to meet, and the manufacturers' field personnel inspected, graded, and set prices on the grade. Food manufacturers who were large and powerful enough demanded and got produce on their terms. Sellers (notably fast food chains) who were big enough often had leverage to demand both quality and price from manufacturers, or the manufacturers ran the risk of losing contracts.

Primary producers, it must be remembered, are also at the mercy of seed, feed, and chemical manufacturers to whom they often owe large sums of money that cannot be paid until harvest.

c.   New or Future Times

I see a new relationship emerging (Table 12.1) that puts the retailer and seller in the most influential and dominating position within the food microcosm. Their desire to attain a more commanding position not only by making each running meter of their store shelves not only profitable but by maximizing profitability this tends to lead them to avoid unproven products; new products *per se* are not in their picture unless their introduction means more *guaranteed* profits.

Retailers control sales information; they see what combinations of purchases are made and when these are made. Through information technology programs, this sector obtains the clearest picture of the customer and consumer. By controlling information and its communication, this sector in effect controls a major portion of the food microcosm. Their information is actual; it is real. It is not based on mall surveys or on focus groups composed of "maybe" customers or consumers. It is based on real customers and consumers in real time. They control:

- Where and what products are displayed in the stores to maximize sales
- The number of facings that products receive to provide best exposure
- In-store advertising, promotions, and pricing

They are gatekeepers for the introduction of new products (indeed for all products) onto store shelves, and they charge consumer product manufacturers for the privilege of introducing a product. The retailing sector dictates to manufacturers through JIT (just in time) delivery systems and ECR (efficient consumer response) when and what they want delivered and what products they want developed.

I regard this sales data, part of the market research for product development, as tainted information. Developers should regard such marketplace data as requiring very careful study and interpretation alongside their own data. It is the seller who literally and figuratively stacks this data in favor of what is profitable for the seller. Is such data biased? It reflects what items sold most, brought in the best returns for the store per unit of shelf space (indirectly what the customer wanted or was willing to pay for based on availability and display), and is biased by how the seller has positioned these goods and used the promotional material.

Retailers decide what products are displayed for sale. These decisions are based on what their sales data tell them, and heaven help the customer (or consumer) who does not fit what their data tells them. This does have its amusing aspect.

In Quebec, liquor is sold only in government-run outlets. As a confirmed single malt scotch drinker, I have repeatedly spoken to my local outlet about their poor selection of malt scotches and so-so selection of blended scotches. My local manager told me that I lived in, and his store was situated in, a wine, gin, and vodka district and that to get a better selection of malt scotches I had to drive to a more distant store; this is hardly a convenience. My problem was solved by shopping on the Internet directly from the government warehouse; delivery is free and I can specify when I want delivery.

Store managers do preferentially stock their shelves with products that suit the demands of the major customer group in their district and hence provide products that are most profitable to the store. They ignore other market niches. The situation is excellently and cynically put in the cartoon described Chapter 4 and restated here:

Here at MegaFoodCorp, you're more than just a customer … You're a completely predictable compilation of spending habits and product data.

i.   The Tied No. 2 Position

The food manufacturing and customer and consumer sectors are tied for the second position of influence and dominance. The manufacturing sector is beset by pressures from the retailing sector and both give pressure to and get pressure from the suppliers, below them in the chain of influence. I have received so many mixed reviews and analyses regarding the value of programs such as JIT and ECR in the course of my food work that I cannot judge dispassionately; the systems work for some and have even been described as an aid in new product development. Distributors and members of JIT and ECR associations are generally in favor; most manufacturers and their traffic managers hate them or have serious reservations about them. As a factory-bred food technologist, my opinion is biased: I suggest that it is naïve to think ECR works in anyone's advantage except the retailers. *If* it assists the customer and consumer it is only incidental to the interests of the retailer being satisfied. Retailers exert pressure on manufacturers.

ii.   Efficient Consumer Response

ECR is a more advanced version of the QR (quick response) tools describing close manufacturer and retailer partnerships.

> Efficient Consumer Response incorporates the concept of customer-specific marketing into Quick Response (Morris, 1993).

Morris cites the Food Marketing Institute as defining ECR as "a responsive, consumer-driven system in which distributors and suppliers work together as business allies to maximize consumer satisfaction and minimize cost."

It has been suggested that ECR has improved the success rate for new product introductions by providing an early weeding out of losers. Perhaps this is so, but I suggest that the weeding process strongly favors directly meeting the needs of the retailer, and only indirectly meeting the needs of the customer.

iii.   The Bottom of The Pile: No. 4 Position

I would classify, temporarily at least, the biotechnology companies and many primary producers together, especially the very large primary producers, in this bottom position. Chemical companies (including some pharmaceutical companies) have moved into the more exciting areas of biotechnology and genomics either by forming new companies, buying companies already active in these new fields, or merging through partnerships with other companies. These entities in turn have purchased many seed companies through which to develop new varieties of plant species through genetic research, produce them, and sell them. In effect, these companies have strong influence, if not control, over the primary producer; they supply seed, feed, and chemicals (e.g., herbicides, fertilizers) to primary producers, who are often indebted to these companies until harvest time when they can pay back their loans. In time, this supplying of primary producers will likely include genetically modified animals, poultry, and marine species.

When and if that time comes, this entity, the complex composing the chemical industry and primary producers, will exert greater influence — and power — in the food microcosm. Where this complex might be placed in the hierarchy is anybody's guess.

d.   A Tentative Observation Resulting from Changing Positions of Influence

This shuffling of the relative positions of influence in the food microcosm is a contributing factor in the failures of many new product introductions. Market researchers do not take full cognizance of the impact retailers might have on the market researchers' point of sale data, on their research data, and on their promotional activities during introduction. The techniques used to gather data on the introduction of a product and the interpretation of the results have often failed to fully integrate the role of the retailer. These techniques do not factor in the influence of the retailers and their knowledge of their districts. The data gathered by the marketing people are colored by the seller's in-store activities; this data has the imprint of the outlet.

Retailers have no interest in the introduction itself or empathy for the developer. Introductions are an intrusion into their routines, and no matter how much may be paid to them for this intrusion, it remains that, an intrusion. The introduction for the retailer is an impersonal thing very unlike the feelings of the developer, who has been intimately involved with the product. It is the developer's "baby."

A lack of enthusiasm, perhaps even a lack of cooperation between developer and retailer, requires that marketing work closely with retailers to get their assistance or at least to avoid their antipathy. A successful launch requires the active participation of the retailer.

Two other factors that bear on product failures and involve retailers are:

- Management's decision to pull an introduction too soon and not allow for sufficient buy-in time can upset retailers. Management are too greedy to get their returns; when they fail to get these within their time horizons, they bail out of the introduction. This can upset retailers, who feel they have been overlooked in the decision and hence may be very reluctant in the future to participate in introductory programs or to cooperate with sales data.
- Activity by the competition using any number of tactics in retail outlets can upset interpretation of sales data. Marketing personnel must be wary of data that might not be normal. Introducers of new products must be alert to unusual sales numbers, competitive me-too product introductions, and promotional activities of the competition.

"The buzz" is an element for market research personnel to be aware of during an introduction, indeed to be aware of during any stage of a product's life cycle. It is unrelated to any impact the retailing sector may have in the food microcosm. It *may* represent competitive activity. The buzz, so I am told, refers to something akin to rapidity of communication, indeed basically to instant messaging. It is a tool of activist consumers and some consumer groups. Marketing and advertising people are frightened by the rapidity with which negative reactions to new product introductions or to newly opened food service establishments can spread. Bad buzz, instant feedback, can hurt new food products, especially those aimed at the teens and older groups of consumers who are frequent visitors to chat rooms.

Serious errors can be anticipated in the interpretation of market research data that does not include the role of the retailer if new product developers continue to operate with the social constructs depicted in Table 12.1 within the "modern times" hierarchy of the food microcosm. That structure is changing rapidly. I have not been convinced that ECR is a good or even proper vehicle to bring an understanding of changes to come that will affect developers.

II. WHAT FOOD SCIENCE AND TECHNOLOGY HAVE WROUGHT

A. THE IMPACT OF FOOD SCIENCE AND TECHNOLOGY

> The old-fashioned housewife's menus were carefully thought out; the modern house-wife's menus are carefully thawed out.
>
> *Anonymous*

1. How Food Savvy Are the Customer and Consumer?

Kesterton (2003) credits this cynical aphorism to Harry Balzer, vice president of market research firm NPD Group of Park Ridge, IL, who commented that the majority of Americans eat dinners that come prepared and need only to be reheated. Indeed, less than 50% of diners sat down to dinners in which at least one of the dishes was assembled. The description *assembled* applied to any food preparation activity from putting together cheese and crackers or a jam sandwich to baking pies. For a more extensive discussion of the impact of food science and technology on society, Pyke provides a startling insight in his essay written in 1971.

This reliance on prepared foods stems from many causes:

- Many people do not know how to plan or prepare a nutritious snack or meal because they have never learned. They have no need to learn because they have a multitude of chilled, frozen, canned, or dehydrated prepared food items available for them.
- The busy lifestyles of many people demand foods requiring minimal preparation in the home. The family sit-down dinner has become a thing of fiction except for festive occasions, as family members rush off to planned activities. The social intercourse of the family meal has gone. Progress in food technology has been a socially divisive influence on home life (Pyke, 1971).
- People, women and mothers in particular, no longer feel guilty about not slaving all day in the kitchen or setting before guests a home-prepared meal. They now take pride rather in greater participation in family activities, community affairs, or self-improvement courses.

A personal anecdote — and I fully realize the danger of generalizing from anecdotal observation — illustrates either the first or last of the above causes:

> My wife regularly picked up our own children plus a couple of others after school. For snacks she invariably prepared carrot sticks by the simple method of cleaning, peeling, and slicing carrots (this was before the days of the ubiquitous prepackaged baby-cut carrots). One young boy always devoured these avidly and one day commented to his mother when she arrived to fetch him, "Why can't we have these?" To which she, a university lecturer, replied, "I've never seen them in the stores" and asked my wife where she bought them.

Woolf (2000) wrote in an editorial, "Many children have simply no idea that cereals have to be harvested to produce everything from bread to pasta, so what chance now has the communication of sophisticated concepts ...?" Woolf's comments were not entirely out of context. He was commenting on how food technologists expected people to accept genetically modified foods based on how little they presently knew about the food they ate. Overcoming the ignorance of the general public about their food and about the food microcosm in general will be a price that innovators will have to pay.

To provide the convenience consumers want, retailers use several techniques; they stock a wide selection of ready-to-eat delicatessen items in the chilled foods section of their stores or they operate a hot deli bar with hot, fully cooked items for take-out or for eating in-store. They are thus competing with the fast food chains. In the frozen food sections, easily prepared entrées vie for consumers' attention. Products are displayed to convey ideas for tasty and nutritious meal combinations to hurried consumers. Unfortunately, many of the gatekeepers and consumers do not understand any principles of balanced meal preparation and basic nutrition; hence nutritious meal combinations have no meaning. The August 20, 2003, IFT *Weekly E-Newsletter* provides a headline: "Consumers Want Healthy but Buy Convenience." This article states that consumers are confused about health claims and do not recognize which foods are healthy. Governments are stricken apoplectic at the national obesity problem and the health issues caused by it, yet they have restricted funding for nutrition and household science education in public schools over the years. They now agitate for all kinds of legislation for funds either to educate the public, to (somehow) prevent obesity, or to apply a tax on fatty foods. An excellent article written by Ceci Connolly (2003) described this situation; it appeared in the August 10, 2003, issue of the *Washington Post* (page A01) and was cross-referenced in IFT's *Weekly E-Newsletter* of August 13, 2003.

Wise development team members looking for product ideas should spend some time watching and talking to people as they shop in the many marketplaces that are available. In too many companies, neither senior management nor the new product development team have ever ventured into these marketplaces and questioned their sales forces who do talk with customers, consumers, and retailers. What new food products do they, who see and talk with customers, consumers, and retailers, want to put on the shelves beside their competition's products? It is in these marketplaces that new food product ideas can originate. Developers would see how retailers influence customers to buy with store design for customer traffic control, product layout, in-store food sampling, and cooking demonstrations. For example, the perishable fruit and vegetable section is laid out to resemble a fresh air market designed to channel customers through the displays. It has salad bars for carry-out foods as well as trimmed, cut, ready-to-eat fresh fruit and vegetables (e.g., sliced mushrooms, peeled orange segments, celery and carrot sticks). Bringing an atmosphere of freshness and naturalness to the fore, in-store bakeries waft the delicious aroma of fresh bread throughout the store to complete the picture.

2. Impact of Technology

The origins and causes of the latent fear of technology found in many of the nonscientific community are irrelevant. The nonscientific community is either confused by, frightened of, or suspicious of technology except when this technology is applied to the pharmaceutical and medical devices fields where they are very thankful for the advantages science brings. Newspapers, popular science magazines, radio phone-in shows, and television hosts either quote or interview experts, medical doctors, scientists, nutritionists, and dieticians on controversial issues of food and food safety, health and disease, cooking, and nutrition. Each expert argues for his or her respective position either pro or con; the hosts of the shows or the journalists garner interest and maximum publicity by stirring up controversy between guests or interviewees with opposing views. The result for customers and consumers is often not enlightenment but confusion. No wonder the public harbors a suspicion of science and technology when the so-called experts cannot agree on what appears to the public to be simple and clear-cut questions such as "Is it good or is it not good for me?" This is a question scientists cannot answer, or they waffle on their answers with abstruse, technical jargon.

a.   Big Science: Biotechnology

Biotechnology shows great hope to relieve pain and suffering with its promise of medical advances. The public is not so welcoming when biotechnological techniques are applied to the food supply, where this branch of science promises to have value for farmers, food processors, and the malnourished. Here, suspicion, caution, politics, greed, big business, big science, government, and consumer advocacy groups all seem to cloud the issue of what impact biotechnology could have on agriculture and world food production. Meanwhile consumers are wondering whether they really want hard, bruise-free, or square tomatoes or straighter bananas, both of which would be easier to package and ship. They see no perceived advantages for themselves in such products.

Bovine somatotropin (BST), derived from biotechnology, has been used to increase milk production in cows, yet the general public reads of gluts of milk on markets and wonders why prices of milk and milk products are not coming down. If marketing boards control both the supply of milk and the availability of industrial milk, where is the need for more milk production from fewer cows? Who benefits? Consumer advocacy groups are suspicious of the possible effects of BST's long-term use on people consuming products derived from milk of treated animals. Small dairy farmers are concerned that large dairy factories that can afford to use BST to increase milk production will force them out of business. Their dairy associations lobby the government for protection.

Leaf culture techniques now permit scientists to grow many plant products, for example, "natural" vanilla, in vats in factories situated far from the natural geographic sources of vanilla bean production (see, e.g., Knorr et al., 1990). The impact on the economies of nations such as Madagascar, the Comoro Islands, the Island of Réunion, Tahiti, and Mexico, for whom the vanilla bean is an important export commodity, is a concern both for these countries and those activists against such technologies.

Product development and developers responsible for the technology should be aware of public reaction to their use of genetically modified foods or ingredients and should proceed cautiously to overcome fear, suspicion, and politics before technology, and especially biotechnology, makes a major contribution. Controlling the gene that controls senescence in a tomato is of little interest to the general public. Tomatoes are not meant for the salvation of shippers and distributors; they are meant for consumers. Will there be a perceived advantage for the consumer? That is the key issue: convincing the customer and consumer with tangible evidence that the technology will benefit them. So far the biotechnologists have failed to put this message across successfully; simply proclaiming that the genetically modified products are safe is not enough.

Breeding and gene selection programs for sturdier, more disease-resistant banana trees that have straighter bananas that pack better and resist transit damage are fine advantages for growers and distributors. But again, these are not advantages easily perceived by the consumer. "Will my bananas be different?" laments the consumer.

The introduction of manufactured microorganisms into the environment for nitrogen fixation or as a cryoprotectant for plants is an issue understood by scientists but not by the public. Any advantage to the public is seen dimly. And it is grist for the mills of the environmentalists, who can readily shake the confidence of the public respecting the hazards, real or imagined, that such an introduction may bring. Customers and consumers remember the introduction of DDT into the environment; they also remember the assurances of scientists from the chemical industry who proclaimed its safety and the attacks on the naysayers. They also remember chlorofluoro hydrocarbons (CFCs) that were touted as harmless, biologically inert, noncombustible substances. Then the world, including consumers and scientists, discovered the effect CFCs had on the ozone layer. All of a sudden something considered harmless, when used in large quantities, became dangerous. And it was leading experts on government appointed committees again who claimed lead in gasoline was safe until accumulations in the body fluids of children proved otherwise.

Food biotechnologists bear the onus of providing tangible evidence not only of the safety of their discoveries but also of their advantages to the general public; too often these advantages are seen as advantages to the scientists and their research institutes through recognition, patents and royalties, and the private companies they establish on the basis of their research findings. Product developers must realize that unproven technologies require careful investigation before they are used in new products, for if they fail to be approved or accepted by the public, the work and effort, as well as the good name of their company, may be forfeit.

Harlander (1992) stated the problem succinctly:

> If biotechnology is to be used to ensure a safe, abundant and affordable food supply, it must be accepted by the public; therefore it is critical for us to come to grips with the scientific, as well as the social, moral and ethical issues that influence our thinking about the food supply.

For biotechnology to be accepted in new consumer food products, the products must be perceived as having benefits for the customer and consumer and as being environmentally safe or neutral over the long term and not merely as an advantage to the grower, shipper, distributor, manufacturer, or discoverer. (For a fuller discussion, see Harlander, 1989.)

b. Consumer Reaction

Molly O'Neill summed up consumer reaction in an article headlined "Geneticists' Latest Discovery: Public Fear of 'Frankenfood.'" Frankenfood is, of course, genetically altered food, a name coined by Paul Lewis in a letter to the editor of *The New York Times*. The newspaper article featured developments at the exhibition of the Institute of Food Technologists (1992), in particular, a frost-resistant tomato incorporating a fish gene.

This may very well have advantages for a grower, but where is the advantage for consumers? It requires a bit of a stretch of the imagination for consumers to see benefits. This difficulty may in part be due to consumers' lack of knowledge of the food chain from grower to retailer. If the consumers cannot see how the modification benefits them or appreciate the advantage the modification gives to them, it is, in their thinking, an unnecessary alteration of an already accepted and successful product. No amount of education will convince them to accept this product *at this time*. Already many companies have backed away from the application of some of their genetic research; they do not want the job of educating the consumer for fear of reprisals.

Some parallels in the early stages of the introduction of irradiation techniques to those of the introduction of genetically modified foods are apparent. It will be remembered that irradiation as a food preservation technique was originally touted as prolonging the shelf life of foods. This benefit was not perceived by customers and consumers as an advantage for them, but rather as an advantage for the processor (Best, 1989a) and the distributor. Customers thought they would get staler (i.e., older) product because it would last longer on the shelves. Furthermore, they were convinced that processors would be able to skimp on sanitary procedures because they could zap the microbial contamination under which the product was handled or prepared. Is this not logical reasoning on the parts of consumer and customer alike? Irradiation for food products and the acceptance of genetically modified foods and food ingredients are both clouded with highly polarized views for or against. Some of the rhetoric and the extent to which apologists for irradiation will have to go in educating the consumer is reviewed by Pszczola (1990). It is good reading for proponents of the use of genetically modified foods.

c. Factory Farming and Agricultural Practices

Factory farming techniques are perceived as cruelty to animals by many in the general public, by animal ethicists, by vegetarians, and by empathetic meat eaters. Odors from farms annoy people, and run-off from pig and poultry factory farms pollutes streams, rivers, and lakes and angers environmentalists. Even a traditional practice such as force feeding of geese to produce *foie gras* has been declaimed.

There have been violent demonstrations against both food retailers displaying factory-farmed poultry and veal and against farmers employing factory farming techniques. These incidents plus bumper stickers, T-shirts, and placards deploring certain agricultural practices (often displayed by those lacking any knowledge of farm animals and their habits) clearly indicate the existence of social constraints to the technologies and crafts applied to food production and processing. Spokespersons for the food microcosm are needed to teach the public about food, its production, manufacture, nutrition, and proper kitchen preparation lest many new product ideas using the promised benefits of biotechnology will be aborning. Developers alert to the cachet that such novel product introductions carry will have ready an effective educational campaign to accompany them.

Old food crops are being rediscovered, and developers have created new products from them to satisfy the consumer, for example, quinoa-based cereals, flaxseed bread, and amaranth flours. Such crops as millet, quinoa, and amaranth have long been favored in other cultures. Brücher (1983) gives a brief history and description of amaranth ("the poor man's high-protein cereal"), plus information on its cultivation. Teutonico and Knorr (1985) reviewed the properties, composition, and applications of both the grain and the vegetable amaranths.

Organic farming and organic products have surpassed the multimillion dollar level. They occupy expanding sections of the produce and prepared food departments in supermarkets. Many customers now contract with market gardeners and farmers for weekly deliveries of organic produce to their doors. Consumer products based on organic produce will not be far behind; indeed, some major food manufacturers are now going mainstream with such products.

3. Trends as Social History

Hindsight is a wonderful sense. One should learn from hindsight, but one unfortunately does not always do so. The predictions made many years ago and some made more recently concerning what we would be eating today have taught me not to make predictions.

a. The Past and What Can or Cannot Be Learned

Reviewing earlier predictions of what was to have happened in food and agricultural areas is a sobering, educational exercise. Why have some forecasts come to fruition rapidly? Why are some still pie-in-the-sky dreams? Why have some been slow to come? Lawrie and Symons (2001) reviewed predictions made by Dr. J. G. Davis in a lecture entitled "The Food Industry in AD 2000" given in 1965 to the Royal Society of Arts when he was president of the IFST (U.K.). Davis's predictions are presented in Table 12.2. The right-hand column has Lawrie and Symons's comments and some of mine.

Whitehead (1976) wrote enthusiastically of the changes that might be seen in 1999. Most of the following are technically feasible, albeit the technology for some, for example, controlling machinery by thought waves, is still very much in its infancy:

- Meals would include algae-fed oysters from sea farms, with an entree of mock chicken from spun soy protein, accompanied by a mixed vegetable casserole in a base of single cell protein, topped with a cheese analogue. Spun protein meat analogues have struggled for acceptance as such but

TABLE 12.2
Then and 35 Years Later According to U.K. Data Based on Excerpts from a Speech by Dr. J. G. Davis

1965	2000[a]
Over 25% of personal income is spent on food, and the food microcosm employs more people than any industry.	19% of personal income is spent on food, and the number of people employed in the agricultural sector of the food microcosm has declined sharply.
The most important nutritional problem in 1965 and into the future is a sufficient supply of good quality protein.	FAO/WHO had recommended the amount of good quality protein to be 70 g daily. This has since been reduced to 53 to 55 g per adult male and ~45 to 47 g per female per day.
The form of malnutrition in developed countries is overnutrition.	Obesity is a major health issue (A.O.A., 2002; Birmingham et al., 1999; Lachance, 1994; MacAulay, 2003; NCHS, 1999).
It will be necessary for a fourfold increase in the world's food supply by 2000.	There has been more than a twofold increase in world population. Population growth shows signs of slowing. Food supplies have increased, but availability is uneven.
Qualified food technologists will control every aspect of food from production to selling.	There has been a great increase in the numbers of trained personnel within the food microcosm. Education of the public in nutrition and food hygiene has lagged (Fuller, 2001).
Protein extracted from green leaves could be used for human and animal nutrition (Pirie, 1987).	This has never been developed to the potential that Pirie and Davis anticipated, partly because of the Green Revolution; the lack of a convenient, economical, and acceptable vehicle to use the protein in; and costs of extraction.
By 2000, waste and food poisoning will be a thing of the past.	1965: the innocence of the age of antibiotics! (author's comment). Not only has the number of cases of food poisoning increased, but the pattern of pathogens has changed. Waste and environmental problems still confound the food industry.
Convenience foods, apartment living, increasing numbers of the elderly, and working women will lead to communal feeding.	Instead of communal eating, there has been a burgeoning growth of convenience foods, a wide gamut of eating establishments and take-out food. Home cooking from scratch has declined to be replaced by so-called speed-scratch cookery.
Freezing as a means of preservation is the "least objectionable method" of preservation, and most people would own a refrigerator and freezer by 2000.	The availability of frozen foods has grown enormously, but so have many other means of food preservation.
Desalination would be economically feasible by 2000.	The availability of fresh water is a major world concern, and cheap power is as elusive as ever.
Africa will become a food-producing country.	This has not occurred because of political instability, disease, and drought.

[a] Critiqued by Lawrie and Symons (2001), with added comments by the author.

have been popular in the sophistication of canned stews, dried soup preparations, and pet foods. Cake baked using triticale flour would be dessert. Triticale flour has never been wholeheartedly accepted by the public.

- The superfarm would be laid out in long narrow strips spanned by moving bridges in which sits the farmer, who from his position on this bridge can program all activities from cultivation, seeding, irrigation, weeding, and harvesting of most crops.
- By 1999, atomic energy plants would supply cheap electricity and heat to farm communities and "farmers wearing cybernetic equipment may be able to control their machinery by thinking about what they want it to do." Most atomic power plants have been shut down or mothballed, and the public has been particularly resentful of where these have been sited. Thought control of machinery has a long way to go for commercialization.
- Aquaculture farming, particularly for salmon and catfish, would grow in importance. Lobster and oyster farming would gain greater prominence. This has occurred, but this technology has led to serious problems of water pollution, introduction of diseases to the wild stock, and concern for the genetic stability of the wild stock.
- The soft drink industry would develop a market for syrups and powders that are carbonated at home.
- Insects, cattle manure, poultry droppings, and municipal sewage would provide the basis for protein for animal feed. Animal ethicists frown on these as sources for animal feed; obviously they have never lived on a farm and watched the free range chickens peck away at insects or observed pigs follow hungrily behind cows.

Social, practical, ethical, environmental, and economic reasons have slowed or stopped the acceptance of many of the above developments.

The American Society of Agricultural Engineers dared to make some very long-range forecasts of changes in farming practices to the year 2076 (Anon., 1977). These include the following food predictions:

- By 2076, 75% of meals will be prepared at large food service commissaries and not in the home. This is an interesting prediction. Some of today's more radical architects have argued that the kitchen in a home is a wasted space. People eat out, buy take-out food, or buy frozen prepared foods. Kitchen preparation is minimal, consisting largely of putting together meal components to thaw or reheat. Putting together components does not require a separate room dedicated to food preparation with its built-in appliances for frozen and refrigerated storage, electric and microwave ovens, and cupboards. However, kitchens are still built into houses, and to this day there still appears to be some appeal for homestyle prepared foods.
- Fruits and vegetables will be grown in solar-heated greenhouses near, or in, urban centers. Fresh fruits and vegetables will be delivered to consumers daily.

- Poultry, as well as other livestock, will be raised in environmentally controlled high-rise buildings as part of urban areas, with feeding and waste management automatically controlled. Organic wastes will be extracted for their energy.
- Harvesting and in-field processing of field crops will result in the separation of the edible material into protein, carbohydrates, fats, and other useful food ingredients. Cellulosic and ligninic material in the waste will be converted into plastics.

These predictions are an interesting hodgepodge of already here, those yet to come, and those not likely to come. The majority of eating occasions already are removed from the home; this prediction was fulfilled within the first 20 years, but will it still be true in 2076 or will communal eating in commissaries be the vogue? Devotees of sci-fi writings by authors such as Asimov, Clarke, and others read that these authors frequently view communal eating in commissaries as the way of the future; that is, eating out will prevail.

On the other hand, the home office has become a reality thanks to improvements in telecommunications, the personal computer, and networking facilities; this suggests eating in will prevail. What will be its impact on meals away from home? The home office often has meant more work and less of a distinction between home life and work life. Therefore, with less time for food preparation, a greater impetus for home delivery of prepared foods may result. Already many restaurants take part in meal programs whereby menu items can be ordered by telephone and delivered home, with each member eating from a different restaurant's menu. Perhaps this is where developers should be looking for product development ideas.

Animal rearing in urban settings is simply too expensive to contemplate; land is too valuable, and objections from social activist groups and ethicists would surely be disruptive if urban housing was used for animals when housing for the homeless is desperately needed. The same criticism would be expected of urban space being used for indoor culture of vegetables. Some hydroponic culture of crops has developed into major industries but not in urban centers (excepting marijuana growing). The amount of hydroponic culture of herbs and salad greens by chefs on the roofs of their hotels or establishments is insignificant and primarily a promotional ploy. The introduction of fresh herbs and herb blends in the produce department has been a remarkably successful market development.

Predictions are usually more accurate when the projected time frame over which the predictions are made is short. The unpredictability of technological breakthroughs, the legislative activity of government in the food microcosm, and the volatility of both customers and consumers do not distort the data on which these shorter-term forecasts are based. A study, released by Frost & Sullivan (Anon., 1980) gave the following predictions for the 1980s:

- The use of analogues, especially for microwave-designed new food products, will expand.
- The use of whey-based ingredients, high fructose corn syrups, encapsulated flavors, and savory flavors will expand.

- Food service menus will broaden their menu selection. Salad bars will be more prominent in food service establishments.
- Government influence on nutritional labeling and dietary claims will continue to grow.
- Nutritionally positioned products will displace naturally positioned products. Nutrition will be the "turn-on" for consumers in the future.
- The over-65s is one of the fastest growing segments of the population. The geriatric food market will present a challenge to food product developers and marketers in the future.
- Ethnic foods will grow and cause segmentation of the market.

These predictions, covering a shorter forecasting period, describe the 1980s and early 1990s very well. Fast food restaurants are certainly going upscale with white tablecloth seating and broader menu selection; expanded deli salad bars are found in fast food restaurants and even in supermarkets. McDonald's and Burger King have both experimented with special dinner menus (*The Gazette, Montreal*, August 16, 1992).

Nutritional products are prominently positioned now, but one might argue whether the natural foods (no additives, or organic) or minimally processed foods have been displaced to any extent. In addition, some believe the bloom is fading from nutritionally positioned products. Nutraceutical additions or adjuncts to foods, directed to prevention of disease conditions, are increasingly popular.

Ryval (1981) made the following predictions but wisely put no time frame on them:

- Meat consumption, particularly red meat, will drop and with it animal fat consumption because meat's replacement, foods such as lentils, contain little fat — but there was no suggestion that one contributing cause might be BSE (bovine spongiform encephalopathy), the public's aversion to meat at this time, and health concerns. More fresh fruits and vegetables will be consumed, as well as more minimally processed foods.
- Food selection will broaden, with a greater acceptance of ethnic foods. This happened and is continuing on a large scale. The cause has been massive migrations of peoples caused by political and economic instability. Both causes were unpredictable.
- Seafood consumption will rise. There will be a greater reliance on aquaculture and fish farming. Underutilized species will receive greater recognition as food in products such as fish sticks, fish sausage, and surimi. Again this prediction has come to be, with the result that many wild fish stocks have been overfished. Fish farming, now a big business, has been condemned in many areas as a pollutant.
- A more diverse variety of plant protein sources will be used to complement the traditional sources such as soybean, peanut, and cereal grains. The variety of plants being utilized has increased partly from people bringing their traditional foods to new countries and partly because people are searching for new foods with health benefits. There may also be a backlash from those opposed to genetic modification of plants and animals who

have raised concerns that biotechnology may lead to monoculture, with the result that many valuable and diverse food crops will be lost.

- Analogues will continue to gain favor, not necessarily as total meat replacements but rather as extenders of other products.
- Waste recovery techniques will undergo intensive research to develop new feed sources by fermentation with microorganisms or to extract valuable food ingredients from the waste.
- The retort pouch "spells the end of the tin can" because of the energy savings the pouch offers, plus the opportunity for improved quality.
- Sterilized milk will move milk out of the refrigerator cabinet and onto the store shelves.

In the two decades since Ryval collated these prognostications from experts, some have come to pass, while others have sputtered out or continue to arouse yawns in customers and consumers. What has made the difference?

Aquaculture has proven to be successful but has not proven the panacea it was once thought it would be in making seafood cheap and readily available. There are still disease problems to be overcome. There are pollution problems caused by the aquaculture industry, which are threatening its expansion. Aquaculture has not quelled the fears that many wild fish stocks have been seriously depleted.

Analogues from various protein sources enjoy some success in many products where they can be blended in with or incorporated structurally with their natural counterpart. Single cell protein, particularly yeast protein, has had some success as a base for food ingredients. Triticale flour has never become popular; I have seen it only in some health food stores or occasionally in the health food section of supermarkets.

Insects as food or feed require a great deal more consumer education than most companies will risk undertaking to overcome cultural taboos. In Western culture insects are no more than a novelty food, for example, chocolate covered ants; before consumers eat insects or before consumer activists knowingly allow animals to be fed insect-derived protein or manure-derived feed, another cultural taboo, much education of the public is required. Taboos exist; consumers have not fully accepted atomic power, especially in their backyards, nor are they ready to accept farmyard droppings or insects as foods and they want only quality cuts of meat ground into their hot dogs.

b. Barriers to Predicting Food Development Events

There are curbs both to an unlimited or unchecked growth in new food products, new processes, and to consumers' acceptance of new food products. The following observations sum up the situation:

- New food product development is expensive, and technologies leading to new processing and ultimately to new equipment are all expensive. Local, national, and world economic conditions dictate the availability of monies for development. Processors cannot purchase new and untried equipment technologies unless that technology promises huge benefits. The sciences

related to customer and consumer research are not fully understood, and
food, nutrition, and their relation to health and disease prevention are
undergoing rapid rethinking and reinterpretation. Because of this trial and
error nature of development, it is costly. Yet it is necessary for growth.

- Legislation pertaining to food and its labeling, packaging, advertising,
safety, and manufacturing and to new foods will continue to be a thorn
in the side of new food product developers. Governmental agricultural
policies, international trade agreements and affiliations, and world politics
will influence sources and availability of ingredients. Hence their costs.

- Customers, the gatekeepers, and consumers are fickle. They have not been
educated about, and indeed, may resist, some of the newer technologies
associated with foods, for example, genetically modified foods, irradiation, and scientists as well as food companies have been loath to undertake
this education. Scientists do not consider it their concern, and food manufacturers fear reprisals and hence economic ruin.

All these barriers make the senior management and financial managers of food
businesses nervous.

The sciences and technologies, including the sciences related to consumer studies, on which food product development is based are unpredictable and are still
undergoing extensive development. Food technology has outstripped the food and
nutritional sciences upon which it should be based. The result is that technology has
had to rely on trial and error and craftsmanship in equipment design, product
formulation, preservative technology, and ingredient technology. Consider, for example, the technologies of stabilizing with high pressure, irradiation, water control, and
pulsed electric fields, which still lack a theoretical basis for their effectiveness.
Where, for example, is the mathematical basis for irradiation or high-pressure processing, as there is for thermal processing? According to Coppock (1978):

> Man had always been concerned with the technology of food, because he needed food
> to survive and that his understanding of the basic principles underlying the art, or craft,
> had always been slower than its development (see also Taylor, 1969).

Without the science, applying technology for new product development will
always be costly in time and money, and it will be hazardous.

c.   Consumer Responses

Consumers paid no heed to the prognostication of some food technologists that
freezing as a process was going to replace canning (thermal processing of food);
products employing both technologies coexist successfully. Glass containers were
to replace metal cans as the container of the future; then thin-profile containers (the
retortable flexible pouch and its cognate, the semirigid container) were to replace
both as the packaging of the future. But again consumers fooled the predictors:
North American consumers have never really adapted to thin-profile containers
despite their huge success in Japan and modest success in Europe. None of these
predictions came to pass. Foods packaged in glass and metal are on the shelves side

by side with plastic-packed and laminated paper board– and cello-wrapped and metal-packed foods. A variety of packaging materials reigns.

Sloan (1999, 2001, 2003b) has more conservatively opted to follow consumer trends (or are they more accurately customer trends?). These are presented in Table 12.3. Here the volatility of both customer and consumer responses to happenings in the food microcosm are highlighted better than with a study of past predictions. The fulfillment of predictions is always thwarted by unexpected and unpredictable events

---

TABLE 12.3
Top Ten Trends to Watch for in the Future

Food Technology August 1999	*Food Technology* April 2001	*Food Technology* April 2003
"Americanization of flavor": plain American with just a twist of foreign	"Do-it-for-me foods": the use of premade or take-out meals	"Heat and eat": tasty, quick-to-prepare, no mess, no clean-up foods
"Super simple": simple to prepare foods with taste appeal for busy living	"Super savory and sophisticated": demand for more upscale and cultivated taste in foods	"Retro nutrition": renewed interest in sugar- and fat-containing foods but with reduced amounts of calories
"Street foods": fast, portable, tasty, and accessible	"Balance": extremism in food selection is vanishing	"Casual indulgence": upscale food for casual comfort
"Living foods": fresh, natural, organic foods	"Form follows function": growth of the "minis," e.g., appetizers	"Country charisma": ethnic and regional foods with an American twist
"Deliver me": carry-out purchases and Internet food shopping	"A new kind of 'homespun'": developing foods for regional tastes; communal (one-dish) meals; homey, comfy living room style restaurants	"Table talk": upscale, fashionability (more flavor) in favorite foods
"Eatertainment": family style, communal "one-plate" eating; restaurant samplers	"Kid-influenced": catering to the influence children have on food purchases	"Simple solutions": single serve, combined (multiingredient, multinutrient foods); mini-packs
"Freestyle": grazing, foods for all day, mealtime anytime eating	"Light and lively": fresher, healthier, and more attractive looking items	"Custom catering": catering to children as decision makers in food purchases
"In-dull-gence": eating with moderation in all varieties of food	"Crossover meal patterns": mealtime is grazing time; out with traditional meal patterns	"Correcting conditions": foods to prevent, ward off, or relieve unhealthy conditions
"Self-treatment & trial": do-it-yourself health; eating foods to ward off something	"'Do-it-yourself health'": selecting food to improve or prevent a health problem from occurring	"Exceptionally pure": organic, natural, pure, no artificial additives foods
"Trusting technology": consumer willingness to try "genetically modified that tastes better" (cf. European attitudes)	"Clean, pure, natural, and safe": all natural, "green," non-factory-raised, organic foods	"Snacks & mini meals": snacking, grazing, eat-all-day foods

From Sloan, E.A., *Food Technol.*, 53, 40, 1999; Sloan, E.A., *Food Technol.*, 55, 38, 2001; Sloan, E.A., *Food Technol.*, 57, 30, 2003b.

occurring over the longer time range; trends are short range. Sloan follows and acknowledges the volatility and unpredictability of the customer by observing what is occurring in the various marketplaces in much shorter time intervals.

There is, nevertheless, a danger associated with following trends; they, too, are historical. The buying statistics on which they are calculated are based on the past month's or past 6 month's or even past year's purchases. When one reflects on the brevity of life cycles of products, one may be jumping in at the wrong time. At best, a study of trends emphasizes the volatility of the customer. They change their food buying habits to jump on the next food fad, shattering the previous market niche only to introduce new niches. If, indeed, as Sloan claims, the statistics show that children dominate food choice decision making, who could be more faddish and volatile in food choices than children, tweenies, or teens?

The family unit with its traditional ways has evolved, and with it many traditional food habits have changed; the daily family dinner is a rarity. The proportion of elderly in the population has grown such that it has now become a recognizable pressure group within society, as have the teens, the tweenies, and other groupings such as yuppies, baby boomers, home cookers, and so on. Within these groups there are subdivisions. They all represent separate market niches.

The sciences associated with understanding the buying habits of customers and what influences consumers choices are still inadequate, but the basis for an accurate understanding of what motivates both these groups is beginning to be clarified. Clearer identification of the characteristics of consumers certainly will help developers formulate new products and help marketers effectively promote these to customers.

d. Politics and Government

The most recent factors accelerating the intervention of governments into the safety of the food microcosm within their respective countries are:

- The potential for food bioterrorism directed against crops and animals: Strict agricultural polities and policies on emerging animal and plant diseases have emerged.
- Political instability in many food exporting countries: This has increased border food inspection.

More benign factors are an increasing awareness of the importance of food in the prevention of many human diseases and consumer demands for minimally processed foods. The results have been:

- An increasing body of legislation regulating the growing, harvesting, production, and retailing of food and increasing support for national farm, agricultural, and seafood policies directed at, as near as is possible, self-sufficiency. These directly influence regional environmental policies.
- A growing body of safety, health, and nutrition policies. Many such policies are being and will be used in the future as nontariff trade barriers.

- A new consumer attitude toward foods and nutrition, especially in developed countries. No longer are consumers eating sensibly for good health, but are eating to prevent disease. This has alerted governments to the need to curb overhyped advertising.
- Newer technologies, genetically modified foods for one, have rightly or wrongly aroused passions in customers and consumers. Despite the best efforts of governments and scientists to alleviate these fears, many in the general public fear the heavy involvement of the chemical and biochemical companies in agribusiness.

All of these arouse a profound interest by government in the food microcosm and confound prognostication in the food business.

B. FACTORS SHAPING FUTURE NEW PRODUCT DEVELOPMENT

1. The Influences

Predicting with certainty which new products will emerge as successes is impossible. One can, nevertheless, certainly forecast those elements that will influence whatever new products do successfully emerge. These elements are derived from the factors listed above that confound prognosticators; they sort themselves into four broad areas:

- Social, cultural, religious, and ethical concerns of customers and consumers, which are hard to overcome through education.
- Customer and consumer skepticism (perhaps even growing apathy) about scientific and technological advances. Announcement of these advances excite customers and consumers with the benefits they promise. At the same time customers and consumers are frightened because of counter opinions and fears expressed by equally prominent scientists, well-meaning environmentalists opposed to the developments, and other groups with some axe to grind.
- The lack of good communication between scientists in the food microcosm and the general public. Scientists must learn to communicate concisely and in an attention grabbing manner with the public. Their status reports, position papers, and information briefs are too often wordy, pedantic, and dull, appearing to have been written for other scientists and certainly not for journalists working on a deadline for their newspapers, radio newscasts, or television shows. Journalists want something eye catching that will sell newspapers and advertising.
- The changing marketplaces in which customers and consumers shop.

Pervading each of these elements to a degree is government's reaction to the new developments and all the pressures on government (see, e.g., Figure 6.1).

a. Social, Cultural, Religious, and Ethical Concerns

The education of customers and consumers in North America was sparked by the writings of Catherine Parr Traill (1855) describing good household practices; the beginnings of consumer-oriented magazines such as *Good Housekeeping Magazine,* which launched in 1885; the growth of associations such as women's institutes; and writings by, for example, the Womens Alliance of the First Church of Deerfield (Anon., 1897). Perhaps indirectly these also spawned the growth of consumerism.

Today, the Internet plays an increasingly strong role in informing and educating, for better or worse, customers and consumers. This tool has certainly let customers and consumers comparison shop for the products they want and shop in a worldwide variety of electronic food marketplaces without leaving their homes. Information about food, nutrition, health, and cookery can be found on the Internet. However, in searching for information, Web surfers often get well-meaning but unsubstantiated and perhaps dangerous quackery. For a lay person to distinguish truths from half truths from arrant mistruths is difficult.

The "greening" movement (distinct from the "green revolution" of the 1950s) means many things to many people. Some issues involved are anti–free trade agreement attitudes to third world nations and the resultant exploitation of cheap labor in these countries (e.g., fair trade coffee), agricultural practices (actively pro organic farming, anti factory farming and pro animal rights), and health issues (pure food, water, and air; additive-, pesticide-, herbicide-free food). Generally, environmental issues are lumped in with the greening movement; environmental issues are themselves a complex mixture of issues and emotions. It has even reached into the processing industry (Mattson and Soneson, 2003). Greening becomes a mushy confusion of issues to which, nevertheless, companies had better pay attention.

Perhaps associated with greening, there has developed a growing consumer mistrust of bigness within the food microcosm, spawned by books such as Rachel Carson's *Silent Spring* and a consumer movement known as Naderism. This antipathy toward bigness, whether it be business, science, or government, has its unsettling effects within the food microcosm.

b.   Eating One's Way to Health: Nutrition and Pharmafoods

Knowledge of the interrelationships between nutrition and disease has progressed well beyond knowing, as in the 1930s, the need for vitamins in the diet lest scurvy or rickets or night blindness develop. Today there is a growing awareness that the foods people eat may play an important role in preventing the onset of certain diseases (Taylor, 1980; Jenkins, 1980; Maugh, 1982; Ames, 1983; NCI, 1984; USDHHS, 1984; USDHHS, 1985; Cohen, 1987; Berner et al., 1990). Diet may also play a role in how people behave (Kolata, 1982; Wurtman, 1989; Barinaga, 1990; Erickson, 1991).

Most nutrition information available to the general public comes from newspapers, magazines, or the Internet. Very little information comes through recognized educational sources. Most of the readily available information is not directly concerned with healthy eating habits but is directed at disease avoidance: for example, fiber and prevention of bowel disorders and cancer; soy protein–containing foods as a preventive therapy against cancer; saturated fats and trans fatty acids and heart disease, and so on.

The implications of disease-avoidance foods for governments are powerful incentives to promote such foods in the diet. If medical expenses and hospital stays can be cut, absenteeism from work reduced, and if people live healthy, more productive lives, the reduced costs of health care and the increased productivity would benefit the government. These benefits would surely come if consumers could be encouraged to:

- Eat foods containing the prebiotics and probiotics able to prevent disease
- Eat consumer products enriched with these disease-preventing foods
- Dose themselves with supplements containing the responsible component extracted from these beneficial foods
- Avoid certain foods considered to be bad for one and thereby reduce one's risk of developing certain diseases

The rush has been for consumers to self-medicate with herbal preparations with purported disease-preventing properties and for consumer product manufacturers to provide products enhanced with the responsible nutraceuticals. The onus is then on governments to educate the public and at the same time to verify the efficacy, safety, and purity of power bars, sports drinks, teas, beverages, and snacks fortified with various nutraceuticals and other nonnutritive substances. Such foods and verifiable knowledge about them must be available in addition to support policies to assure their availability at prices customers can afford.

The result will be an even greater growth in the numbers of the elderly, with the potential for a healthier, more productive lifestyle and a longer life if they eat the "right" foods. This has implications for the work force, retirement policies, pension plans, medical plans, the health care system, agricultural policies, food policies, and companies wanting to capitalize on people's health concerns. The key, of course, is whether all the promises of the nutraceuticals are scientifically true, as opposed to anecdotally true; scientists are finding much contrary evidence.

Current evidence clearly suggests that certain diets containing foods with these nonnutritives added can reduce the risks of some diseases:

- Kritchevsky (1991) reviews the effect of garlic on cardiovascular disease and discusses its hypolipidemic, hypotensive, anticoagulant, and fibrinolytic properties. This is certainly a good bet for an enterprising developer to come up with a pharmafood.
- Mills and coworkers (1992) discuss work on the biochemical interactions of peptides derived from the breakdown of foods and the possible influence these may have on food intolerances — work of immense value to consumers with food allergies. Two possible spin-offs from this work are the screening of foods for precursors causing intolerance and the possibility for developers to remove such precursors from new foods.
- Bioactive peptides derived from milk proteins, e.g., opioid peptides, immunopeptides, and mineral-binding peptides, are discussed by Meisel and Schlimme (1990). These authors speculate on the role of these derived dietary products as food hormones and "natural" drugs.
- Bifidobacteria have received prominence lately, and bifidus milk is on the market as a treatment for various intestinal ailments. O'Sullivan et al.

(1992) review the subject of probiotic bacteria, describe some of the products on the market, and suggest areas for further work. Hughes and Hoover (1991) discuss bifidobacteria in dairy products and describe products already on the market shelves in various countries.

- *Aphrodisiacs* is the provocative title of an article by O'Donnell (1992), who reports on the ability of certain foods to affect human reactions; that is, they can act on sensory perception (cf. alcohol or hallucinogenic mushrooms); they can irritate, or they can stimulate. These "mood foods" offer opportunities for new products.

- The antioxidant hypothesis of cardiovascular disease, in which oxidized cholesterol and not cholesterol is the risk factor, is reviewed by Duthie (1991). According to this hypothesis, insufficient antioxidant intake to prevent the oxidation of cholesterol and associated free radical activity is the problem. This clearly suggests a tie-in with vitamins E and C, beta-carotene, and selenium, all food nutrients.

The literature and concerns about the use of nutraceuticals are discussed in Chapter 10 of this volume and in Fuller (2001). As reliable data accumulate, the evidence for these health benefits will mount and new food product development will be influenced. People will demand these products. Certainly there should be a large demand for the "deep-sea protein" food supplement made by a European company and claimed to banish wrinkles (Cremers, 1993). Developers will want to produce these products, and governments will have to provide the legislation and inspection services to protect consumers with respect to the safety of the products and the truth of the claims made for them.

The findings cited in the list above concerning nutraceuticals and the products that could be made based on them, coupled with the promotional activities to support them, would blur the fine line between responsible ("good for you") and irresponsible marketing claims ("increases life expectancy"). Most certainly the food industry can expect, and is currently getting, government intervention in the form of advertising guidelines. Those developing such products for export need to be cautious, as there is not unanimity on safety issues, claims, or permitted dosage levels.

The abundance of propaganda for the use of foods for disease avoidance coupled with reports of contrary evidence on the value of these foods has confused customers and consumers alike. The confusion has spawned its own black humor, of which the following e-mail message sent to me is evidence, with apologies to the national groups mentioned:

"Here's the final word on nutrition and health. It's a relief to know the truth after all those conflicting medical studies.

1. The Japanese eat very little fat and suffer fewer heart attacks than the British or Americans.
2. The Mexicans do eat fat and suffer fewer heart attacks than the British or Americans.
3. The Japanese drink very little red wine and suffer fewer heart attacks than the British or Americans.

4. The Italians and French do drink red wine and suffer fewer heart attacks than the British or Americans.
5. The Germans drink beer and eat lots of sausages and fats and suffer fewer heart attacks than the British or Americans.

CONCLUSION: Eat and drink what you like. Speaking English is apparently what kills you."

Such satire shows that many within the general public have developed a skepticism and cynicism toward health claims.

c. Marketplace Influences

The many elements that influence the various marketplaces and the customers and consumers in them have been discussed in the foregoing chapters. A major point to be made is this: the introduction of new products onto store shelves cannot be a continuous and unchecked growth in numbers. There are physical limitations. To introduce new products, there must be an attrition of older established products if these do not satisfy the many new market niches that customers and consumers can be subdivided into. Supermarkets cannot accommodate all the newly introduced products; consequently, old, established products that do not pay the retailer's expected return per running foot of shelf space must be culled, or developers must maintain different versions of them that satisfy customers and consumers in these new niches. A classic example might be instant coffee, which emerged as a liquid concentrate, then a powder, then freeze dried, then with versions with and without caffeine, and now has liqueur-flavored versions and has managed to stay on the shelves.

Without maintenance to satisfy new customers and consumers, old products will die because the needs, that is, criteria, of the retailer have not been fulfilled. Likewise new products will succeed if the retailer, the customer, and the consumer are satisfied.

i. Marketplace Changes

Changes in the marketplace are legion. The warehouse outlet is climbing in popularity as customers buy in bulk to control food costs. Whether such purchasing really helps customers control costs or not is immaterial if customers believe it does. Which products could be adapted (and how?) to a bulk market is something developers must determine.

The field of communication is expanding at an enormous rate, in particular interactive television. The impact of teleboutique shopping on the food retail marketplace and ultimately on the introduction of new food products is as yet unknown. Developers with a heavy expenditure of television advertising monies can initiate a demand among customers and consumers for new products that retailers may have to stock. This is costly for developers.

These new communication vehicles present a problem for the introduction of products to elusive market niches. How does one target an elusive population with such a broad-based approach? Snow (1992) suggests that "instead of broadcasting adverts to the old 'admass' the new buzz word is 'narrowcasting.'" Marketers must learn to find and target customer niches and blitz these niches with their highly targeted promotions and advertising. How this is to be done on the Internet, for example, is anyone's guess.

Conceptually, the supermarket is changing. It is still a marketplace all under one roof, but now various departments may be privately owned or leased to specialist tradespeople, that is, the meat department is owned and operated by a professional butcher, the in-store bakery run by bakers, the fruits and vegetables might be run by a knowledgeable greengrocer who cares about the produce. There is professionalism. One no longer stands bewildered, desperately seeking help from a teenager stocking shelves. In the new supermarket there are staff knowledgeable about the breads they sell, the sausages they make, the coffees they are roasting, and the tea blends available; demonstrations provide customers and consumers an opportunity to taste products. The whole effect is to produce a carnival-like atmosphere typical of a marketplace where people can eat, meet, socialize, and shop. This is the environment within which developers must fit their product introductions.

d. The Impact of Legislation and Other Government Intervention

Government intervention will become an increasing burden on the food industry and especially on food companies heavily committed to new product and new process development. This will be felt in areas as diverse as:

- Patent protection for genetically altered foods
- Toxicological testing of new ingredients and foods derived through biotechnology and submission of safety data to regulatory authorities for these and other new products such as degradable and edible films
- Development of a nomenclature for biotechnologically derived products
- Labeling regulations and guidelines to define or clarify such concepts as *natural, nature-identical, organic,* or *minimally processed*
- Advertising claims for the new ingredients and products with health benefits

In a good example of the latter, the consumer protection department of the French government has recently cracked down on special diet foods for athletes that contain large amounts of carnitine and for which claims are made that carnitine increases the amount of energy to cells, enhances athletic performance, and reduces the amount of fat in cells (Patel, 1993).

Concern for the safety of products derived from biotechnology will be particularly suspect when the microorganisms used in their production are hazardous.

Political and economic factors will influence the growth of new food product development. The awareness of the influence of food and nutrition on human behavior and disease will force governments to adopt agricultural and food policies that promote healthy food production and reduce the risk of nutritionally related diseases. An attempt by the National Cancer Institute of the U.S. Department of Health and Human Services (1984) to publish a guide to healthy food choices to reduce the risk of cancer caused a furor from vested interest groups, mainly meat producers but also some vegetable growers.

There are new trade patterns emerging that will have an impact on new food product development. If nothing else, the new markets that these agreements will create will usher in new competitors with new products to flood new niches in the various markets. There will be new customers and consumers to market products to. The impact on the movement of raw materials as well as added-value consumer goods will be immense, not only within the established zones but also between the new trading blocs.

However, new economic alliances and free trade zones also bring their own forms of protectionism and they will most certainly introduce more food legislation. Agricultural groups in dominant powers within free trade alliances often find ways to protect their farmers, while farmers in minor powers find markets closed to them. Vulnerable elements within the agricultural and food manufacturing systems will always clamor for protection, and governments will have to pay them some lip service, perhaps with the application of nontariff trade barriers as impediments to the free flow of products. Free trade zones do open up new markets and present new opportunities for new products and give new life to old established ones. Agricultural marketing and farm board subsidization policies will undergo upheavals. Despite the removal of trade barriers, there will still be impediments to trade that politics must treat.

There will eventually be a global marketplace; few people would argue with this. A global marketplace does not mean global food customs and traditions or a global economy. Food products will have to be styled to satisfy all traditions, habits, tastes, and customs within this world marketplace. The successful food developer will adapt products, their shapes, textures, flavors, and colors to the needs and expectations of consumers in the geographical marketplace the developer wants to penetrate. The truly innovative developer will know that in established markets, taste and flavor preferences of consumers are being challenged by the greater varieties of ethnic foods available.

No single product will be universally successful in all the marketplaces and market niches of the global community. Even the ubiquitous fast food chains realize this and cater to local tastes in the many countries they have penetrated. The success of any product in any given marketplace will always be limited by the social and political upheavals that plague the world, either in cataclysmic fashion or in more subtle fashion as styles and eating habits change as populations shift.

e. Economic Facts of Life

If companies are to survive and grow, the sciences and technologies upon which the development process is based will have to improve. Companies cannot afford to waste money on new product development if those products prove unsuccessful; the odds of success are poor. The success ratio must improve with better screening, forecasting, and consumer research; the management of creativity must be improved.

The costs of development are enormous and will have to be reduced. Even something as basic as the application of statistically based experimental designs in testing, such as rotatable designs (Mullen and Ennis, 1979a,b) would reduce work, reduce the number (and costs) of experimental trials, and assist in the interpretation of data. There are innumerable statistical software packages available to perform this design task.

In short, companies have to get smarter about carrying out product development. Criteria for the success of new products may have to be revised:

- How valid is measuring success by how much market share was obtained or taken from a competitor?
- Would establishment of a growable market niche of a group of faithful customers and consumers be an acceptable alternative?
- Could the company accept a longer time frame within which to recover their development and promotion costs?

These are management decisions, and if management sets its sights only on short-term gains, then new product development may have to remain a crapshoot for this ilk of management.

C. New Food Products of the Future

What the new food products of the future will be is anybody's guess. There are far too many unpredictable factors for any company to make predictions. Equally certain is that new products drive profitability. The pushes and pulls within the food micro-cosm continue and will continue to make new food product development a very risky, but very necessary, business.

New products of the future can perhaps be characterized, without specifying their exact nature or how successful they will be:

- There will continue to be food products nutritionally designed to satisfy some dietary or health or lifestyle goal of consumers. However, consumers will want flavor and convenience, too.
- Foods will be minimally processed, and their final preparation for consumption will be simpler and easier. To do this requires an enormous amount of technology, and developers will be challenged to design stabilizing systems for such foods.
- Grazing, that is, several eating opportunities per day, will become more common. For such grazing, finger foods (snack foods) will become more nutritious and attain a degree of acceptance and sophistication that will blur the distinction between fine dining and "grabbing a bite."
- More ethnic foods will become an accepted part of the North American diet, much as Italian and Chinese cuisine have become adopted by and adapted to North American taste. Throughout the global marketplace, the variability and diversity of food will make consumers more adventuresome.
- Vegetarian main courses (not necessarily vegetarianism) will find wider acceptance. This will be spurred partly through the growth of ethnic foods and partly for health, religious, and social reasons. Vegetarianism will also become more pervasive as a food tradition.

- Organically grown foods, and minimally processed added-value foods based on organic foods, will grow in popularity. Consumers of organic food, presently a minority of consumers, will grow but probably remain a minor segment of the consuming public.

- More meals and finger foods will be eaten away from the home as more leisure time becomes available as the work week decreases in hours and retirement ages continue to fall. Consequently, the food service industry will experience rapid growth if it develops new foods to satisfy the increased demand for meals away from home. Better attention to health concerns because of the backlash of the problems associated with obesity will direct development in the food service industry. The impact of working at home on this is as yet unknown; work-at-homers appear to work ever longer hours and may require home-delivered meals.

- More fragmentation of markets will be found and more characterization of customers and consumers into new recognizable niches will occur. New products will have to be found to fill these niches. Companies will have to be satisfied with smaller markets and longer payback periods.

The marketplace of the future will change and will inevitably require a change in food products. Supermarkets are meeting the challenge of the fast food chains by opening deli counters where customers can purchase prepared lunches, dinners, and even breakfasts. Street vendors with their pushcart finger foods provide a direct challenge to the fast food chains and stand-alone restaurants, each of which are saddled with high real estate costs. Another innovation has been introduced by the station restaurant in Westport, CT, for commuters. Commuters can place their dinner orders with the restaurant in the morning and pick the meal up on their way home at night.

Opportunities for new food products have been opened up by a consumer revolt against bigness (perhaps a sameness of taste, lack of a uniqueness of taste), the growing sophistication of consumers' food tastes, and the more cosmopolitan nature of consumers. Traditional markets have been fragmented and there is more opportunity for niche marketing — placing new products in markets that are either too small for large companies to fill or that provide margins that are too unprofitable for them to be concerned with. Within these niches are very profitable markets for the right products.

A prime example of niche marketing can be seen in the brewing industry. For years bigness was in style and there was a limited number of beers available. Then small, local breweries started up. These microbreweries, as they are called in Canada, or craft breweries as they are called in the U.S., kept away from the so-called popular beer tastes and concentrated more on traditional beers. They have become a huge success and have caused some fragmentation of the beer market by targeting beer lovers rather than beer drinkers. The result has been a takeover or attempted takeover of this market by the big breweries buying microbreweries.

A similar fragmentation has occurred in other markets. Flavored beers, low-alcohol or alcohol-free beers, alcohol-free wines, wine coolers, carbonated and still fruit drinks, and flavored natural waters have all snatched part of a market away from established beverage products and created niches that have matured or are maturing into very profitable markets. Snow (1992) considers this move to niche marketing a result of marketing people's uncertainties regarding consumers and consumers' fragmentation of the retail marketplace. In short, the consumers' volatility has fostered new marketing opportunities that marketers have not quite come to terms with.

In many instances, these new niche products are easier and cheaper to develop and introduce into specialty markets. The developer stands a better chance of getting a return on investment if financial expectations are moderated.

1. On the Future

The statistics presented by Friedman (1990) and Kantor (1991) show that new food product introductions have been increasing year after year for nearly 30 years, and during the latter half of the 1980s the increase in introductions was meteoric (see Chapter 1). If one had extrapolated the data in the mid-1990s to determine the trend, one could have anticipated something approaching 30,000 new products by 2000. I stated "A simple extrapolation is not justified" and proceeded to explain why (Fuller, 1994). One obvious reason was that data at that time, if extrapolated, indicated a phenomenal 25,000 to 30,000 new products each year, which retailers simply could not handle. Data presented in Figure 1.5 suggests that there has been a slackening of the rate of introductions. Indeed, since 1976 declines in new product introductions have outnumbered increases.

The numbers are not of any concern. What does concern companies is what the new products of the future will be. Any attempt to precisely predict the nature of new products in the future is doomed to failure because these predictions depend on:

- The next enactment of food legislation that restricts the use of a food ingredient for whatever reason. For example, Broihier (1999) highlighted beverage products containing functional foods, among which were products containing kava kava; later there was the announcement in newspapers (Branswell, 2002) and on Health Canada's Web site (Health Canada, 2002) indicating kava kava could cause liver damage. Still later (2003) came the announcement that Canada had put a stop sale order on all products containing kava kava.
- The next scientific or technological breakthrough that alters processing technology or uncovers, for example, new nutritional knowledge. These breakthroughs cause a stampede of new products employing that knowledge or technology, but one is cautioned to remember irradiation and the resistance its introduction encountered.
- The next ecological or environmental finding regarding damage to the environment by a food process or agricultural practice.

These are all unknowns, and combined with customers and consumers who are fickle and quixotic and whose activities require careful observation and study, they underscore the pitfalls of development.

About 40 years ago I wrote a paper about what humankind's foods might be in the year 2000. It fell laughably short of meeting any of the actual events that did occur by 2000. It failed for all the reasons expounded in this chapter. The technologies were all there; all my predictions were possible, but those that have come to pass have only met with modest economic success.

What went wrong? The major factor in all new food product development, and in the basic and applied research that leads up to their development, is the ever-changing profile of the customer, the consumer, and the new marketplaces. My predictions overlooked the needs of customers and consumers and the reality of the sellers in the new marketplaces: I satisfied only my needs as a technologist. Food manufacturers must satisfy the needs of real people in highly competitive marketplaces.

The predictions in this chapter were made by technocrats or those applying scientific method to their research and thus they satisfied the needs of technocrats, not the needs of those they were to serve. They said, "Look what we can do." The developers, including the researchers at their laboratory benches working at the so-called cutting edge of science, must realize that for successful new product development, they must satisfy consumers' needs. Sticking one's thumb into a Christmas pie of esoteric research, pulling out some technical plum and saying, "What a good boy am I" will not satisfy tomorrow's customers and consumers. The application of technology *per se* to create new products will not be any guarantee of product success in the marketplace. Customers and consumers will not beat a path to products simply because they are technological marvels. They will beat a path, however, to products that satisfy their hidden needs and desires.

III. WHAT I HAVE LEARNED SO FAR

What I have learned so far:

- All growth is limited. Management cannot expect growth of their present product mix to continue upward forever; one need only compare the life cycles of food products. Eventually one has to stop and go in a different direction, finding something new; for example, instant coffees, beers, and wines have been changed several times over to become new liqueur-flavored coffees, more flavorful beers put out by microbreweries, and wine coolers, ice wines, and fruit wines. These were abrupt changes in thinking. Science, and to a degree human progress, does not proceed in an orderly fashion building on preceding events. They progress by abrupt breaks with previous dogmas and move in new directions. Senior strategists within the food microcosm must themselves be innovative and flexible in their development of strategy and constantly searching for new ways that are not more of the same old ways to ensure survival in the business world.
- New product development is still an art. It probably will never be more than what is referred to as a soft science. Like all practitioners of the arts, product developers must learn the tools of their trade, must learn well the skills that are required in product development, and be skilled in their

application. Just as the painter must learn the properties of paper, canvas, and coloring materials; know how to mix paints; master perspective; study human anatomy; even know how to use a brush for effect; so too must product developers fully know and understand their tools before they can even begin to have any success at product development.

- The skills reflected in the soft and hard sciences of product development must be learned well and used dispassionately. Every available avenue must be used to gain knowledge of the customer, the consumer, the competition, and the retailer and the retailing environment. New product development is no place for gut feel for projects. Bad projects should be cut quickly.
- Trends should be observed closely and cautiously lest they be short-lived fads. Following fads is costly and consumes time. Developers need to think ahead toward what the trends suggest is happening to the customer and consumer in the longer term.
- Greed for short-term profits is dangerous in new product development. Management must be patient in developing markets and educating customers about the value of their products. Time criteria for judging a successful introduction must be reasonable.

The final words shall be left to Sir Isaac Newton, who wrote in *Mathematical Principles of Natural Philosophy*:

Errors are not in the art but in the artificers.

References

Abu-nasr, D., Healthy chips, *The Gazette Montreal*, W7, Oct. 3, 1998.

Adams, J. P., Peterson, W. R., and Otwell, W. S., Processing of seafood in institutional-sized retort pouches, *Food Technol.,* 37, 123, 1983.

Ahmad, J. I., Free radicals and health: is vitamin E the answer?, *Food Sci. Technol. Today,* 10, 147, 1996.

AIC /CIFST joint statement on food irradiation, Ottawa, 1989.

Akre, E., Green politics and industry, *Eur. Food Drink Rev.,* 5, 1991.

A.O.A (American Obesity Association), Obesity in the U.S., http://www.obesity.org/subs/fastfacts/obesity_US.shtml, Nov. 12, 2002.

Ames, B. N., Dietary carcinogens and degenerative diseases, *Science,* 221, 1256, 1983.

Amoriggi, G., The marvellous mango bar, *Ceres,* 24, 25, 1992.

Anderson, J. R., Boyle, C. F., and Reiser, B. J., Intelligent tutoring systems, *Science,* 228, 456, 1985.

Andrieu, J., Stamatopoulos, A., and Zafiropoulos, M., Equation for fitting desorption isotherms of durum wheat pasta, *J. Food Technol.,* 20, 651, 1985.

Ang, J. F. and Miller, W. B., The case for cellulose powder, *Cereal Foods World,* 36, 562, 1991.

Anon., *The Pocumtuc Housewife,* The Women's Alliance of the First Church of Deerfield, Deerfield, MA (first published in 1805), rev. ed., 1897.

Anon., R & D lame ducks, *Chem. Ind.,* 913, 1971.

Anon., The future as seen by agricultural engineers, *Food Eng.,* 49, 120, 1977.

Anon., Surgeon General says get healthy, eat less meat, *Science,* 205, 1112, 1979.

Anon., Analogs, ethnic and geriatric products offer top growth potential: "natural" to wane, *Food Prod. Dev.,* 16, 52, 1980.

Anon., Irradiation for fruits & vegetables, *Food Eng.,* 53, 152, 1981.

Anon., Introducing the hamburger, *The Gazette, Montreal,* Sept. 16, 1986.

Anon., The creative approach, *R. Bank Lett.,* 69, 1988a.

Anon., Fatal reaction, *Can. Con.,* 18, 6, 1988b.

Anon., Showcase: fiber ingredients, *Prep. Foods,* 157, 151, 1988c.

Anon., Modified Atmosphere Packaging: A. An Extended Shelf Life Packaging Technology, B. Investment Decisions, C. The Consumer Perspective, Report Series, Food Development Division, Agriculture Canada, Ottawa, 1990.

Anon., Psyllium stabilizer: label friendly and functional, *Prep. Foods,* 159, 127, 1990a.

Anon., Natural oxidation inhibitor breathes shelf life into meats, *Prep. Foods,* 159, 71, 1990b.

Anon., Food Industry Source Book™, *Prep. Foods,* 160, 1991.

Anon., Nutraguide, *Food Process.,* 59, 38, 1998.

Anscombe, A., Regulating botanicals in food, *Food Technol.,* 57, 18, 2003.

Anthony, S., Clearing the air about MAP, *Prep. Foods,* 158, 176, 1989.

Aram, J. D., Innovation via the R & D underground, *Res. Manage.,* 24, 1973.

Archer, D. L., The true impact of foodborne infections, *Food Technol.,* 42, 53, 1988.

Archer, D. L., The need for flexibility in HACCP, *Food Technol.,* 44, 174, 1990.

Arendt, J., Regulating the body's internal clock, *Food Technol. Int. Eur.,* 25, 1989.

Argote, L. and Epple, D., Learning curves in manufacturing, *Science,* 247, 920, 1990.

Astorg, P., Food carotenoids and cancer prevention: an overview of current research, *Trends Food Sci. Technol.*, 8, 406, 1997.

Babayan, V. K. and Rosenau, J. R., Medium-chain-triglyceride cheese, *Food Technol.*, 45, 111, 1991.

Babic, I., Hilbert, G., Nguyen-The, C., and Guirard, J., The yeast flora of stored ready-to-use carrots and their role in spoilage, *Int. J. Food Sci. Technol.*, 27, 473, 1992.

Bailey, C., Scaling up in R & D for production, *Food Technol. Int. Eur.*, 115, 1988.

Baird, B., SteiGenics to build new irradiation plant, *Food Technol.*, 53, 12, 1999.

Banks, G. and Collison, R., Food waste in catering, *Inst. Food Sci. Technol. Proc.*, 14, 181, 1981.

Barinaga, M., Amino acids: how much excitement is too much?, *Science*, 242, 20, 1990.

Barnby-Smith, F. M., Bacteriocins: applications in food preservation, *Trends Food Sci. Technol.*, 3, 132, 1992.

Bastos, D. H. M., Domenech, C. H., and Arêas, J. A. G., Optimization of extrusion cooking of lung proteins by response surface methodology, *Int. J. Food Sci. Technol.*, 26, 403, 1991.

Bauman, H., HACCP: concept, development, and application, *Food Technol.*, 44, 156, 1990.

Baxter, J., Blood, R. M., and Gibbs, P. A., Assessment of antimicrobial effects of lactic acid bacteria, *Br. Food Manuf. Ind. Res. Assoc. Res. Rep.*, No. 425, 1983.

Beauchamp, G. K., Research in chemosensation related to flavor and fragrance perception, *Food Technol.*, 44, 98, 1990.

Beckers, H. J., Incidence of foodborne diseases in The Netherlands: annual summary 1982 and an overview from 1979 to 1982, *J. Food Prot.*, 51, 327, 1988.

Bender, F. E., Douglas, L. W., and Kramer, A., *Statistical Methods for Food and Agriculture*, AVI Publishing, Westport, CT, 1982, chap. 3.

Bengtsson, N., Contact grilling, in *Proc. IUFoST Int. Symp. Progress in Food Preparation Processes*, SIK — Swedish Food Institute, Göteborg, 1986, 129.

Berger, K. G., Tropical oils in the U.S.A.: situation report, *Food Sci. Technol. Today*, 3, 232, 1989.

Berner, L. A., McBean, L. D., and Lofgren, P. A., Calcium and chronic disease prevention: challenges to the food industry, *Food Technol.*, 44, 50, 1990.

Bertin, O., Labatt stumbled on to low-cal "fat," *The Globe Mail*, Toronto, B18, Aug. 3, 1991.

Best, D., Marketing technology through the looking glass, *Activities Rep. Res. Dev. Assoc.*, 41, 86, 1989a.

Best, D., Analogues restructure their market, *Prep. Foods*, 158, 72, 1989b.

Best, D., New perspectives on water's role in formulation, *Prep. Foods*, 161, 59, 1992.

Best, D. and O'Donnell, C. D., Food products for the next millennium, *Prep. Foods*, 161, 48, 1992.

Beuchat, L. and Golden, D., Antimicrobials occurring naturally in foods, *Food Technol.*, 43, 134, 1989.

Beuchat, L. R., Sensitivity of *Vibrio parahaemolyticus* to spices and organic acid, *J. Food Sci.*, 41, 899, 1976.

Binkerd, E. F., The luxury of new product development, *Food Can.*, 35, 31, 1975.

Birmingham, C. L., Muller, J. L., Palepu, A., Spinelli, J. L., and Anis, A. H., The cost of obesity in Canada, *Can. Med. Assoc. J.*, 160, 483, 1999.

Bishop, D. G., Spratt, W. A., and Paton, D., Computer plotting in 3-dimensions: a program designed for food science applications, *J. Food Sci.*, 46, 1938, 1981.

Biss, C. H., Coombes, S. A., and Skudder, P. J., The development and application of ohmic heating for the continuous heating of particulate foodstuffs, in *Process Engineering in the Food Industry*, Field, R. W. and Howell, J. A., Eds., Elsevier, London, 1989, 17.

Blades, M., Food allergy and food intolerance, *Food Sci. Technol. Today*, 10, 82, 1996.

Blanchfield, J. R., How the new food product is designed, *Food Sci. Technol. Today*, 2, 54, 1988.

Block, E., The chemistry of garlic and onions, *Sci. Am.*, 252, 114, 1985.

Bogaty, H., Development of new consumer products: ways to improve your chances, *Res. Manage.*, 26, 1974.

Boileau, T. W.-M., Liao, Z., Kim, S., Lemeshow, S., Erdman J. W., Jr., and Clinton, S. K., Prostate carcinogenesis in N-methyl-N-nitrosourea (NMU)-testosteron-treated rats fed tomato powder, lycopene, or energy-restricted diets, *J. Nat. Cancer Inst.*, 95, 1578, 2003.

Bolaffi, A. and Lulay, D., The foodservice industry: continuing into the future with an old friend, *Food Technol.*, 43, 258, 1989.

Bond, S., New products in the market place, *Food Technol. Int. Eur.*, 109, 1992.

Bone, B., The importance of consumer language in developing product concepts, *Food Technol.*, 41, 58, 1987.

Boudouropoulos, I. D. and Arvanitoyannis, I. S., Current state and advances in the implementation of ISO 14000 by the food industry. Comparison of ISO 14000 to ISO 9000 to other environmental programs, *Trends Food Sci. Technol.*, 9, 395, 1998.

Boxer, A. and Back, P., *The Herb Book*, Octopus Books, London, 1983.

Brackett, R. E., Microbiological safety of chilled foods: current issues, *Trends Food Sci. Technol.*, 3, 81, 1992.

Bradbury, F. R., McCarthy, M. C., and Suckling, C. W., Patterns of innovation: Part I, *Chem. Ind.*, 22, 1972.

Bramsnaes, F., Maintaining the quality of frozen foods during distribution, *Food Technol.*, 35, 38, 1981.

Branswell, H., Kava linked to liver failure: Health Canada bans supplement, *The Gazette, Montreal*, A12, Aug. 22, 2002.

Brassart, D. and Schiffrin, E. J., The use of probiotics to reinforce mucosal defence mechanisms, *Trends Food Sci. Technol.*, 8, 321, 1997.

Bremness, L., *Herbs*, Stoddart, Toronto, 1996.

BrightHouse Institute, BrightHouse Institute for Thought Sciences launches first "neuromarketing" research company: company uses neuroimaging to unlock the consumer mind, available from http://www.thoughtsciences.com/pressrelease060302.htm, accessed 12/12/02.

Bristol, P., Packaging freshness, *Food Can.*, 50, 30, 1990.

Brody, A. L., Chilled foods distribution needs improvement, *Food Technol.*, 51, 120, 1997.

Brody, A. L., The return of the retort pouch, *Food Technol.* 57, 76, 2003.

Broihier, K., A thirst for nutraceuticals, *Food Process.*, 60, 42, 1999.

Bronowski, J., The creative process, in *Scientific American*, Sept. 1958, in *Readings from Scientific American: Scientific Genius and Creativity*, W. H. Freeman, New York, 1987, 2–9.

Brooker, J. R. and Nordstrom, R. D., Developments in engineered seafoods for commercial and military markets, *Act. Rep. Res. Dev. Assoc.*, 39, 56, 1987.

Brouns, F. and Kovacs, E., Functional drinks for athletes, *Trends Food Sci. Technol.*, 8, 414, 1997.

Brown, G., Is it safe to eat out?, *Food Sci. Technol.*, 16, 50, 2002.

Brücher, H., Amaranth, an old Amerindian crop, *DRAGOCO Rep.*, 35, No. 2, 1983.

Bruhn, C. and Schutz, H. G., Consumer awareness and outlook for acceptance of food irradiation, *Food Technol.*, 43, 93, 1989.

Buchanan, R. L. Predictive food microbiology, *Trends Food Sci. Technol.*, 4, 6, 1993.

Buchanan, R. L. and Doyle, M. P., Foodborne disease significance of *Escherichia coli* 0157:H7 and other enterohemorrhagic *E. coli*, *Food Technol.*, 51, 69, 1997.

Burch, N. L., and Sawyer, C., Hospital foodservice requirements: special diet convenience foods, *Food Technol.*, 40, 131, 1986.

Busch, L., Biotechnology: consumer concerns about risks and values, *Food Technol.*, 45, 96, 1991.

Bush, P., Expert systems: a coalition of minds, *Prep. Foods*, 158, 162, 1989.

Buxton, A., Going online, *Trends Food Sci. Technol.*, 2, 266, 1991.

Campbell, I., Consumer attitudes to food safety, *Visions*, 2, 1991.

Campbell, L. B. and Pavlasek, S. J., Dairy products as ingredients in chocolate and confections, *Food Technol.*, 41, 78, 1987.

Caragay, A. B., Cancer-preventive foods and ingredients, *Food Technol.*, 46, 65, 1992.

Cardoso, G. and Labuza, T. P., Prediction of moisture gain and loss for packaged pasta subjected to a sine wave temperature/humidity environment, *J. Food Technol.*, 18, 587, 1983.

Carlin, F., Nguyen-The, C., Chambroy, Y., and Reich, M., Effects of controlled atmospheres on microbial spoilage, electrolyte leakage and sugar content of fresh "ready-to-use" grated carrots, *Int. J. Food Sci. Technol.*, 25, 110, 1991.

Cauvain, S. P., Improving the control of staling in frozen bakery products, *Trends Food Sci. Technol.*, 9, 56, 1998.

Cerf, O., Davey, K. R.. and Sadoudi, A. K., Thermal inactivation of bacteria: a new predictive model for the combined effect of three environmental factors: temperature, pH and water activity, *Food Res. Int.*, 29, 219, 1996.

Chapman, S. and McKernan, B. J., Heat conduction into plastic food containers, *Food Technol.*, 17, 79, 1963.

Chinachoti, P., Water mobility and its relation to functionality of sucrose-containing food systems, *Food Technol.*, 47, 134, 1993.

Chirife, J., and Favetto, G. J., Some physico-chemical basis of food preservation by combined methods, *Food Res. Int.*, 25, 389, 1992.

Chung, K.-T., Wei, C.-I., and Johnson, M. G., Are tannins a double-edged sword in biology and health?, *Trends Food Sci. Technol.*, 9, 168, 1998.

Church, N., Developments in modified-atmosphere packaging and related technologies, *Trends Food Sci. Technol.*, 5, 345, 1994.

Chuzel, G., and Zakhia, N., Adsorption isotherms of gari for estimation of packaged shelf-life, *Int. J. Food Sci. Technol.*, 26, 583, 1991.

Clark, J. P., Processing equipment covered at Food Expo®, *Food Technol.*, 56, 102, 2002.

Clarke, D., Chilled foods: the caterer's viewpoint, *Food Sci. Technol. Today*, 4, 227, 1990.

Clausi, A. S., The Story of Dry Cereal and Freeze Dried Fruit, presented at School of Food Science, Cornell University, Ithaca, New York, Apr. 21, 1971.

Clausi, A. S., The Role of Technical Research in Product Development Programs, presented at Institute of Food Technologists' Short Course on Ingredient Technology for Product Development, New Orleans, May 12–15, 1974.

Cleveland, W. and McGill, R., The proper display of data, reported in Kolata, G., *Science*, 226, 156, 1984.

Clifford, M., Are polyphenols good for you?, *Food Sci. Technol.*, 15, 24, 2003.

Cocup, R. O. and Sanderson, W. B., Functionality of dairy ingredients in bakery products, *Food Technol.*, 41, 86, 1987.

Coghlan, A., Out of the frying pan, *New Scientist*, 157, 14, 1998.

Cohen, J. C., Applications of qualitative research for sensory analysis and product development, *Food Technol.*, 44, 164, 1990a.

Cohen, J., What I have learned (so far), *Am. Psychol.*, 45, 1304, 1990b.

Cohen, L. A., Diet and cancer, *Sci. Am.*, 257, 42, 1987.

Colby, M. and Savagian, J., Irradiation: progress or peril? Con: consumers say no!, *Prep. Foods*, 158, 62, 1989.

Cole, K. C., Is there such a thing as scientific objectivity? *Discover*, 6, 98, 1985.

Cole, M. B., Databases in modern food microbiology, *Trends Food Sci. Technol.*, 2, 293, 1991.

Converse, P. E. and Traugott, M. W., Assessing the accuracy of polls and surveys, *Science*, 234, 1094, 1986.

Cooper, L., A computer system for consumer complaints, *Food Technol. Int. Eur.*, 247, 1990.

Coppock, J. B. M., Has food technology outstripped food science?, *Inst. Food Sci. Technol. Proc.*, 11, 193, 1978.

Cremers, H. C., Fishy cure for craggy features, *New Sci.*, 137, 8, 1993.

Crockett, J. G., The new products manager, *Food Technol.*, 23, 25, 1969.

Crone, G., Welcome to the war room, *Financ. Post*, C15, Feb. 11, 1999.

Culhane, C. T., Resounding roar, *Food Can.*, 59, 16, 1999.

Cullinane, D., In the know, *Food Sci. Technol.*, 16, 48, 2002.

Culpeper, N., *Culpeper's Complete Herbal, and English Physician*, Harvey Sales, 1981 (reproduced from 1826 edition).

Curiale, M. S., Shelf-life evaluation analysis, *Dairy, Food Environ. Sanit.*, 11, 364, 1991.

Daeschel, M. A., Antimicrobial substances from lactic acid bacteria for use as food preservatives, *Food Technol.*, 43, 164, 1989.

Daniel, S. R., How to develop a customer complaint feedback system, *Food Technol.*, 38, 41, 1984.

Daniels, R. W., Home food safety, *Food Technol.*, 52, 54, 1998.

Datta, A. K. and Hu, W., Optimization of quality in microwave heating, *Food Technol.*, 46, 53, 1992.

Davey, K. R. and Daughtry, B. J., Validation of a model for predicting the combined effect of three environmental factors on both exponential and lag phases of bacterial growth: temperature, salt concentration and pH, *Food Res. Int.*, 28, 233, 1995.

Davidson, C., Drinking by numbers, *New Sci.*, 158, 36, 1998.

Davies, R., Birch, G. G., and Parker, K. J., Eds., *Intermediate Moisture Foods*, Applied Science Publishers, Essex, U.K., 1976.

Davies, A., Kochar, S. P., and Weir, G. S., Studies on the efficacy of natural antioxidants, *Leatherhead Food Res. Assoc. Tech. Circ.*, No. 695, 1979.

Day, B., Extension of shelf-life of chilled foods, *Eur. Food Drink Rev.*, 47, 1989.

Day, B., A perspective of modified atmosphere packaging of fresh produce in Western Europe, *Food Sci. Technol. Today*, 4, 215, 1990.

Dean, R. C., The temporal mismatch: innovation's pace vs management's time horizon, *Res. Manage.*, 12, 1974.

Decareau, R. V., Cooking by microwaves, in *Proc. IUFoST Int. Symp. Progress in Food Preparation Processes*, SIK — Swedish Food Institute, Göteborg, 1986, 173.

Deis, R. C., The facts on functional foods, *Food Prod. Design*, 13, 41, 2003.

Demetrakakes, P., Zap!, *Food Processing*, 59, 20, 1998a.

Demetrakakes, P., Information overload, *Food Process.*, 59, 17, 1998b.

Dempster, J. F., Hawrysh, Z. J., Shand, P., Lahola-Chomiak, L., and Corletto, L., Effect of low-dose irradiation (radurization) on the shelf life of beefburgers stored at 3°C, *J. Food Technol.*, 20, 145, 1985.

Denton, D. K., Four steps to resolving conflicts, *Qual. Prog.*, 29, 1989.

Dethmers, A. E., Utilizing sensory evaluation to determine product shelf life, *Food Technol.*, 33, 40, 1979.

De Vries, U., Velthuis, H., and Koster, K., Baking ovens and product quality: a computer model, *Food Sci. Technol. Today*, 9, 232, 1995.

Dörnenburg, H. and Knorr, D., Monitoring the impact of high-pressure processing on the biosynthesis of plant metabolites using plant cell cultures, *Trends Food Sci. Technol.*, 9, 355, 1998.

Downer, L., *Japanese Vegetarian Cooking*, Pantheon, New York, 1986.

Drewnowski, A., Henderson, S., Hann, C., Berg, W., and Ruffin, M., Genetic taste markers and preferences for vegetables and fruit of female breast care patients, *J. Am. Diet. Assoc.*, 100, 191, 2000.

Duffy, V. and Bartoshuk, L., Food acceptance and genetic variation in taste, *J. Am. Diet Assoc.*, 100, 637, 2000.

Duthie, G. G., Antioxidant hypothesis of cardiovascular disease, *Trends Food Sci. Technol.*, 2, 205, 1991.

Duxbury, D., New products, flavors, and uses of under-utilized species, *Act. Rep. Res. Dev. Assoc.*, 39, 51, 1987.

Dziezak, J. D., Fats, oils and fat substitutes, *Food Technol.*, 43, 66, 1989.

Dziezak, J. D., Taking the gamble out of product development, *Food Technol.*, 44, 110, 1990.

Dziezak, J. D., A focus on gums, *Food Technol.*, 45, 116, 1991.

Eagle, S., Pushing the envelope, *Food Can.*, 59, 23, 1999.

Earle, M. D., Changes in the food product development process, *Trends Food Sci. Technol.*, 8, 19, 1997a.

Earle, M. D., Innovation in the food industry, *Trends Food Sci. Technol.*, 8, 166, 1997b.

Eilperin, J., Sugar industry aims to block WHO report, *The Gazette, Montreal*, A21, Apr. 25, 2003.

Eisner, M., *Introduction into the Technique and Technology of Rotary Sterilization*, 2nd ed., private author's edition, M. Eisner, Brockdorf, Germany, 1988.

Elliott, J. G., Application of antioxidant vitamins in foods and beverages, *Food Technol.*, 53, 46, 1999.

Engel, C., Natural colours: their stability and application in food, *Br. Food Manuf. Ind. Res. Assoc. Sci. Tech. Surveys*, No. 117, 1979.

Erhard, D., Nutrition education for the 'now' generation, *J. Nutr. Educ.*, Spring, 135, 1971.

Erickson, D., Brain food, *Sci. Am.*, 265, 124, 1991.

Erkmen, O., Predictive modelling of *Listeria monocytogenes* inactivation under high pressure carbon dioxide, *Lebensm.-Wiss. Technol.*, 33, 514, 2000.

European Union (EU), Regulation (EC) No 258/97 of the European Parliament and of the Council of 27 January 1997 concerning novel foods and novel food ingredients, *Official Journal*, L043, 14/02/1997 P0001-0006.

Ewaidah, E. H. and Hassan, B. H., Prickly pear sheets: a new fruit product, *Int. J. Food Sci. Technol.*, 27, 353, 1992.

Farber, J. M., Foodborne pathogenic microorganisms: characteristics of the organisms and their associated diseases. I. Bacteria, *Can. Inst. Food Sci. Technol. J.*, 22, AT/311, 1989.

Farr, D., High pressure technology in the food industry, *Trends Food Sci. Technol.*, 1, 14, 1990.

FDD (Food Development Division), Ag. Canada, Modified atmosphere packaging: (A) An extended shelf life technology; (B) Investment decisions; (C) The consumer perspective, Food Development Division, Agricultural Development Branch, Agriculture Canada, Ottawa, 1990.

Fernandez de Tonella, M. L., Taylor, R. R., and Stull, J. W., Properties of a chocolate-flavored beverage from chick-pea, *Cereal Foods World*, 26, 528, 1981.

Figueiredo, A. A., Mesquite: history, composition and food uses, *Food Technol.*, 44, 118, 1990.

Floros, J. D. and Chinnan, M. S., Computer graphics-assisted optimization for product and process development, *Food Technol.*, 42, 72, 1988.

Foot, D. K., and Stoffman, D., *Boom, Bust & Echo: Profiting from the Demographic Shift in the 21st Century*, Stoddart, Toronto, 2001.

Francis, F. J., Natural food colorants, *Cereal Foods World*, 26, 565, 1981.

Francis, F. J., A new group of food colorants, *Trends Food Sci. Technol.*, 3, 27, 1992.

Frenzen, P. D., Majchrowicz, A., Buzby, J. C., and Imhoff, B., Consumer acceptance of irradiated meat and poultry products, USDA, Economic Research Service, *Agric. Res. Bull.*, No. 757, 2000.

Friedman, M., Twenty-five years and 98,900 new products later…, *Prep. Foods*, 159, 23, 1990.

Fuchs, C. S., Giovannucci, E. L., Colditz, G. A., Hunter, D. J., Stampfer, M. J., Rosner, B., Speizer, F. E., and Willett, W. C., Dietary fiber and the risk of colorectal cancer and adenoma in women, *New Engl. J. Med.*, 340, 169, 1999.

Fuller, G. W., *New Food Product Development: From Concept to Marketplace*, CRC Press, Boca Raton, FL, 1994.

Fuller, G. W., *Getting the Most Out of Your Consultant: A Guide to Implementation*, CRC Press, Boca Raton, FL, 1999.

Fuller, G. W., *Food, Consumers, and the Food Industry: Catastrophe or Opportunity?*, CRC Press, Boca Raton, FL, 2001.

Fuller, G. W., Additives, in *Encyclopedia of Food and Culture*, Charles Scribners Sons/Thomson Gale, New York, 2003, 7–14.

Gabriel, S. L., Separation and Identification of the Anthocyanins in the Sweet Potato (*Ipomoea batatas*) Using High Pressure Liquid Chromatographic Methods, M.S. thesis, Food Science and Nutrition, University of Massachusetts, 1989.

Gains, N. and Thomson, D., Contextual evaluation of canned lagers using repertory grid method, *Int. J. Food Sci. Technol.*, 25, 699, 1990.

Gaisford, S. E., Information, databases and the food technologist, *Food Technol. Int., Eur.*, 33, 1989.

Garcia, D. J., Omega-3 long-chain PUFA nutraceuticals, *Food Technol.*, 52, 44, 1998.

Garetto, J. M., Protecting trademarks and servicemarks, *Food Technol.*, 57, 42, 2003.

Gaunt, I. F., Food irradiations: safety aspects, *Inst. Food Sci. Technol. Proc.*, 19, 171, 1985.

Geake, E. and Coghlan, A., Industry "does not need research," *New Sci.*, 133, 15, 1992.

Geeson, J. D., Genge, P., Smith, S. M., and Sharples, R. O., The response of unripe conference pears to modified atmosphere retail packaging, *Int. J. Food Sci. Technol.*, 26, 215, 1991.

Geeson, J. D., Smith, S. M., Everson, H. P., George, P. M., and Browne, K. M., Modified atmosphere packaging to extend the shelf life of tomatoes, *Int. J. Food Sci. Technol.*, 22, 659, 1987.

Gehm, B. D., McAndrews, J. M., Pei-Yu, C., and Jameson, J. L., Resveratrol, a polyphenolic compound found in grapes and wine, is an agonist for the estrogen receptor, *Proc. Natl. Acad. Sci.*, 94, 14138, 1997.

Gélinas, P., Freezing and fresh bread, *Alimentech*, 4, 12, 1991.

George, K. C., Hebbar, S. A., Kale, S. P., and Kesavan, P. C., Caffeine protects mice against whole-body lethal dose of γ-irradiation, *J. Radiol. Prot.*, 19, 171, 1999.

Gershman. M., *Getting It Right the Second Time: How American Ingenuity Transformed Forty-Nine Failures into Some of Our Most Successful Products*, Addison-Wesley, Reading, MA, 1990.

Gibbons, M., Greer, J. R., Jevons, F. R., Langrish, J., and Watkins, D. S., Value of curiosity-oriented research, *Nature*, 225, 1005, 1970.

Gibbs, P. A. and Williams, A. P., Using mathematics for shelf life prediction, *Food Technol. Int. Eur.*, 287, 1990.

Gibson, L. D., The psychology of food: why we eat what we eat when we eat it, *Food Technol.*, 35,54, 1981.

Giddings, G. G., Sterilization of spices: irradiation vs gaseous sterilization, *Act. Rep. Res. Dev. Assoc.*, 36, 20, 1984.

Giddings, G. G., Irradiation: progress or peril? Pro: Safety is no longer an issue, *Prep. Foods*, 158, 62, 1989.

Giese, J., University centers ease product development, *Food Technol.*, 53, 98, 1999.

Giese, J., Additional university research centers, *Food Technol.*, 54, 63, 2000.

Giese, J., Selecting an outside food testing laboratory, *Food Technol.*, 55, 70, 2001.

Giese, J. H., Alternative sweeteners and bulking agents, *Food Technol.*, 47, 114, 1993.

Giovannucci, E., Tomatoes, tomato-based products, lycopene and cancer: review of the epidemiologic literature, *J. Nat. Cancer Inst.*, 91, 317, 1999.

Gitelman, P., Opportunities in the food industry, *Can. Inst. Food Sci. Technol. J.*, 19, xix, 1986.

Glew, G., Introduction to catering production planning, in *Proc. IUFoST Int. Symp. Progress in Food Preparation Processes*, SIK — Swedish Food Institute, Göteborg, 1986, 203.

Glyer, J., Diet healing: a case study in the sociology of health, *J. Nutr. Educ.*, 4, No. 4, 163, 1972.

Godfrey, W., A retailing perspective, *Food Sci. Technol. Today*, 2, 56, 1988.

Goff, H. D., Low-temperature stability and the glassy state in frozen foods, *Food Res. Int.*, 25, 317, 1992.

Goldenfield, I., The regulations affecting product development: industry view, *Food Technol.*, 31, 80, 1977.

Goldman, A., A Study of Product Development Management Practice among Food Manufacturing Companies Located in Southern Ontario, Working Paper No. 84-303, Dep. Consumer Studies, University of Guelph, Guelph, Canada, 1983.

Goldman, A., personal communication, 1993.

Gordon, M. H., Finding a role for natural antioxidants, *Food Technol. Int. Eur.*, 187, 1989.

Gould, G., Predictive mathematical modelling of microbial growth and survival in foods, *Food Sci. Technol. Today*, 3, 89, 1989.

Gowland, H., Allergic customers and food suppliers: bridging the information gap, *Food Sci. Technol.*, 16, 44, 2002.

Graf, E. and Saguy, I. S., *Food Product Development: From Concept to the Marketplace*, Graf, E., and Saguy, I. S., Eds., Van Nostrand Reinhold, New York, 1991, chap. 3.

Graham, D., Quality programs and consumer complaints, *Food Technol. Int. Eur.*, 245, 1990.

Green, C., Getting the best value from outsourcing, *Food Can.*, 56, 8, 1996.

Grodner, R. and Hinton, A., Jr., Low dose gamma irradiation of *Vibrio cholerae* in crabmeat (*Callinectes sapidus*), in *Proc. Eleventh Annu. Tropical and Subtropical Fisheries Conf. of the Americas*, Grodner, R. and Hinton, A., Jr., Eds., Texas A & M Univ., College Station, 1986, 219.

Gutteridge, C. S., New methods for finding the right market niche, *Food Technol. Int. Eur.*, 127, 1990.

Halden, K., De Alwis, A. A. P., and Fryer, P. J., Changes in the electrical conductivity of foods during ohmic heating, *Int. J. Food Sci. Technol.*, 25, 9, 1990.

Hamm, D. J., Preparation and evaluation of trialkoxytricarbalkylate, trioxycitrate, trialkoxyglycerylether, jojoba oil and sucrose polyester as low calorie replacements of edible fats and oils, *J. Food Sci.*, 49, 419, 1984.

Hammer, M., Why projects fail..., *Ceres*, 26, 32, 1994.

Han, J. H., Antimicrobial food packaging, *Food Technol.*, 54, 56, 2000.

Hang, Y. D. and Woodams, E. E., Enzymatic production of soluble sugars from corn husks, *Lebensm.-Wiss. Technol.*, 32, 208, 1999.

Hang, Y. D. and Woodams, E. E., Corn husks: a potential substrate for production of citric acid by *Aspergillus niger, Lebensm.-Wiss. Technol.*, 33, 520, 2000.

Hang, Y. D. and Woodams, E. E., Enzymatic production of reducing sugars from corn cobs, *Lebensm.-Wiss. Technol.*, 34, 140, 2001.

Hannigan, K. J., Dried citrus juice sacs add moisture to food products, *Food Eng.*, 54, 88, 1982.

Hardas, N., Danviriyakul, S., Foley, J. L., Nawar, W. W., and Chinachoti, P., Accelerated stability studies of microencapsulated anhydrous milk fat, *Lebensm.-Wiss. Technol.*, 33, 506, 2000.

Hardy, K. G., Fickle tastebuds, *Bus. Q.*, Spring, 40, 1991.

Harlander, S., Food technology: yesterday, today, and tomorrow, *Food Technol.*, 43, 196, 1989.

Harlander, S., Social, moral and ethical issues in food biotechnology, *Food Sci. Technol. Today*, 6, 66, 1992.

Harlfinger, L., Microwave sterilization, *Food Technol.*, 46, 57, 1992.

Harris, J. M., Food product introductions continue to decline in 2000, *Food Rev.: Consum.-Driven Agric.*, 25, 24, 2002.

Harvey, P., Joint ventures: problems and opportunities, *Chem. Ind.*, 949, Dec. 3, 1977.

Hasler, C. M., Functional foods: their role in disease prevention and health promotion, *Food Technol.*, 52, 63, 1998.

Hauck, K., Alone on the range, *Prep. Foods*, 161, 32, 1992.

Hayashi, R., Application of high pressure to food processing and preservation: philosophy and development, in *Engineering and Food*, Vol. 2, Spiess, W. E. L. and Schubert, H., Eds., Elsevier Applied Science, London, 1989, 815.

Head, A. W., The technical manager: present and future role, *Chem. Ind.*, 716, 1971.

Headford, L., The environmental policies of the food industries in Europe and USA, *Food Sci. Technol. Today*, 10, 99, 1996.

Health Canada, Health Canada issues a stop-sale order for all products containing Kava, Aug. 21, 2002, http://www.hc-sc.gc.ca/english/protection/warnings/2002/2002_56e.htm.

Heath, H., *Flavor Technology: Profiles, Products, Applications*, AVI Publishing, Westport, CT, 1978.

Hegenbart, S., The R&D and marketing melee, *Prep. Foods*, 159, 117, 1990.

Helander, I. M., von Wright, A., and Mattila-Sandholm, T-M., Potential of lactic acid bacteria and novel antimicrobials against Gram-negative bacteria, *Trends Food Sci. Technol.*, 8, 146, 1997.

Heldman, D. R. and Newsome, R. L., Kinetic models for microbial survival during processing, *Food Technol.*, 57, 40, 2003

Henika, R. G., Simple and effective system for use with response surface methodology, *Cereal Sci. Today*, 17, 309, 1972.

Herrod, R. A., Industrial applications of expert systems and the role of the knowledge engineer, *Food Technol.*, 43, 130, 1989.

Hibler, M., Fast foods, *Can. Consum.*, 18, 19, 1988.

Hill, C. G., Jr. and Grieger-Block, R. A., Kinetic data: generation, interpretation, and use, *Food Technol.*, 34, 56, 1980.

Hill, S., The IFIS food science and technology bibliographic databases, *Trends Food Sci. Technol.*, 2, 269, 1991.

Hollingsworth, P., The perils of product development, *Food Technol.*, 48, 80, 1994.

Hollingsworth, P., Slotting fees under fire, *Food Technol.*, 54, 2000.

Hollingsworth, P., Federal R&D: a boon to food companies, *Food Technol.*, 55, 45, 2001.

Hollingsworth, P., The "self-care shopper" emerges, *Food Technol.*, 57, 20, 2003.

Holmes, A. W., The control of research for profit, *Br. Food Manuf. Ind. Res. Assoc. Tech. Circ.*, No. 412, 1968.

Holmes, A. W., Securing innovation in the food industry, *Br. Food Manuf. Ind. Res. Assoc. Tech. Circ.*, No. 636, 1977.

Hoover, D. G., Bifidobacteia: activity and potential benefits, *Food Technol.*, 47, 120, 1993.

Hoover, D. G., Metrick, C., Papineau, A. M., Farkas, D., and Knorr, D., Biological effects of high hydrostatic pressure on food microorganisms, *Food Technol.*, 43, 99, 1989.

Hsieh, Y. P. C., Pearson, A. M., and Magee, W. T., Development of a synthetic meat flavor mixture by using surface response methodology, *J. Food Sci.*, 45, 1125, 1980.

Huggett, A. C. and Conzelmann, C., EU regulation on novel foods: consequences for the food industry, *Trends Food Sci. Technol.*, 8, 133, 1997.

Hughes, D. B. and Hoover, D. G., Bifidobacteria: their potential for use in American dairy products, *Food Technol.*, 45, 74, 1991.

Huizenga, T. P., Liepins, K., and Pisano, D. J., Jr., Early involvement, *Qual. Prog.*, 81, 1987.

Hunt, J. T., Moving target, *Natl. Post Bus. Mag.*, 48, Oct. 2000.

Idziak, E. S., personal communication, 1993.

IFST(UK), *Guidelines for the Handling of Chilled Foods*, 2nd ed., Institute of Food Science & Technology, London, 1990.

IFST(UK), Cyclospora: an IFST Information Statement, *Food Sci. Technol.*, 17, 8, 2003.

Ishibashi, N. and Shimamura, S., Bifidobacteria: research and development in Japan, *Food Technol.*, 47, 126, 1993.

James, W., *The Principles of Psychology*, 1890, chap. 2.

Jay, J. M., Do background microoganisms play a role in the safety of fresh foods?, *Trends Food Sci. Technol.*, 8, 421, 1997.

Jenkins, D., Dietary fibre and its relation to nutrition and health, *Inst. Food Sci. Technol. Proc.*, 13, 51, 1980.

Jeremiah, L. E., Penney, N., and Gill, C. O., The effects of prolonged storage under vacuum or CO on the flavor and texture profiles of chilled pork, *Food Res. Int.*, 25, 9, 1992.

Jezek, E. and Smyrl, T. G., Volatile changes accompanying dehydration of apples by the Osmovac process, *Can. Inst. Food Sci. Technol. J.*, 13, 43, 1980.

Johnston, W. A., Surimi: an introduction, *Eur. Food Drink Rev.*, 21, 1989.

Jolly, D. A., Schutz, H. G., Diaz-Knauf, K., V., and Johal, J., Organic foods: consumer attitudes and use, *Food Technol.*, 43, 60, 1989.

Jones, K., The EEC flavoring and food labelling directives, *DRAGOCO Rep.*, 103, No. 3, 1992.

Josephson, E. S., Military benefits of food irradiation, *Act. Rep. Res. Dev. Assoc.*, 36, 30, 1984.

Juven, B. J., Schved, F., and Linder, P., Antagonistic compounds produced by a chicken intestinal strain of *Lactobacillus acidophilus*, *J. Food Prot.*, 55, 157, 1992.

Kantor, D., New product proliferation: are the benefits worth the cost?, *Prep. Foods*, 160, 28, 1991.

Kardinaal, A. F. M., Waalkens-Berendsen, D. H., and Arts, C. J. M., Pseudo-oestrogens in the diet: health benefits and safety concerns, *Trends Food Sci. Technol.*, 8, 327, 1997.

Katz, F., The changing role of water binding, *Food Technol.*, 51, 64, 1997.

Katz, F., That's using the old bean, *Food Technol.*, 52, 42, 1998.

Katz, F., Top product development trend in Europe, *Food Technol.*, 53, 38, 1999.

Katzenstein, A. W., The Food Update Delphi Survey: forecasting the food industry 10 years from now, *Food Prod. Dev.*, 11, 1975.

Kennedy, J. P., Structured lipids: fats of the future, *Food Technol.*, 45, 76, 1991.

Kernon, J. M., The Foodline scientific and technical, marketing and legislation databases, *Trends Food Sci. Technol.*, 2, 276, 1991.

Kesterton, M., Cooking: a minority art?, *The Globe Mail*, A18, Jan. 15, 2003.

King, V. A.-E., and Zall, R. R., A response surface methodology approach to the optimization of controlled low-temperature vacuum dehydration, *Food Res. Int.*, 25, 1, 1992.

Kinsella, J. E., Frankel, E., German, B., and Kanner, J., Possible mechanisms for the protective role of antioxidants in wine and plant foods, *Food Technol.*, 47, 85, 1993.

Kirk, D. and Osner, R. C., Collection of data on food waste from catering outlets in a University and a Polytechnic, *Inst. Food Sci. Technol. Proc.*, 14, 190, 1981.

Kirkpatrick, K. J. and Fenwick, R. M., Manufacture and general properties of dairy ingredients, *Food Technol.*, 41, 58, 1987.

Kläui, H., Naturally occurring antioxidants, *Inst. Food Sci. Technol. Proc.*, 6, 195, 1973.

Klensin, J. C., Information technology and food composition databases, *Trends Food Sci. Technol.*, 2, 279, 1991.

Knorr, D., Recovery and utilization of chitin and chitosan in food processing waste management, *Food Technol.*, 45, 114, 1991.

Knorr, D., Plant cell and tissue cultures as a model system for monitoring the impact of unit operations on plant food, *Trends Food Sci. Technol.*, 5, 328, 1994.

Knorr, D., Technology aspects related to microorganisms in functional foods, *Trends Food Sci. Technol.*, 9, 295, 1998.

Knorr, D., Beaumont, M. D., Caster, C. S., Dörnenburg, H., Gross, B., Pandya, Y., and Romagnoli, L. G., Plant tissue culture for the production of naturally derived food ingredients, *Food Technol.*, 44, 71, 1990.

Knorr, D, Schlueter, O., and Heinz, V., Impact of high hydrostatic pressure on phase transitions of foods, *Food Technol.*, 52, 42, 1998.

Kolata, G., Food affects human behaviour, *Science*, 218, 1209, 1982.

Konuma, H., Shinagawa, K., Tokumaru, M., Onove, Y., Konno, S., Fujino, N., Shigehisa, T., Kurata, H., Kuwabara, Y., and Lopes C. A. M., Occurrence of *Bacillus cereus* in meat products, raw meat and meat product additives, *J. Food Prot.*, 51, 324, 1988.

Kraushar, P. M., *New Products and Diversification*, Business Books, London, 1969.

Kritchevsky, D., The effect of dietary garlic on the development of cardiovascular disease, *Trends Food Sci. Technol.*, 2, 141, 1991.

Krizmanic, J., Here's who we are!, *Veg. Times*, 182, 72, 1992.

Kroll, B. J., Evaluating rating scales for sensory testing with children, *Food Technol.*, 44, 78, 1990.

Kuhn, M. E., Functional foods overdose?, *Food Process.*, 59, 21, 1998a.

Kuhn, M. E., Partner power, *Food Process.*, 59, 67, 1998b.

LaBarge, R. G., The search for a low-caloric oil, *Food Technol.*, 42, 84, 1988.

Labuza, T. P., The effect of water activity on reaction kinetics of food deterioration, *Food Technol.*, 34, 36, 1980.

Labuza, T. P. and Hyman, C. R., Moisture migration and control in multi-domain foods, *Trends Food Sci. Technol.*, 9, 47, 1998.

Labuza, T. P. and Riboh, D., Theory and application of Arrhenius kinetics to the prediction of nutrient losses in foods, *Food Technol.*, 36, 66, 1982.

Labuza, T. P. and Schmidl, M. K., Accelerated shelf-life testing of foods, *Food Technol.*, 39, 57, 1985.

Lachance, P. A., Human obesity, *Food Technol.*, 48, 127, 1994.

Lampert, A., Montreal schoolchildren speak in many tongues, *The Gazette, Montreal*, A1, Oct. 9, 2002.

Land, E., The Second Great Product of Industry: The Rewarding Working Life, presented at Science and Human Progress: 50th Anniversary of Mellon Institute, Pittsburgh, 1963, 107.

Lanier, T. C., Functional properties of surimi, *Food Technol.*, 40, 107, 1986.

Lawrie, R. A. and Symons, H., The food industry: changes foreseen and unseen 1965–2000, *Food Sci. Technol.*, 15, 50, 2001.

Leadley, C., Developments in non-thermal processing, *Food Sci. Technol.*, 17, 40, 2003.

Lechowich, R. V., Microbiological challenges of refrigerated foods, *Food Technol.*, 42, 84, 1988.

Lee, C. M., Surimi process technology, *Food Technol.*, 38, 69, 1984.

Lee, J., The development of new products for the European Market, *Food Sci. Technol. Today*, 5, 155, 1991.

Lee, K., Food neophobia: major causes and treatment, *Food Technol.*, 43, 62, 1989.

Lee, Y.-K. and Salminen, S., The coming of age of probiotics, *Trends in Food Sci. Technol.*, 6, 241, 1995.

Leistner, L., Hurdle technology applied to meat products of the shelf stable product and intermediate moisture food types, in *Properties of Water in Foods*, Simatos, D. and Multon, J. L., Eds., Martinus Nijhoff, Dordrect, The Netherlands, 1985, 309.

Leistner, L., Shelf stable products and intermediate moisture foods based on meat, IFT-IUFoST Basic Symp. Water Activity: Theory and Applications, Dallas, June 13–14, 1986, chap. 13.

Leistner, L., Food preservation by combined methods, *Food Res. Int.*, 25, 151, 1992.

Leistner, L. and Rödel, W., Inhibition of micro-organisms in food by water activity, in *Inhibition and Inactivation of Vegetative Microbes*, Skinner, F. A. and Hugo, W. B., Eds., Academic, London, 1976a, 219.

Leistner, L. and Rödel, W., The stability of intermediate moisture foods with respect to micro-organisms, in *Intermediate Moisture Foods*, Davies, R., Birch, G. G., and Parker, K. J., Eds., Applied Science, London, 1976b, chap. 10.

Leistner, L., Rödel, W., and Krispien, K., Microbiology of meat and meat products in high- and intermediate-moisture ranges, in *Water Activity: Influences on Food Quality*, Rockland, L. B. and Stewart, G. F., Eds., Academic, New York, 1981, 855.

Lenz, M. K. and Lund, D. B., Experimental procedures for determining destruction kinetics of food components, *Food Technol.*, 34, 51, 1980.

Levitt, T., Marketing myopia, *Harv. Bus. Rev.*, 38, 45, 1960.

Levitt, T., Marketing myopia, *Harv. Bus. Rev.*, 53, 26, 1975.

Lewandowski, R., Corporate confidential, *Financ. Post.*, Mar. 18, 1999.

Lieber, H., Preserved Radio-active Organic Matter and Food, U.S. Patent 788.480, 1905.

Light, N., Young, H., and Youngs, A., Operating temperatures in chilled food vending machines and risk of growth of food poisoning organisms, *Food Sci. Technol. Today*, 1, 252, 1987.

Lightbody, M. S., New technological approaches to reducing uniformity in processed foods, *Food Sci. Technol. Today Proc.*, 4, 37, 1990.

Lingle, R., AmeriQual Foods: at ease with MRE's, *Prep. Foods*, 158, 144, 1989.

Lingle, R., Degradable plastics: all sizzle and no steak?, *Prep. Foods*, 159, 144, 1990.

Lingle, R., Streamlining package design through computers, *Prep. Foods*, 160, 86, 1991.

Lingle, R., A sign of changing times, *Prep. Foods*, 161, 52, 1992.

Linnemann, A. R., Meerdink, G., Meulenberg, M. T. G., and Jongen, W. M. F., Consumer-oriented technology development, *Trends Food Sci. Technol.*, 9, 409, 1998.

Livingston, G. E., Foodservice: older than Methuselah, *Food Technol.*, 44, 54, 1990.

Loaharanu, P., International trade in irradiated foods: regional status and outlook, *Food Technol.*, 43, 77, 1989.

Loaharanu, P., Cost/benefit aspects of food irradiation, *Food Technol.*, 48, 104, 1994.

Looney, J. W., Crandall, P. G., and Poole, A. K., The matrix of food safety regulations, *Food Technol.*, 55, 60, 2001.

Lord, J. B., Launching the new product, in *Developing New Food Products for a Changing Marketplace,* Brody, A. L. and Lord, J. B., Eds., Technomic, Lancaster, PA, 2000, chap. 17.

Lovel, J., 'Neuromarketing' firm launched by Atlanta ad veteran, *Atlanta Bus. Chron.,* June 14, 2002, available from http://www.thoughtsciences.com/abc061402.htm, accessed 12/12/02.

Lund, B., *Cyclospora cayetanensis* update, *Food Sci. Technol.,* 16, 48, 2002.

Lund, D. B., Considerations in modeling food processes, *Food Technol.,* 37, 92, 1983.

MacAulay, J., Obesity: an issue of national importance, *Food Technol.,* 57, 20, 2003.

MacFie, H., Factors affecting consumers' choice of food, *Food Technol. Int. Eur.,* 123, 1990.

MacDonald, G. A. and Lanier, T., Carbohydrates as cryoprotectants for meats and surimi, *Food Technol.,* 45, 150, 1991.

MacNulty, C. A. R., Food products and the future, *Food Technol. Int. Eur.,* 19, 1989.

Malpas, R., Chemical technology: scaling greater heights in the next ten years?, *Chem. Ind.,* 111, 1977.

Mans, J., Kyotaru's bridge across the Pacific, *Prep. Foods,* 161, 85, 1992.

Marcotte, M., Irradiated strawberries enter the U.S. market, *Food Technol.,* 46, 80, 1992.

Mardon, J., Cripps, W. C., and Matthews, G. T., The organisation and administration of technical departments in large multi-plant companies, *Chem. Ind.,* 450, 1970.

Marechal, P. A., Martínez de Marnañón, I., Poirier, I., and Gervais, P., The importance of the kinetics of application of physical stresses on the viability of microorganisms: significance for minimal food processing, *Trends Food Sci. Technol.,* 10, 15, 1999.

Marlow, P., Qualitative research as a tool for product development, *Food Technol.,* 41, 74, 1987.

Martin, D., The impact of branding and marketing on perception of sensory qualities, *Food Sci. Technol. Today Proc.,* 4, 44, 1990.

Mason, L. H., Church, I. J., Ledward, D. A., and Parsons, A. L., The sensory quality of foods produced by conventional and enhanced cook-chill method, *Int. J. Food Sci. Technol.,* 25, 247, 1990.

Matthews, M. E., Foodservice in health care facilities, *Food Technol.,* 36, 53, 1982.

Mattson, P., Eleven steps to low cost product development, *Food Prod. Dev.,* 6, 106, 1970.

Mattson, B. and Soneson, U., Eds., *Environmentally-Friendly Food Processing,* Woodhead Publishing, Abington, U.K., 2003.

Maugh, T. H., II, Cancer is not inevitable, *Science,* 212, 36, 1982.

Mayo, J. S., AT&T: management questions for leadership in quality, *Qual. Prog.,* 34, 1986.

Mazza, G., Development and consumer evaluation of a native fruit product, *Can. Inst. Food Sci. Technol. J.,* 12, 166, 1979.

McCaskill, D. R. and Zhang, F., Use of rice bran oil in foods, *Food Technol.,* 53, 50, 1999.

McDermott, B. J., Identifying consumers and consumer test subjects, *Food Technol.,* 44, 154, 1990.

McDermott, R. L., Functionality of dairy ingredients in infant formula and nutritional specialty products, *Food Technol.,* 41, 91, 1987.

McGinn, C. J. P., Evaluation of shelf life, *Inst. Food Sci. Technol. Proc.,* 15, 153, 1982.

McLellan, M. R., An introduction to artificial intelligence and expert systems, *Food Technol.,* 43, 120, 1989.

McLellan, M. R., Hoo, A. F., and Peck, V., A low-cost computerized card system for the collection of sensory data, *Food Technol.,* 41, 66, 1987.

McPhee, M., Organically grown-up?, *Prep. Foods,* 161, 17, 1992.

McProud, L. M. and Lund, D. B., Thermal properties of beef loaf produced in foodservice systems, *J. Food Sci.,* 48, 677, 1983.

McWatters, K. H., Enwere, N. J., and Fletcher, S. M., Consumer response to akara (fried cowpea paste) served plain or with various sauces, *Food Technol.*, 46, 111, 1992.

McWatters, K. H., Resurreccion, A. V. A., and Fletcher, S. M., Response of American consumers to akara, a traditonal West African food made from cowpea paste, *Int. J. Food Sci. Technol.*, 25, 551, 1990.

McWeeny, D. J., Long term storage of some dry foods: a discussion of some of the principles involved, *J. Food Technol.*, 15, 195, 1980.

Megremis, C. J., Medium-chain triglycerides: a nonconventional fat, *Food Technol.*, 45, 108, 1991.

Meheriuk, M., Girard, B., Moyls, L., Beveridge, H. J. T., MCKenzie, D.-L., Harrison, J., Weintraub, S., and Hocking, R., Modified atmosphere packaging of 'Lapins' sweet cherry, *Food Res. Int.,* 28, 239, 1995.

Meisel, H. and Schlimme, E., Milk proteins: precursors of bioactive peptides, *Trends Food Sci. Technol.*, 1, 41, 1990.

Meltzer, R., Value added products: a noteworthy niche, *Visions*, 2, 1991.

Mermelstein, N. H., Retort pouch earns 1978 IFT Food Technology Industrial Achievement Award, *Food Technol.*, 32, 22, 1978.

Mermelstein, N. H., Software for food processing, *Food Technol.*, 54, 56, 2000.

Mermelstein, N. H., Military and humanitarian rations, *Food Technol.*, 55, 73, 2001.

Mertens, B. and Knorr, D., Developments of nonthermal processes for food preservation, *Food Technol.*, 46, 124, 1992.

Messens, W., Van Camp, J., and Huyghebaert, A., The use of high pressure to modify the functionality of food proteins, *Trends Food Sci. Technol.*, 8, 107, 1997.

Metrick, C., Hoover, D. G., and Farkas, D. F., Effects of high hydrostatic pressure on heat-resistant and heat-sensitive strains of *Salmonella*, *J. Food Sci.*, 54, 1547, 1989.

Meyer, R. S., Eleven stages of successful new product development, *Food Technol.*, 38, 71, 1984.

Mills, E. N. C., Alcocer, M. J. C., and Morgan, M. R. A., Biochemical interactions of food-derived peptides, *Trends Food Sci. Technol.*, 3, 64, 1992.

Morris, C. E., Why new products fail, *Food Eng.*, 65, 130, 1993.

Moskowitz, H. R., Benzaquen, I., and Ritacco, G., What do consumers really think about your product?, *Food Eng.*, 53, 80, 1981.

Mossel, D. A. A. and Ingram, M., The physiology of the microbial spoilage of food, *J. App. Bacteriol.*, 18, 232, 1955.

Mullen, K. and Ennis, D. M., Rotatable designs in product development, *Food Technol.*, 33, 74, 1979a.

Mullen, K. and Ennis, D. M., Mathematical system enters development realm as aid in achieving optimum ingredient levels, *Food Prod. Dev.*, 15, 50, 1979b.

Mullen, K. and Ennis, D., Fractional factorials in product development, *Food Technol.*, 39, 90, 1985.

Muller, R. A., Innovation and scientific funding, *Science*, 209, 880, 1980.

Mundy, C. C., Accessing the literature of food science, *Trends Food Sci. Technol.*, 2, 272, 1991.

NAS, *Underexploited Tropical Plants with Promising Economic Value*, Report of an Ad Hoc Panel of the Advisory Committee on Technology Innovation, National Academy of Sciences, Washington, DC, 1975.

Nathan, I., Hackett, A., and Kirby, S., Vegetarianism and health: is a vegetarian diet adequate for the growing child?, *Food Sci. Technol. Today*, 8, 13, 1994.

Nazario, S. L., Big firms get high on organic farming, *Wall St. J.*, B1, Mar. 21, 1989.

NCHS, Prevalence of overweight and obesity among adults: United States, 1999–2000, http://www.cdc.gov/nchs/products/pubs/pubd/hestats/obese/obse99.htm, 1999.

NCI (National Cancer Institute/Office of Cancer Communications), Cancer prevention research: chemoprevention and diet, *Backgrounder*, Dec. 1984.

Neaves, P., Gibbs, P. A., and Patel, M., Inhibition of *Clostridium botulinum*, by preservative interactions, *Br. Food Manuf. Ind. Res. Assoc. Res. Rep.*, No. 378, 1982.

Neff, J., Timid steps, *Food Process.*, 59, 33, 1998.

Neugebauer, W., "Electronic noses": possibilities and limitations of chemical sensor systems, *DRAGOCO Rep.*, 257, Jun. 1998.

Newiss, H., The patenting of genetically modified foods, *Trends Food Sci. Technol.*, 9, 368, 1998.

Newsome, R., Organically grown foods, *Food Technol.*, 44, 26, 1990.

NHANES, Overweight among U.S. children and adolescents, http:www.cdc.Gov/nchs/nhanes.htm, undated.

Noel, T. R., Ring, S. G., and Whittam, M. A., Glass transitions in low-moisture foods, *Trends Food Sci. Technol.*, 1, 62, 1990.

Norback, J., Techniques for optimization of food processes, *Food Technol.*, 34, 86, 1980.

Normand, F. L., Ory, R. L., and Mod, R. R., Binding of bile acids and trace minerals by soluble hemicelluloses of rice, *Food Technol.*, 41, 86, 1987.

Norwig, J. F. and Thompson, D. R., Making accurate energy efficiency measurements in foodservice equipment, *Act. Rep. Res. Dev. Assoc.*, 36, 37, 1984.

O'Brien, J., An overview of online information resources for food research, *Trends Food Sci. Technol.*, 2, 301, 1991.

O'Donnell, C. D., Computers: what's the use?, *Prep. Foods*, 160, 62, 1991.

O'Donnell, C. D., Aphrodisiacs, *Prep. Foods*, 162, 69, 1992.

O'Donnell, C. D., The formulation challenge: formulating for the big freeze, *Prep. Foods*, No. 6, 55, 1993.

Ohlsson, T., Boiling in water, in *Proc. IUFoST Int. Symp. Progress in Food Preparation Processes*, SIK — Swedish Food Institute, Göteborg, 1986, 89.

Ohlsson, T., Minimal processing: preservation methods of the future: an overview, *Trends Food Sci. Technol.*, 5, 341, 1994.

Ohlsson, T. and Bengtsson, N., *Minimal Processing Technologies in the Food Industries*, Ohlsson, T. and Bengtsson, N., Eds., CRC Press, Boca Raton, FL, 2002.

Ohr, L. M., Fats for healthy living, *Food Technol.*, 57, 91, 2003.

Ohshima, T., Recovery and use of nutraceutical products from marine resources, *Food Technol.*, 52, 50, 1998.

Oickle, J. G., *New Product Development and Value Added*, Food Development Division, Agriculture Canada, Ottowa, 1990.

Okamoto, M., Kawamura, Y., and Hayashi, R., Application of high pressure to food processing: textural comparison of pressure- and heat-induced gels of food proteins, *Agric. Biol. Chem.*, 54, 183, 1990.

Okoli, E. C. and Ezenweke, L. O., Formulation and shelf-life of a bottled pawpaw juice beverage, *Int. J. Food Sci. Technol.*, 25, 706, 1990.

Olson, A., Gray, G. M., and Chiu, M., Chemistry and analysis of soluble dietary fiber, *Food Technol.*, 41, 71, 1987.

Olson Zaltman Associates, Overview of ZMET Research Process, Olson Zaltman Associates, State College, PA, undated.

O'Neill, M., Geneticists' latest discovery: public fear of "Frankenfood," *NY Times*, 1, Jun. 28, 1992.

O'Sullivan, M. G., Thornton, G., O'Sullivan, G. C., and Collins, J. K., Probiotic bacteria: myth or reality?, *Trends Food Sci. Technol.*, 3, 309, 1992.

Palmer, G. M., Wine production from cheese whey, *EPA-600/2-79-189*. Industrial Environmental Research Laboratory, Office of Research and Development, U.S. Environmental Protection Agency, Cincinnati, OH, 1979.

Park, C. E., Szabo, R., and Jean, A., A survey of wet pasta packaged under a CO:N (20:80) mixture for Staphylococci and their enteriotoxins, *Can. Inst. Food Sci. Technol. J.*, 21, 109, 1988.

Parrott, D. L., Use of ohmic heating for aseptic processing of food particulates, *Food Technol.*, 46, 68, 1992.

Parsons, R., *Statistical Analysis: A Decision-Making Approach*, 2nd ed., Harper and Row, New York, 1978, chap. 5–6.

Paster, N., Juven, B. J., Gagel, S., Saguy, I., and Padova, R., Preservation of a perishable pomegranate product by radiation pasteurization, *J. Food Technol.*, 20, 367, 1985.

Patel, T., German syringes turn up in French quarry, *New Sci.*, 135, 7, 1992.

Patel, T., Energy booster "a con" says French food watchdog, *New Sci.*, 137, 8, 1993.

Patterson, J. and Haas, S., A rationale for outsourcing, *Food Technol.*, 53,52, 1999.

Pearce, F., Silence of the experts, *New Sci.*, 150, 50, 1996.

Pearce, S. J., Quality assurance involvement in new product design, *Food Technol.*, 41, 104, 1987.

Peck, M. W., *Clostridium botulinum* and the safety of refrigerated processed foods of extended durability, *Trends Food Sci. Technol.*, 8, 186, 1997.

Pehanich, M., Top tips for working with chains, in *Foodservice Annual*, in *Food Product Design*, 45, Apr. 2003.

Pennington, J. A. T. and Butrum, R. R., Food descriptions using taxonomy and the "Langual" system, *Trends Food Sci. Technol.*, 2, 285, 1991.

Penny, C., Detailing dietary fibre, *Food Ingred. Process. Int.*, 14, Nov. 1992.

Peryam, D. R., Sensory evaluation: the early days, *Food Technol.*, 44, 86, 1990.

Peters, D., Power, influence, and group dynamics in sensory evaluation, *Food Technol.*, 41, 62, 1987.

Peters, J. W., Foodservice R & D at Gino's: spotting the winners faster, *Food Prod. Dev.*, 16, 28, 1980.

Petersen, K., Nielsen, P. V., Berteses, G., Lawther, M., Olsen, M. B., Nilsson, N. H., and Mortensen, G., Potential of biobased materials for food packaging, *Trends Food Sci. Technol.*, 10, 52, 1999.

Petesch, B. L. and Sumiyoshi, H., Recent advances on the nutritional benefits accompanying the uses of garlic as a supplement, *Trends Food Sci. Technol.*, 9, 415, 1999.

Piggott, J. R., Simpson, S. J., and Williams, S. A. R., Sensory analysis, *Int. J. Food Sci. Technol.*, 33, 7, 1998.

Pine, R. and Ball, S., Productivity and technology in catering operations, *Food Sci. Technol. Today*, 1, 174, 1987.

Pirie, N. W., *Leaf Protein and Its By-products in Human and Animal Nutrition*, 2nd ed., Cambridge University Press, London, 1987.

Pokorný, J., Natural antioxidants for food use, *Trends Food Sci. Technol.*, 2, 223, 1991.

Port, O. and Carey, J., Getting to 'Eureka!': researchers are tracking how breakthroughs are made, *Bus. Week*, Nov. 10, 1997, available from http://www.businessweek.com.

Porter, W. L., Storage life prediction under noncontrolled environmental temperatures: product-sensitive environmental call-out, in *Proc. Food Processors' Institute Shelf-life: A Key to Sharpening Your Competitive Edge*, Food Processors' Institute, Washington, DC, 1981, 1.

Poste, L. M., Mackie, D. B., Butler, G., and Larmond, E., *Laboratory Methods for Sensory Analysis of Food*, Publication 1864/E, Research Branch, Agriculture Canada, Ottawa, 1991.

Pothakamury, U. R., Barbosa-Cánovas, G. V., and Swanson, B. G., Magnetic-field inactivation of microorganisms and generation of biological changes, *Food Technol.*, 47, 85, 1993.

Powers, J. J., Uses of multivariate methods in screening and training sensory panelists, *Food Technol.*, 42, 123, 1988.

Pszczola, D. E., Food irradiation: countering the tactics and claims of opponents, *Food Technol.*, 44, 92, 1990.

Pszczola, D. E., Irradiated poultry makes U.S. debut in Midwest and Florida markets, *Food Technol.*, 47, 89, 1993.

Puzo, D. P., First irradiated fruit on market sells quickly, *Los Angeles Times*, 1986.

Pyke, M., A food scientist's reflections, *Food Sci.*, 9, 10, 1971.

Ramarathnam, N. and Osawa, T., International conference on food factors: chemistry and cancer prevention. Report of a conference held in Hamamatsu, Japan, Dec. 10–15, 1995, *Trends Food Sci. Technol.*, 7, 64, 1996.

Reiser, S., Metabolic effects of dietary pectins related to human health, *Food Technol.*, 41, 91, 1987.

Reuters, Worldwide increase in obesity could become disastrous, doctors say, *The Gazette, Montreal*, B15, May 17, 1996.

Righelato, R. C., Biotechnology and food manufacturing, *Food Technol. Int. Eur.*, 155, 1987.

Ripsin, C. M. and Keenan, J. M., The effects of dietary oat products on blood cholesterol, *Trends Food Sci. Technol.*, 3, 137, 1992.

Rizvi, S. S. H. and Acton, J. C., Nutrient enhancement of thermostabilized foods in retort pouches, *Food Technol.*, 36, 105, 1982.

Robbins, M. P., Ed., *The Cook's Quotation Book*, Robert Hale, London, 1987.

Roberts, T. A., Combinations of antimicrobials and processing methods, *Food Technol.*, 43, 156, 1989.

Roberts, T. A., Predictive modelling of microbial growth, *Food Technol. Int. Eur.*, 231, 1990.

Robertson, G. L., Shelf life of packaged foods, its measurement and prediction, in *Developing New Food Products for a Changing Marketplace*, Brody, A. L. and Lord, J. B., Eds., Technomic, Lancaster, PA, 2000, chap. 13, p. 329.

Robinson, D. S., Irradiation of foods, *Inst. Food Sci. Technol. Proc.*, 19, 165, 1985.

Rockland, L. B. and Nishi, S., Influence of water activity on food product quality and stability, *Food Technol.*, 34, 42, 1980.

Roller, S., Ingredients as preservatives and the role of biotechnology, *Food Sci. Technol. Today*, 9, 116, 1995.

Roos, Y. and Karel, M., Applying state diagrams to food processing and development, *Food Technol.*, 45, 66, 1991a.

Roos, Y. and Karel, M., Amorphous state and delayed ice formation in sucrose solutions, *Int. J. Food Sci. Technol.*, 26, 553, 1991b.

Roser, B., Trehalose, a new approach to premium dried foods, *Trends Food Sci. Technol.*, 2, 166, 1991.

Ross, C., personal communication, 1980.

Rowley, D. B., Significance of sublethal injury of foodborne pathogenic and spoilage bacteria during processing, *Act. Rep. Res. Dev. Assoc.*, 36, 41, 1984.

Rusoff, I., Nutrition and dietary fiber, *Cereal Foods World*, 29, 668, 1984.

Rutledge, K. P. and Hudson, J. M., Sensory evaluation: method for establishing and training a descriptive flavor analysis panel, *Food Technol.*, 44, 78, 1990.

Rutledge, K. R., Accelerated training of sensory descriptive flavor analysis panelists, *Food Technol.*, 46, 114, 1992.

Ryan, J. K., Jelen, P., and Sauer, W. C., Alkaline extraction of protein from spent honey bees, *J. Food Sci.*, 48, 886, 1983.

Ryley, J., The impact of fast foods on U.K. nutrition, *Inst. Food Sci. Technol. Proc.*, 16, 58, 1983.

Ryu, C. H. and West, A., Development of kimchi recipes suitable for British consumers and determination of vitamin C changes in kimchi during storage, *Food Sci. Technol. Today*, 14, 76, 2000.

Ryval, M., The shape of food to come, *Financ. Post Magazine*, 38, Apr. 15, 1981.

Saca, S. A. and Lozano, J. E., Explosion puffing of bananas, *Int. J. Food Sci. Technol.*, 27, 419, 1992.

Saguy, I. and Karel, M., Models of quality deterioration during food processing and storage, *Food Technol.*, 34, 78, 1980.

Saito, Y., The retort pouch is a way of life in Japan, *Act. Rep. Res. Dev. Assoc.*, 35, 8, 1983.

Saldana, G., Meyer, R., and Lime, B. J., A potential processed carrot product, *J. Food Sci.*, 45, 1444, 1980.

Salminen, S., Ouwehand, A., Benno, Y., and Lee, Y.-K., Probiotics: how should they be defined?, *Trends Food Sci. Technol.*, 10, 107, 1999.

Sanders, M. E., Probiotics, *Food Technol.*, 53, 67, 1999.

Sastry, S. K. and Palaniappan, S., Ohmic heating of liquid-particle mixtures, *Food Technol.*, 46, 64, 1992.

Schaller, E., Bosset, J. O., and Escher, F., "Electronic noses" and their application to food, *Lebensm.-Wiss. Technol.*, 31, 305, 1998.

Schiffmann, R. F., Microwave processing in the U.S. food industry, *Food Technol.*, 46, 50, 1992.

Schlegel, W., Commercial pasteurization and sterilization of food products using microwave technology, *Food Technol.*, 46, 62, 1992.

Schmidl, M. K., Massaro, S. S., and Labuza, T. P., Parenteral and enteral food systems, *Food Technol.*, 42, 77, 1988.

Schneeman, B. O., Soluble vs insoluble fiber: different physiological responses, *Food Technol.*, 41, 81, 1987.

Schutz, H. G., Multivariate analyses and the measurement of consumer attitudes and perceptions, *Food Technol.*, 42, 141, 1988.

Schwarcz, J., 'Natural' remedies come under microscope, *The Gazette, Montreal*, A1, Jan. 6, 2002.

Scott, V. N., Control and prevention of microbial problems in new generation refrigerated foods, *Act. Rep. Res. Dev. Assoc.*, 39, 22, 1987.

Scriven, F. M., Gains, N., Green, S. R., and Thomson, D. M. H., A contextual evaluation of alcoholic beverages using the repertory grid method, *Int. J. Food Sci. Technol.*, 24, 173, 1989.

Selman, J. D., Trends in food processing and food processes, *Food Technol. Int. Eur.*, 55, 1991.

Sensory Evaluation Division, IFT, *Sensory Evaluation Guide for Testing Food and Beverage Products*, Institute of Food Technologists, Chicago, 1981.

Settlemyer, K., A systematic approach to foodservice new product development, *Food Technol.*, 40, 120, 1986.

Shahidi, F., Arachchi, J. K. V., and Jeon, Y.-J., Food applications of chitin and chitosans, *Trends Food Sci. Technol.*, 10, 37, 1999.

Shahin, J., They're heeeere, *Am. Way*, 42, Dec. 1, 1995.

Shapton, D. A. and Shapton, N. F., *Principles and Practices for the Safe Processing of Foods*, Shapton, D. A. and Shapton, N. F., Eds., Butterworth-Heinemann, Oxford, 1991.

Shelef, L. A., Naglik, O. A., and Bogen, D. W., Sensitivity of some common food-borne bacteria to the spices sage, rosemary, and allspice, *J. Food Sci.*, 45, 1042, 1980.

Sherman, H. C., *Food Products*, MacMillan, New York, 1916, 397.

Shi, Z., Bassa, I. A., Gabriel, S. L., and Francis, F. J., Anthocyanin pigments of sweet potatoes: *Ipomoea batatas*, *J. Food Sci.*, 57, 755, 1992a.

Shi, Z., Francis, F. J., and Daun, H., Quantitative comparison of the stability of anthocyanins from *Brassica oleracea* and *Tradescantia pallida* in non-sugar drink model and protein model systems, *J. Food Sci.*, 57, 768, 1992b.

Silveira, E. T. F., Rahman, M. S., and Buckle, K. A., Osmotic dehydration of pineapple: kinetics and product quality, *Food Res. Int.*, 29, 227, 1996.

Silver, D., The seven deadly sins of product development, *Food Prod. Design,* 12, 93, 2003a.

Silver, D., Vegetarian dishes go mainstream, *Food Prod. Design*, 13, 99, 2003b.

Singhal, R. S., Gupta, A. K., and Kulharni, P. R., Low-calorie fat substitutes, *Trends Food Sci. Technol.*, 2, 241, 1991.

Sinki, G., Technical myopia, *Food Technol.*, 40, 86, 1986.

Skarra, L., Rollout roulette, *Prep. Foods*, 167, Aug. 1998.

Skinner, R. H. and Debling, G. B., Food industry applications of linear programming, *Food Manuf.*, 44, 35, 1969.

Skjöldebrand, C., Cooking by infrared radiation, in *Proc. IUFoST Int. Symp. Progress in Food Preparation Processes*, SIK — Swedish Food Institute, Göteborg, 1986, 157.

Slade, L. and Levine, H., Beyond water activity: recent advances based on an alternative approach to the assessment of food quality and safety, *Crit. Rev. Food Sci. Nutr.*, 30, 115, 1991.

Slade, P. J., Monitoring *Listeria* in the food production environment. I. Detection of *Listeria* in processing plants and isolation methodology, *Food Res. Int.*, 25, 45, 1992.

Slight, H., The storage and transport of chilled foods: a temperature survey, *Leatherhead Food Res. Assoc. Res. Rep.*, No. 340, 1980.

Sloan, E. A., Top 10 trends to watch and work on for the millennium, *Food Technol.,* 53, 40, 1999.

Sloan, E. A., Top 10 trends to watch and work on: 3rd biannual report, *Food Technol.*, 55, 38, 2001.

Sloan, E.A., Weighing in on the weight-control market, *Food Technol.*, 57, 18, 2003a.

Sloan, E. A., Top 10 trends to watch and work on: 2003, *Food Technol.*, 57, 30, 2003b.

Snow, C. P., *Science and Government*, Harvard University Press, Cambridge, MA, 1961.

Snow, P., The shape of shopping to come, *Oxford Today*, 5, 55, 1992.

Snyder, O. P., A model food service quality assurance system, *Food Technol.*, 35, 70, 1981.

Snyder, O. P., Jr., Minimizing foodservice energy consumption through improved recipe engineering and kitchen design, *Act. Rep. Res. Dev. Assoc.*, 36, 49, 1984.

Snyder, O. P., Jr., Microbiological quality assurance in foodservice operations, *Food Technol.*, 40, 122, 1986.

Solberg, M., Buckalew, J. J., Chen, C. M., Schaffner, D. W., O'Neill, K., McDowell, J., Post, L. S., and Boderck, M., Microbiological safety assurance system for foodservice facilities, *Food Technol.*, 44, 68,1990.

Solhjoo, K., Food technology and the Baháí faith, *Food Sci. Technol. Today*, 8, 2, 1994.

Speck, M. L., Dobrogosz, W. J., and Casas, I. A., *Lactobacillus reuteri* in food supplementation, *Food Technol.*, 47, 90, 1993.

Spitz, P., The public granary, *Ceres,* 12, 16, 1979.

Stanton, J. L., Diffusion of slotting allowances: a marketing practice going global, www.johnlstanton.com, 2003.

Stavric, S. and Speirs, J. I., *Escherichia coli* associated with hemorrhagic colitis, *Can. Inst. Food Sci. Technol. J.*, 22, AT/205, 1989.

Steinbock, E., Knowledge is power, *Food Sci. Technol.*, 16, 49, 2002.

Stent, G. S., Prematurity and uniqueness in scientific discovery, *Scientific American*, Dec. 1972, in *Readings from Scientific American: Scientific Genius and Creativity*, W. H. Freeman, New York, 1987, 95–104.

Stevenson, K. E., Implementing HACCP in the food industry, *Food Technol.*, 44, 179, 1990.

Stiles, M. E. and Hastings, J. W., Bacteriocin production by lactic acid bacteria: potential for use in meat preservation, *Trends Food Sci. Technol.*, 2, 247, 1991.

Strauss, S., Save stomachs: zap food, Canada advised, *Globe Mail*, Jun. 17, A1, 1998.

Stuller, J., The nature and process of creativity, *Sky*, 11, 37, 1982.

Sudman, S., Sirken, M., and Cowan C., Sampling rare and elusive populations, *Science*, 240, 991, 1988.

Tamime, A. Y., Khaskheli, M., Barclay, M. N., and Muir, D. D., Effect of processing conditions and raw materials on the properties of kishk. 1. Compositional and microbiological qualities, *Lebensm.-Wiss. Technol.*, 33, 444, 2000

Taylor, A. W., Scaling-up of process operations in the food industry, *Inst. Food Sci. Technol. Proc.*, 2, 86, 1969.

Taylor, B., Sandwiches in Britain: some observations, *Food Sci. Technol. Today*, 10, 78, 1996.

Taylor, G., A côte of many colours, *The Gazette, Montreal*, A1, A17, Aug. 10, 2002.

Taylor, L. and Leith, S., TQM, *Food Can.*, 51, 18, 1991.

Taylor, T. G., Diet and coronary heart disease, *Inst. Food Sci. Technol. Proc.*, 13, 45, 1980.

Teutonico, R. A. and Knorr, D., Amaranth: composition, properties, and applications of a rediscovered food crop, *Food Technol.*, 39, 49, 1985.

Thakur, S. and Saxena, D., Formation of extruded snack food (gum based cereal-pulse blend): optimization of ingredients' levels using response surface methodology, *Lebensm.-Wiss. Technol.*, 33, 354, 2000.

Thibault, J.-F., Asther, M., Ceccaldi, B., Couteau, D., Delattre, M., Duarte, J., Faulds, C., Heldt-Hansen, H.-P., Kroon, P., Lesage-Meessen, L., Micard, V., Renard, C., Tuohy, M., Van Hulle, S., and Williamson, G., Fungal bioconversion of agricultural by-products to vanillin, *Lebensm.-Wiss. Technol.*, No. 6, 530, 1998.

Thomson, D., Recent advances in sensory and affective methods, *Food Sci. Technol. Today*, 3, 83, 1989.

Thorne, S., What temperature?, *Inst. Food Sci. Technol. Proc.*, 11, 207, 1978.

Tiwari, N. P. and Aldenrath, S. G., Occurrence of *Listeria* species in food and environmental samples in Alberta, *Can. Inst. Food Sci. Technol. J.*, 23, 109, 1990.

Toma, R. B., Curtis, D. J., and Sobotar, C., Sucrose polyester: its metabolic role and possible future applications, *Food Technol.*, 42, 93, 1988.

Traill, C. P., *The Canadian Settlers Guide*, McClelland and Stewart, 1969 (first published in 1855).

Tregunno, N. B. and Goff, H. D., Osmodehydrofreezing of apples: structural and textural effects, *Food Res. Int.*, 29, 471, 1996.

Tsuji, K., Low-dose Cobalt 60 irradiation for reduction of microbial contamination in raw materials for animal health products, *Food Technol.*, 37, 48, 1983.

Tung, M. A., Garland, M. R., and Campbell, W. E., Quality comparison of cream style corn processed in rigid and flexible containers, *Can. Inst. Food Sci. Technol. J.*, 8, 211, 1975.

Tuomy, J. M. and Young, R., Retort-pouch packaging of muscle foods for the Armed Forces, *Food Technol.*, 36, 68, 1982.

Turner, M., and Glew, G., Home-delivered meals for the elderly, *Food Technol.*, 36, 46, 1982.

Tye, R. J., Konjac flour: properties and applications, *Food Technol.*, 45, 81, 1991.

Tyler, V. E., *The Honest Herbal: A Sensible Guide to the Use of Herbs and Related Remedies*, 3rd ed., Pharmaceutical Products Press/Haworth Press, Inc., New York, 1993.

USDA, Food and Nutrient Intakes of Individuals in 1 Day in the United States, Spring 1977, Nationwide Food Consumption Survey 1977–78, Preliminary Report No.2, U.S. Dept. of Agriculture, 1980.

USDHHS (U.S. Dept. of Health and Human Services), Nutrition and Cancer Prevention: A Guide to Food Choices, draft publication, 1984.

USDHHS (U.S. Dept. of Health and Human Services), Cancer Prevention Research Summary: Nutrition, National Institute of Health Publication No. 85-2616, 1985.

Van den Hoven, M., Functionality of dairy ingredients in meat products, *Food Technol.*, 41, 72, 1987.

Vega-Mercado, H., Pothakamury, U. R., Fu-Jung Chang, Barbosa-Cánovas, G. V., and Swanson, B. G., Inactivation of *Escherichia coli* by combining pH, ionic strength and pulsed electric fields hurdles, *Food Res. Int.*, 29, 117, 1996.

Vega-Mercado, H., Martín-Bellos, O., Bai-Lin Qin, Fu Jung Chang, Góngora-Nieto, M. M., Barbosa-Cánovas, G. V., and Swanson, B. G., Non-thermal food preservation: pulsed electric fields, *Trends Food Sci. Technol.*, 8, 151, 1997.

Vegetarian Resource Group, http://www.vrg.org/nutshell/faq.htm#poll, 2003.

Vegetarian Society, http://www.vegsoc.org/info/statve.g.,html, 2003.

Wachsmuth, I. K., *Escherichia coli* 0157:H7: Harbinger of change in food safety and tradition in the industrialized world, *Food Technol.*, 51, 26, 1997.

Waite-Wright, M., Chilled foods: the manufacturer's responsibility, *Food Sci. Technol. Today*, 4, 223, 1990.

Walker, S. J. and Jones, J. E., Predictive microbiology: data and model bases, *Food Technol. Int. Eur.*, 209, 1992.

Walston, J., C.O.D.E.X. spells controversy, *Ceres*, 24, 28, 1992.

Wang, C., Food business: seven ways brand management systems kill innovation, *Food Process.*, 60, 35, 1999.

Warburton, J., The electronic 'NOSE': the technology and applications, *Food Sci. Technol. Today*, 10, 91, 1996.

Webster, S. N., Fowler, D. R., and Cooke, R. D., Control of a range of food related microorganisms by a multi-parameter preservation system, *J. Food Technol.*, 20, 311, 1985.

West, A., Good catering practice: use of manufactured meals in catering, *Food Sci. Technol. Today*, 8, 172, 1994.

Whitehead, R., What'll we eat in 1999?, *Ind. Week*, May 17, 30, 1976.

Whitney, L. F., What expert systems can do for the food industry, *Food Technol.*, 43, 135, 1989.

Wiley, H. W., *Foods and Their Adulteration*, 3rd ed., P. Blakiston's Sons, Philadelphia, 1917, 455.

Wilhelmi, F., Product safety and how to ensure it, *DRAGOCO Rep.*, No. 1, 14 1988.

Williams, A., Another challenge for the food industry?, *Food Sci. Technol.*, 17, 41, 2003.

Williams, A. P., Blackburn, C., and Gibbs, P., Advances in the use of predictive techniques to improve the safety and extend the shelf-life of foods, *Food Sci. Technol. Today*, 6, 148, 1992.

Williams, D., Partnering for successful product development, *Food Technol.*, 56, 28, 2002.

Williams, E. F., Weak links in the cool and cold chain, *Inst. Food Sci. Technol. Proc.*, 11, 211, 1978.

Williams, M., Electronic databases, *Science*, 228, 445, 1985.

Williams, R., Vending in the Canadian foodservice industry, *Visions*, No. 5, 2, 1991.

Williams, R., A profile of the Canadian fast-food sector, *Visions*, No. 2, 3, 1992.

Williamson, M., Tasting tests carried out at the Leatherhead Food Research Association, *Leatherhead Food Res. Assoc. Tech. Circ.*, No. 749, 1981.

Wilson, C., Ethical shopping, *Food Can.*, 52, 7, 1992.

Wilson, D., Marketing mycoprotein: the Quorn Foods story, *Food Technol.*, 55, 48, 2001.

Wolfe, K. A., Use of reference standards for sensory evaluation of product quality, *Food Technol.*, 33, 43, 1979.

Wood, P. J., Oat β-glucan: physicochemical properties and physiological effects, *Trends Food Sci. Technol.*, 2, 311, 1991.

Wood, S., Food law enforcement: where next? An industry viewpoint, *Inst. Food Sci. Technol. Proc.*, 18, 89, 1985.

Woodward, M., Ed., *Gerard's Herbal*, Senate, Twickenham, U.K., 1998.

Worsfold, D. and Griffith, C., Food safety behaviour in the home, *Br. Food J.*, 99, 97, 1997.

Woolf, G., 2002, Editorial, *Eur. Food Drink Rev.*, 4, 2002.

Wrick, K. L., Development opportunities for functional foods, *Food Prod. Design*, 13, 73, 2003.

Wurtman, R. J. and Wurtman, J. J., Carbohydrates and depression, *Sci. Am.*, 260, 68, 1989.

Yankelovich Partners, Today's market?, *The Globe Mail*, A14, Aug. 3, 2000.

Young, J., A perspective on functional foods, *Food Sci. Technol. Today* 10, 18, 1996.

Zaika, L. L. and Kissinger, J. C., Inhibitory and stimulatory effects of oregano on *Lactobacillus plantarum* and *Pediococcus cerevisiae*, *J. Food Sci.*, 46, 1205, 1981.

Zaika, L., Kissinger, J. C., and Wasserman, A. E., Inhibition of lactic acid bacteria by herbs, *J. Food Sci.*, 48, 1455, 1983.

Zaltman, G., The Dimensions of Brand Equity for Nestlé Crunch Bar: A Research Case, Harvard Business School, N9-500-083, Jan. 27, 2000a.

Zaltman, G., Consumer researchers: take a hike!, *J. Consum. Res.*, 26, 423, 2000b.

Zhou, Y. and Lee, A., Mechanism for the suppression of the mammalian stress response by genistein, and anticancer phytoestrogen from soy, *J. Natl. Cancer Inst.*, 90, 381, 1998.

Zind, T., Phytochemicals: the new vitamins?, *Food Process.*, 59, 29, 1998.

Zind, T., The functional foods frontier, *Food Process.*, 60, 45, 1999.

Index